Text

2-

DISPLAY ENGINEERING
Conditioning, Technologies, Applications

DISPLAY ENGINEERING
Conditioning, Technologies, Applications

Edited by

D. BOSMAN
University of Twente
Enschede, The Netherlands

1989

NORTH-HOLLAND
AMSTERDAM ● NEW YORK ● OXFORD ● TOKYO

pp. 141-184, 219-250: copyright not transferred.

ISBN: 0 444 87319 8

Published by:

ELSEVIER SCIENCE PUBLISHERS B.V.
P.O. Box 1991
1000 BZ Amsterdam
The Netherlands

Sole distributors for the U.S.A. and Canada:

ELSEVIER SCIENCE PUBLISHING COMPANY, INC.
655 Avenue of the Americas
New York, N.Y. 10010
U.S.A.

Library of Congress Cataloging-in-Publication Data

Display engineering.

    Includes bibliographies and indexes.
    1. Information display systems. I. Bosman, D.
TK7882.I6D565 1989   621.3815'42   88–33608
ISBN 0-444-87319-8 (U.S.)

PRINTED IN THE NETHERLANDS

ABOUT THE BOOK

Designers and endusers of systems with electronic displays are now confronted
with a choice of available electro-optic technologies and a number of image
generation possibilities. All applications have in common that, given specific
functional and environments, the visual stimuli must be perceived reliably
without undue strain and at acceptable cost.
To make a well founded selection of a display requires in the first place
thorough analysis of its function in its environment, in combination with
sufficient understanding of the capabilities and constraints of the triple
involved in the visual system-technology interface:
- image generation techniques,
- electro-optic conversion technologies and
- the visual system.
This book provides general information about these subjects, together with
examples of their application. The aim is to acquant the reader with the
necessary background to appreciate the importance of visual display factors,
which normally are beyond the traditional discipline boundaries but vital to
successful system operation. Neglect will be felt strongly once the display is
in operation.

The organisation of the subject matter is in three parts. In part I theories
and facts behind image generation, conditioning and addressing, and about the
visual system are broadly explained. The line of reasoning is underpinned by
mathematical symbolism but mathematical proficiency is not necessary to
understand the message given in the text.
In part II the present state of development of display technologies is
reported. As the selection of a display technology in professional applications
is not always driven by the cost factor alone, attention is devoted to specific
advantages owing to the principle of operation and physical characteristics.
Part III discussed the appliccation of display techniques and technologies in
the domains of the office, industrial control and vehicle instrumentation.
These examples are typical for the type of considerations involved in the
selection process.

Each chapter has beeen written by specialists in the field concerned; many are
updated and improved versions from other, similar presentations. This is the
case with Part I, where the original sources either are borrowed from course
material or from work carries out in an ESPRIT project partly financed by the
European Community; and with Part II which in its original version was first
published in AGARD AR-169 by the Advisory Group for Aerospace Research and
Development, North Atlantic Treaty Organisation (AGARD/NATO) in October 1982.

The preparation of this book was done in camera ready form with the help of
numerous persons; in particular the effort by mrs. A.G.M. van Essen is
gratefully acknowledged.

Enschede, October 1988
D. Bosman.

CONTENTS

PART II   **TECHNOLOGIES**

Chapter 5 **Electroluminescent displays**

PART III   **APPLICATIONS**

PART I

CONDITIONING

DISPLAY ENGINEERING: D. Bosman (Editor)
© Elsevier Science Publishers B.V. (North-Holland), 1989

I.1: IMAGE CHARACTERISATION AND FORMATION

D. BOSMAN
University of Twente, The Netherlands

1.1. Introduction

Display image conditioning is necessary to format the signals to the display device such, that the human visual and cognitive systems can extract the desired informations from the presented image. With the wide choice of options available, one should develop a thorough understanding of image features, of the visual channel, of the electro-optic technologies and of the way the operator digests and uses the data presented at the display face. We do not have to acquire detail specialist knowledge in each of these disciplines, but the multi-disciplinary overview is necessary. And it certainly helps if we can express the variables involved in appropriate numbers, and describe certain relations in an exact fashion, e.g. mathematically. Although one would like to be able to do so in relation to the cognitive capability of the display observer as suggested in figure 1.1, the subject matter of this book is restricted to image characterisation, the visual channel and electro-optic technologies. The first two chapters of the book are addressed to the readers who are determined to know sufficient to design themselves a proper VDU workplace; the remaining chapters provide information enabling one to ask the necessary and correct questions with respect to the available display technologies. These have different electrical, optical and appearance characteristics. The characteristics must fit the primary function of a display: to provide a window on the state and behaviour of the process which needs to be observed and/or operated. This function must be maintained over the whole range of environmental conditions, with good to reasonable viewing and posture comfort. To appreciate the meaning of this important but rather vague statement, it is necessary to analyse the visual system/technology interface (VSTI) in order to be able put numbers to the factors involved in display quality.

1.2. The tesselated image or message

Any set of data with two-dimensional correlations can be represented graphically in a plane were locally the emission or reflection of light is modulated according to the existence/magnitude of said correlations. The so obtained picture is considered to consist of many picture elements (pels, pixels) of finite size. For continuous shade pictures this size can be made arbitrarily small. In electronic displays the cost is a function of the number of pels; thus it is necessary to perform an analysis of the images to be displayed in order to minimise the pelcount for a given image quality or legibility. Early examples of such analysis, but carried out for artistic purposes, are found in the mosaics we inherited from ancient cultures. In those mosaics the stones (tessera) each support over their area one colour of uniform intensity. They are the elementary display elements (dels) which either can align with features in the picture or cut through features with resulting jaggedness and discontinuities. The visual agreement depends on the local alignment between display elements and important features of the picture. In artistic mosaics, contours and gradient directions in the image are emphasised by appropriate choice of tessera locations and orientations, as shown in figure 1.2.

This mosaic structure thus depends on the main features of the image to be shown; the display elements have become the pels of the picture as visualised by the artist. By using this technique the artist was able to suggest structural image quality despite the fact that the image is strongly undersampled.

Fig. 1.1 To the concept of mental model.

Fig. 1.2 Mosaic at Katellorizo, Dodecaneses, Greece.

The line drawing of figure 1.1 and the half-tone picture of figure 1.2 demonstrate that every picture provides the observer with two kinds of data: structural, associated with the spatial distribution of pels with correlated luminous intensities; and metrical, associated with the luminous intensity of each pel. The display is required to reproduce the correct shades of grey of every picture element at the correspondingly located display elements (dels).

Electronic displays are programmable, i.e. they are able to show many different images and messages which may be totally unrelated as to their spatial structures. Therefore it is not possible to make the display element dependent on the structure of the image: pels and dels do not coincide. The

del density is uniform over the entire display in order to maintain a constant picture reproduction accuracy independent of the image/message features. In pictures which consist of a limited number of recurrent features, such as the strokes of some classes of stylised alpha-numeric symbols, cost and alignment advantages can be gained from dedicated display element organisation.
These artificial symbols can be constructed such, that low spatial sampling rates do not impair their legibility; compare figure 2.3 [1.1,1.2].

Most natural images are continuous; graphic and text images are either continuous (stroke writing mode on the CRT screen) or discrete (character segments, addressed dots in a matrix, rasterscan). Both matrix addressing and rasterscan lend themselves to reconstruct discrete images stored in computer memory. Rasterscan is well known for its application in television. Using the (monochrome) CRT with its continuous luminophore, digitised video can be presented by sweeping the electron beam continuously across the face, writing successive lines. The image reconstruction then is performed in X direction by continuous convolution of the CRT spot with the video signal, in Y direction by discrete convolution, causing different spatial filtering characteristics in X and Y directions. The radiance distribution of the spot is the "point spread function" (PSF) of the reconstruction filter. In figure 1.3 the monochrome CRT PSF; $F_r(x,y)$ is shown in a).

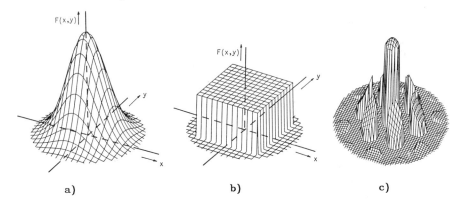

Fig. 1.3 Point spread functions of display technologies a) monochrome CRT, b) LCD, EL, c) colour CRT.

The light generating capacity of the CRT screen is only used at the locations of the written lines. These lines are blurred by the PSF of the spot. The radiance distribution being approximately normal or Gaussian, the cross section of the line (line spread function) is also Gaussian distributed. By choosing the separation or line pitch equal to 2σ (σ being the standard deviation of the line distribution) the screen radiates uniformly (merging raster). See also section 1.3.2, example 1.1.

The screen of the colour CRT is made up of triads of red (R), green (G) and blue (B) phosphors which are selected by proper modulation of the electron beams with respect to a shadow-mask, a metal foil with holes (chapter II.1). The electron beam covers several triads, the resulting PSF becomes multi-modal and the image reconstruction is no longer continuous in the horizontal direction, see figure 1.3c.

In flat panel technologies the continuous convolution also is not possible. In both the X and Y directions discrete convolution takes place. The PSF of the dels is a (usually square) block as shown in figure 1.3b. The display elements are separated by narrow gaps (figure 1.4a) resulting from the need to provide isolation between the electrodes which are deposited on both sides of the electro-optic medium (electroluminescence, liquid crystal, electrochromic materials). With plasma displays and light emitting diodes the existence of separation gaps also is natural. Even when every del would reflect or radiate the same luminance, the grid of separation gaps still is disturbingly visible; the only method to obtain merging of dels is to make the viewing distance so large that the visual angle of the del diagonal is too small for the resolving power of the eye (section 2.3.2). Slightly blurring the displayed image does not help much because the square dels then appear rounded at the corners, making the gap intersections relatively more visible, see figure 1.4b.

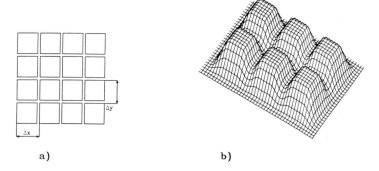

a)   b)

Fig. 1.4 a) Square del matrix arrangement. b) 3-D view of square dels after optical low pass filtering (blurring).

1.3. Description of the displayed picture

1.3.1. The sampling operation

In fully programmable matrix or rasterscan displays the dels are located at fixed points on a lattice; also their PSF is (or should be) independent of the del location on the lattice (shift invariant). The lattice best suited to computer addressing is orthogonal, i.e. the dels are arranged in horizontal lines and vertical rows with pitches Δx and Δy as shown in figure 1.4a.
The luminous fluxes $F_d(m,n)$ to be reflected or radiated at the lattice nodes $(m\Delta x, n\Delta y)$ shown in figure 1.5 are stored in memory in the form of numbers, i.e. metrical information. The total flux distribution F_D thus is given by

$$F_D = \sum_m \sum_n F_d(m,n) \tag{1.1}$$

Each display element $d(m,n)$ responds to $F_d(m,n)$; it spreads this flux according to its point spread distribution $PSF:F_r(x,y)$, see figure 1.3, thus making the numbers set (1.1) visible by providing spatial extent to it.
The resulting image should be a "good enough" mapping of the image $F_0(x,y)$ intended to be displayed.

Taking $F_0(x,y)$ as the input to the system, the members of (1.1) are obtained by multiplication of $F_0(x,y)$ by the two-dimensional Dirac comb $D(x,y)$ depicted in figure 1.5. $D(x,y)$ consists of Dirac functions of intensity 1 located at the nodes m,n of the lattice:

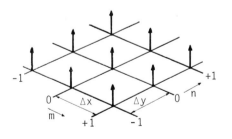

$$D(x,y) = \delta(x-m\Delta x, y-n\Delta y) \tag{1.2a}$$
$$D(x,y) = 1 \ : \ x=m\Delta x, y=n\Delta y \tag{1.2b}$$
$$D(x,y) = 0 \ : \ x\neq m\Delta x, y\neq n\Delta y \tag{1.2c}$$

The result is that

$$F_d(m,n) = F_0(x,y).D(x,y) \tag{1.3}$$

Fig. 1.5 Sampling matrix $D(x,y)$.

which basically describes the spatial sampling operation. The area a around each node (m,n) equals $a = \Delta x.\Delta y$.

The spreading of the flux by the $PSF:F_r(x,y)$ of the del or CRT spot causes the flux magnitude $F_d(m,n)$ destined for location $m\Delta x, n\Delta y$ to contribute to the flux

radiance of neighbouring locations at distances x-mΔx,y-nΔy; according to the
weighing law F_r(x-mΔx,y-nΔy). If F_r(x-mΔx,y-nΔy) is zero for $|$x-mΔx$|$ $>$ 0.5Δx
and $|$y-nΔy$|$ $>$ 0.5 Δy, there is no overlap; otherwise the resulting local
luminance at each location (x,y) equals the sum of the weighed neighbourhood
contributions:

$$F_D(x,y) = \sum_m \sum_n F_0(m\Delta x, n\Delta y) * F_r(x,y)$$

$$= \sum_m \sum_n \{F_d(m,n)\delta(x-m\Delta x, y-n\Delta y)\} * F_r(x,y)$$

$$= \sum_m \sum_n F_d(m,n)\{\delta(x-m\Delta x, y-n\Delta y) * F_r(x,y)\}$$

$$= \sum_m \sum_n F_d(m,n).F_r(x-m\Delta x, y-n\Delta y) \qquad (1.4a)$$

The operator $*$ signifies convolution.

The image is <u>convolved</u> with the point spread function, as shown in figure 1.6a
which depicts the convolution of the Dirac comb D(x,y) with a normally
distributed PSF; the shift distances Δx = Δy are twice the standard deviation
δ of the PSF distribution. For non-overlapping PSFs (liquid crystal, electro-

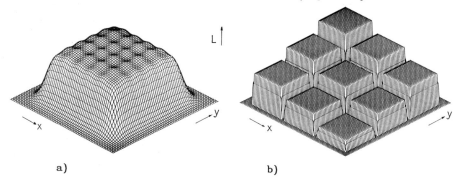

a)   b)

Fig. 1.6 Reconstruction of: a) uniform luminance region by monochrome CRT
b) luminance gradient by square del technology.

luminescence displays), spatial luminance gradients in the image to be
displayed produce sharp edges between dels as depicted in figure 1.6b.
When F_r(x,y) is not continuous, but defined as a matrix of weight factors
(e.g. display processing by computer) we deal with discrete convolution.
The spreading of the flux F_d(m,n) over the area a of the PSF (a = Δx.Δy for
square dels) yields a flux density or luminous exitance M = dF/da; the
resulting luminance L = π^{-1}.M (expression 1.20). The luminance distribution of
the displayed image at the face of a square del display thus is given by

$$L_D(x,y) = (\pi.\Delta x.\Delta y)^{-1} \sum_m \sum_n F_d(m,n).F_r(x-m\Delta x,y-n\Delta y) \qquad (1.4b)$$

1.3.2. Analysis in the spatial frequency domain; filtering

If the displayed picture does not vary during the time required to make the analysis it can be decomposed [1.3] into an infinite number of harmonic wave-planes like the one shown in figure 1.7, of different periods λ, shifted with respect to the origin over angles $\phi(\lambda)$ and rotated over angles $\theta(\lambda)$. The period is measured in the direction of the wave; the intersection of the wave with the x-L plane, yields a period $u^{-1} = \lambda \cos^{-1}\theta$, with the y-L plane $v^{-1} = \lambda \sin^{-1}\theta$. The inverse of the periods are the spatial frequencies of the waveplanes·

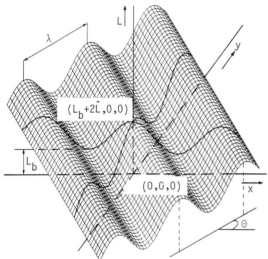

$$\rho = \lambda^{-1} = (u^2+v^2)^{\frac{1}{2}} \qquad (1.5a)$$
$$u = \rho \cos \theta \qquad (1.5b)$$
$$v = \rho \sin \theta \qquad (1.5c)$$

The waveplane thus can be defined by the parameters λ,ϕ and θ or u,v,ϕ. Assume that the foreground luminance is L_f, the background luminance L_b, as in figure 1.7. The mean luminance equals $(L_f+L_b)/2$; the peak amplitude $\hat{L} = (L_f-L_b)/2$.

Fig. 1.7 Two-dimensional harmonic function.

The description of the waveplane is

$$g(x,y) = \hat{L}[\cos\{2\pi(xu+yv)+\phi\}+]+L_b \qquad (1.6a)$$

or, with $\phi = -2\pi(x_0u+y_0v)$

$$g(x,y) = \hat{L}[\cos\{2\pi[(x-x_0)u+(y-y_0)v]\}+1]+L_b \qquad (1.6b)$$

which shows that the phase shift merely shifts the waveplane in the distance domain over respectively x_0 and y_0.

The complete picture is the sum of all the contributing waveplanes; when

8   *Part I: Conditioning*

accounting for possible distance shifts by using both cosine and sine wave-planes in our description:

$$g(x,y) = \int\limits_{+\infty}^{+\infty} \int\limits_{\infty}^{+\infty} G(u,v).\exp\{2\pi j(xu+yv)\}dudv \qquad (1.7)$$

with $j = \sqrt{-1}$ and $G(u,v)$ representing complex amplitudes of the cosine and sine waveplanes with spatial frequencies u and v:

$$G(u,v) = G_e(u,v)+j\ G_o(u,v) \qquad (1.8a)$$

where the indexes e,o signify "even" or cosine and "odd" or sine, respectively. The modulus of $G(u,v)$ calculated from

$$|G(u,v)|^2 = G_e^{\ 2}(u,v)+G_o^{\ 2}(u,v) = M^2(u,v) \qquad (1.8b)$$

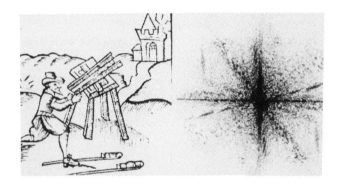

Fig. 1.8 Line drawing and its MSD (shown log density).

defines the magnitude spectral density (MSD) distribution $M(u,v)$ which describes the sum of the mean luminance and the contrast distribution as function of spatial frequency. An example is depicted in figure 1.8 where density represents log M; mean luminance is the density at the origin. The spatial phase spectral density (PSD) distribution, given by:

$$\phi(u,v) = \tan^{-1} \frac{G_o(u,v)}{G_e(u,v)} \qquad (1.8c)$$

describes the amount of distance shift of the harmonic waveplanes as function of their spatial frequency; as such $\phi(u,v)$ is associated with the structural information of the picture whereas $M(u,v)$ is associated with the metrical information. To illustrate this remark, figure 1.9a shows a natural image in its original form: in figure 1.9b the $\phi(u,v)$ has been retained equal to that of the original picture, but the M(u,v) was replaced by M(u,v) = 1. This modification has made the spatial contrast distribution independent of spatial frequency and thus (on the average!) independent of size/distance: it emphasises the importance of spatial phase.

a)   b)

Fig. 1.9 a) Original image CART (512x512) b) CART image with PSD: $\phi(u,v)$ identical to a), but MSD: M(u,v) = 1.

The reader will have come to the conclusion that G(u,v) is the Fourier transform of g(x,y):

$$G(u,v) = \int_{-\infty}^{+\infty} \int_{-\infty}^{+\infty} g(x,y) . \exp\{-2\pi j(xu+yv)\}dxdy \qquad (1.9a)$$

or $G(u,v) = F\{g(x,y)\}$ (1.9b)

and similarly, from expression (1.7)

$g(x,y) = F^{-1}\{G(u,v)\}$ (1.9c)

The transform pair (1.9 b,c) is quite useful because [1.3] more insight can be obtained in the consequences of convolution, as the transform changes the convolution operation into a multiplication and vice versa. Examples of convolution processes in display technology are: the effect of the spot PSF of a CRT (see example 1.1), diffusing (blurring) by anti-glare filters (example 1.2), impulse response of video amplifiers [1.4], responses of the optics and the retina of the visual system (section I.2.3.2). With some restrictions this is also true for discrete convolution.

Let $g(x,y) = L_D(x,y) = (\pi \Delta x . \Delta y)^{-1} \sum_n \sum_m F_d(m,n) . F_r(x-m\Delta x, y-n\Delta y)$ (1.4b) then $G(u,v) = F\{F_d(m,n)\} \cdot F\{F_r(x,y)\}$ (1.10)

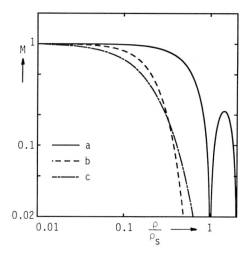

Fig. 1.10   Spatial frequency responses
of three types of low pass filters
a) square, uniform PSF; size = pitch
b) Gaussian PSF, standard deviation
equals to that of a)
c) diffusing anti-glare filter
(modified Bessel function, $\rho\ell$ = 2.5).

Thus the display reconstruction operation (1.4b) can be regarded as a filtering operation on the image data $F_d(m,n)$, as given by expression (1.10). The Fourier transform of $F_r(x,y)$ is called the optical transfer function (OTF). The modulus of the optical transfer function of $F_r(x,y)$ modifies the contrast of the picture as function of spatial frequency: therefore it is called the modulation transfer function (MTF). In the case that the reconstruction filter is continuous with a spatial area wider than the pel area, the MTF has a distinct low pass character as depicted in figure 1.10a. As the del/pel area ratio decreases towards one, the "amount of convolution" decreases and the effects of sampling become more visible.

Example 1.1. The merging raster effect.

Filler operations make use of the overlap of lines (figure 1.6a) written by the spot. Of course the remaining modulation depends on the angle made with respect to the direction of the lines. Along and across the crests, parallel to and perpendicular to the direction of the lines, the maximum contrast is smaller than the maximum contrast in the direction under 45°. To obtain a contrast C_v = 0.036 (see section 1.4.2) the line separation Δy should be about twice the standard deviation σ of the cross section of the line. Often the spot size is quoted as the diameter d at wich the luminance has dropped to 5 % of peak luminance [1.5] (d = 4.9σ).

Given the normal distribution

$$L(h,x-m\Delta x,y-n\Delta y) = L(h,m,n) \cdot \exp - \frac{(x-m\Delta x)^2 + (y-n\Delta y)^2}{2\sigma^2}$$ (1.11a)

$$= L(h,m,n).\exp - \frac{(x-m\Delta x)^2}{2\sigma^2} . \exp - \frac{(y-m\Delta y)^2}{2\sigma^2} \qquad (1.11b)$$

(separable function) the max. perpendicular direction contrast is determined by Gaussians $L(h,m).\exp - \frac{(y-m\Delta y)^2}{2\sigma^2}$ shifted over distances $\Delta y = 2\sigma$. Applying expression (1.4b):

$$L_D(x,y) = (\pi.\Delta x.\Delta y)^{-1} \sum_m \sum_n F_d(m,n).\exp - \{\frac{(x-m\Delta x)^2+(y-n\Delta y)^2}{2\sigma^2}\} \text{ or}$$

$$G(u,v) = F\{F_d(m,n)\}.F\{.\exp - \frac{x^2+y^2}{2\sigma^2}\}$$

$$= F\{F_d(m,n)\}.\exp - 2\pi^2\sigma^2(u^2+v^2) \qquad (1.12)$$

In a uniform filled field, for $u = 0$, the spatial frequency response (MTF) in v direction is given by $\exp -2\pi^2\sigma^2 v^2$ (figure 1.10b). At the Nyquist frequency $v_N = 0.5 v_s$ the MTF (contrast response) still is 0.29 or −11 db.

When in the image to be displayed $F_0(x,y)$ strong components are present with frequencies $v > v_N$, interference (Moiré) patterns can result. Therefore, the MTF in v direction should be limited (by design or by prefiltering) to frequencies below v_N (see section 1.3.3). In the horizontal direction the CRT writes a continuous line; interference does not occur but the resulting MTF is also influenced by the video amplifier response [1.4].

Example 1.2. The diffusing (anti-glare) filter.
In section 1.4.1 the PSF of an optical diffuser is derived. The input to the diffuser is a point source with intensity I, the diffuser transmits a fraction $\sigma < 1$ which is scattered according to a cosine law (Lambertian radiation, section 1.4.1), the remaining part is either diffusely reflected or absorbed.

The luminance distribution at the diffuser surface around the location determined by the normal through the point source, at distance x to that location, is (1.4.1):

$$L(x) = \sigma I \frac{\ell}{\pi(x^2+\ell^2)^{3/2}} \qquad (1.13)$$

with ℓ the distance of the point source to the diffuser surface. Since the PSF is circular symmetric, we take polar coordinates for the spatial frequency domain with radial frequency $\rho^2 = u^2 + v^2$. Transforming expression (1.13) yields

$$\text{MTF} \rightarrow 4\rho\sigma \; I.K_1(\rho\ell) \tag{1.14a}$$

with $K_1(\rho\ell)$ being the modified 1st order Bessel function of height;

$$\lim_{\rho \to 0} 4\rho\sigma \; I.K_1(\rho\ell) = 4\sigma \; I/\ell \tag{1.14b}$$

This MTF is shown in figure 1.10c. Since the argument of the modified Bessel function $K_1(\rho\ell)$ is given by the product $\rho.\ell$, the attenuation of high spatial frequencies (blur) depends strongly on ℓ.

In practice there are no point sources. Placed in front of a CRT, anti-glare filters face normally distributed sources. The CRT spot PSF convolves with the diffuser PSF. In the spatial frequency domain the anti-glare filter/CRT combined response is obtained by multiplication of expressions (1.12) and (1.14a); the spatial distance domain response is obtained by inverse Fourier transform of that product.

The PSF of the anti-glare filter can improve the visual characteristics of a del PSF which has a central dip as is the case with some gas discharge displays. When properly designed, the dip is filled. In the spatial frequency domain the central dip PSF has a MSD with high frequencies enhanced. The low pass characteristic of the anti-glare filter can correct this, making the resultant MSD flat up to the cut-off frequency, with little loss of experienced sharpness.

1.3.3. Analysis in the spatial frequency domain: interference

In section 1.3.1, expression (1.3), it was stated that the sampled version of $F_0(x,y)$ can be described as the result of multiplication of the original picture $F_0(x,y)$ by the two-dimensional Dirac comb $D(x,y)$ shown in figure 1.5. The Fourier transform of a Dirac comb in the (x,y) domain again yields a Dirac comb in the (u,v) domain, with spacings $u_s = (\Delta x)^{-1}$ and $v_s = (\Delta y)^{-1}$.

The multiplication of $F_0(x,y)$ with $D(x,y)$ is in the spatial frequency domain described [1.3] by the convolution of the Fourier transforms of $F_0(x,y)$ and $D(x,y)$ (the operator $*$ again signifies convolution):

$$F\{F_0(x,y).D(x,y)\} = F\{F_0(x,y)\}*F\{D(x,y)\} \qquad (1.15)$$

Thus the MSD of $F_0(x,y)$ is convolved with the Dirac comb in the spatial frequency domain, its centres located at the nodes (iu_s, jv_s), $i,j = 0,1,2,3,$

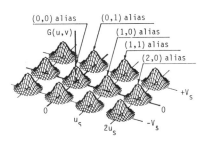

as depicted in figure 1.11. Each shifted representative (alias) of the MSD of $F_0(x,y)$ carries the whole of the information of the image. Only the central representative (0,0 alias) has meaning to the display designer. If the aliases overlap, i.e. exceed the boundaries $(i\pm0.5)u_s$, $(j\pm0.5)v_s$, even an ideal reconstruction with sharp cut-off at $\pm 0.5 u_s$ and $\pm 0.5 v_s$

Fig. 1.11 Frequency domain representation sampled image; u_s and v_s are the sampling frequencies.

cannot prevent corruption of the (0,0) alias by its neighbours (1,0;-1,0;0,1; 0,-1) and (1,1;-1,1;1,-1,;-1,-1) and so on. Aliasing effects are seen in figure 1.12b (the hyperbels of wider sector width, just visible in the result of low pass filtering of figure 1.12a). To avoid such interference, the image to be displayed must be constructed such that its magnitude spectral density remains wi⸱hin the boundaries of the (0,0) alias or, when not, that its spatial frequencies are (sub)multiples of the spatial sampling frequency and also are phase locked: conforming to the condition that line widths are multiples of del size and optimally positioned. That condition can in an orthogonal matrix only be fulfilled in two orthogonal (i.e. the x,y) directions, e.g. for symbols like T and L. In all other directions one must accept the effects of aliasing (jaggedness, figures 1.12a, grey level interference, figure 1.12b; or one may avoid jaggies by local distortion of the orthogonality of the grid in an appropriate manner by the technique of del shifting [1.6] or of apparent centre of radiance shifting [1.7].

In practice the reconstruction PSF:$F_r(x,y)$ of a del is symmetrical, described by even functions. Thus in (1.8): $G_0 = 0$, so that $\phi(u,v) = 0$ (expression 1.8c) and the del response is defined by the MTF:$M_r(u,v)$ only. The situation with square del, non-overlapping PSF reconstruction (filtering) is shown in figure 1.10a.

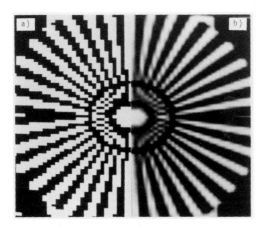

Fig. 1.12 Undersampled spoke pattern
a) reconstruction by square del matrix
showing strong aliasing in distortion
of spokes and jaggedness of edges
b) image of a) after low pass filtering
with grey level representation. Note
the light grey, low contrast,
"hyperbels" due to aliasing.

The MTF associated with a uniformly radiating del is given by a sync function with its first zero located at the sampling frequency. Considerable power of 1st order aliases leaks through; even at higher order aliases the transmission is non-zero. How much power this actually represents depends on the spectral distribution (MSD) of the input image $F_0(x,y)$.
Note: this $M_r(u,v)$ also folds around multiples of the sampling frequency!

The error introduced by the characteristics of the reconstruction filter itself can be easily calculated:

$$P(\epsilon_r) = \int \int F\{F_0(x,y)\}^2 . [1-F\{F_r(x,y)\}^2]dudv \qquad (1.16a)$$

or, normalised to the image input power, the error ϵ_r becomes

$$\epsilon_r = \left| \frac{\int \int |G(u,v)|^2 . \{1-M_r^2(u,v)\}dudv}{\int \int |G(u,v)|^2 \ dudv} \right|^{\frac{1}{2}} \qquad (1.16b)$$

The fraction transmitted by the reconstruction filter is

$$\tau_d = \left| \frac{\int \int |G(u,v)|^2 . M_r^2(u,v)dudv}{\int \int |G(u,v)|^2 \ dudv} \right|^{\frac{1}{2}} \qquad (1.17)$$

A lower cut-off frequency means that less aliasing products (figure 1.13b) are transmitted, but also the filter error (figure 1.13a) increases and vice versa. It turns out that it is possible, given the MSD of the image, to minimise the total error due to both error components caused by sampling (ϵ_s) and by reconstruction filtering (ϵ_r). Evidence is figure 1.13c, which shows

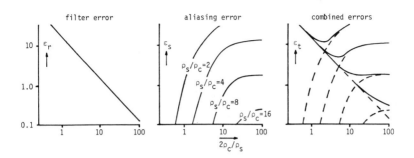

Fig. 1.13 Reconstruction errors in "random" image
a) filter error, b) interference error (aliasing), c) total error
Abcissa: ratio of reconstruction filter cut-off frequency ρ_c over image
(MSD) cut-off frequency ρ_o (bandwith).

the combined error $\epsilon_t = (\epsilon_r^2 + \epsilon_s^2)^{\frac{1}{2}}$ as function of the ratio of the cut-off
frequency ρ_c of the reconstruction filter over the bandwidth ρ_0 of the image
signal, for four sampling frequencies. The magnitude of the optimised error
still is considerable in this case, where the image signal is assumed to be
stochastic with MSD flat up to ρ_0, roll-off 6 db/octave (first order signal).
Improvements are obtained by prefiltering the image signal (i.e. increasing
its order so that roll-off is steeper) and by increasing the sampling rate;
the required system configuration is given in figure 1.14.
The error ϵ_t can also be kept small by construction of synthetic images with
suitably chosen spatial frequency and phase content (graphics, certain text
fonts).

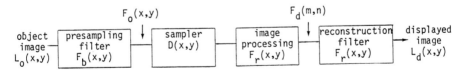

Fig. 1.14 Operations in digital imaging system.

1.3.4. Image quality

Image quality is a subjective judgement of imperfections in the displayed
image. It is generally felt that this subjectivity is a serious drawback when

comparison of the applicability and acceptability of display systems and
electro-optic technologies is necessary. In optics it is possible to judge the
quality of an imaging system by specifying a number of attributes. The spatial
fidelity can be inferred by data on the PSF, internal reflections, geometric
distorsions, chromatic abberations; the radio-metric qualities by F number,
transmission factor, and so on. Although these factors may vary with angle
they are strictly invariant with respect to time and luminous flux. That is
not true in displays: the electro-optical processes are non-linear and, since
perceived image quality involves visual properties, also luminous conditions
and time should be considered.

In spite of this several attempts have been made to relate the qualities of a
display experienced by a standard observer to a number of attributes of the
display in its environment. One may consider e.g. single and multiple
reflections at the display face, its cover or filter(s), and internal
reflections in the display; the contrast available in the "normal"
environmental conditions; and the spatial resolution afforded. Considerations
on reflections and contrast are given in the next section (1.4); the spatial
imaging fidelity attribute is discussed below.

The image is sampled and reconstructed by the display elements (1.3.1) through
the del PSF:$F_r(x,y)$. The grid formed by the separation gaps (figure 1.4a)
multiplies with the image to be displayed. It has two properties: firstly it
reduces the display light output by a fraction determined by the ratio of the
width of the separation gap over the grid period; and secondly the spatial
frequency components of the grid, which are multiples of the spatial sampling
frequency, convolve with the spatial frequency content of the image to be
displayed. Since the spectral distribution of that image either must be
restricted to spatial frequencies below the Nyquist frequencies u_N and v_N, or
must be confined to submultiples of the Nyquist frequencies (in the case of
artificial images like alpha-numerics); no information is lost due to the
existence of the grid. It may be a nuisance if no adequate post filtering
(optical, e.g. by the eye), see figure 1.4b, is provided.

For the square del technologies, the response without anti-aliasing filtering
is depicted in figures 1.10a and 1.12a; interference (Moiré, jaggedness) will
occur. With anti-aliasing filtering, keeping the spatial frequency pass band
well below the Nyquist frequencies, the performance of the square del
technologies becomes comparable to the overlapping PSF technologies like the
CRT. Even jaggedness is reduced because of apparent centre of radiance
shifting. However, the resulting image quality is not only dependent on the

(system) MTF, because for economy the total error ϵ_t due to both the actions of sampling (ϵ_s) and the combination of anti-aliasing and reconstruction filtering (ϵ_f) will be minimised.

In the case that the aliasing error is deliberately kept low the system MTF, in combination with the characteristics of the visual system, can be a good measure of spatial fidelity [1.8] but only in a statistical sense since all **local** effects are averaged into one global property. The system MTF describes the ability of the system to transfer input signal contrast to the luminous contrast of the display as a function of spatial frequency. The standard observers' visual contrast sensitivity (CS) also is a function of spatial frequency and, moreover, has thresholds which must be exceeded (just noticeable differences, JND); both are a function of the luminous conditions (section 1.4.3). Thus, apart from the standard observer, also a standard luminous environment must be defined. A review of several proposed measures of image quality is given in [1.8].

1.4. Analysis in the luminance domain

1.4.1. Luminous variables

The electro-optic conversion process in the display forms the image by producing luminous patterns. The light results from reflection (electro-mechanical, liquid crystal, electrochromic displays); emission (cathode ray tube, gas discharge or electroluminescent, light emitting diode, incandescent displays); transmission (liquid crystal, magneto-optic displays) and scatter (electrowetting display) and combinations thereof. This list of technologies is not complete, see Part II for more details.
The strength of the optical effects is denoted by coefficients which range between 0 and 1: reflection ρ, emission ϵ, transmission τ, scatter σ. Absorption (α) is also involved but a dependent variable since the sum of the coefficients $\alpha + \rho + \tau + \sigma$ equals 1.

The luminous part Φ_v of the total power Φ_e radiated by the display is the desired output. The luminous sensation caused by the physical power [1.9] is converted to a luminous power scale with the unit lumen instead of Watt. Due to the wavelength dependency of the eye sensitivity, see figure 1.15, the conversion factor depends on the wavelength distribution of Φ_v. The wavelengths λ associated with luminous sensation are in the range $400<\lambda<730$

nanometer. These boundaries are somewhat arbitrary; at 400 nm and 730 nm the absolute spectral luminous efficacy $K(\lambda)$ in photopic vision has dropped to 0.33 lumen per Watt or 0.05 % of maximum. Assuming constant radiation density $\Phi_e(\lambda) = C$ for $400<\lambda<730$, the luminous efficacy $K \simeq 200$ ℓm W^{-1}; assuming instead black body radiation distribution, $K = 95$ for $T = 6600$ K(elvin), $K = 20$ for $T = 3000$ K, $K \simeq 1$ for $T = 1900$ K:

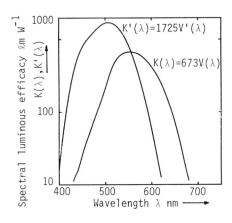

$$K = \frac{\Phi_v}{\Phi_e} = \frac{\int K(\lambda).\Phi_e(\lambda)d\lambda}{\int \Phi_e(\lambda)d\lambda} \qquad (1.18)$$

with Φ_v in lumen, Φ_e in Watt; both are a measure of power.

Fig. 1.15 Luminous efficacies of average healthy young eye
a) in photopic range $K(\lambda)$
b) in scotopic range $K'(\lambda)$.

Definitions

Luminous intensity $I_v = \dfrac{\partial \Phi_v}{\partial \omega}$, (1.19a)

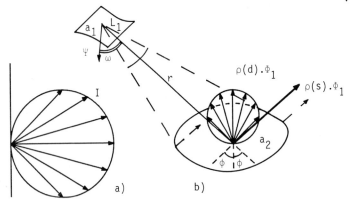

Fig. 1.16 a) intensity distribution of ideal diffusing surface (constant luminance) b) calculation of diffuse $\rho(d)$ and specular $\rho(s)$ reflections.

where ω is the solid angle through which the flux Φ_v from a point source is radiated, I is measured in candela. Unit: lumen per steradian.

Luminance $L = dI_v(da\ \cos\ \theta)^{-1}$ or: (1.19b) the luminance equals the luminous intensity dI_v per projected area $da.\cos\theta$ in the direction of observation. See figure 1.16a for the required intensity distribution associated with a perfectly diffusing surface: L has become independent of θ (Lamberts cosine law). Unit: lumen per steradian per m^2 (or candela per m^2: $cd\ m^{-2}$).

Illuminance, exitance: flux density through a surface area of 1 m^2.

$$\text{Illuminance } E = \frac{\partial\Phi_v}{\partial a} \text{ (lux) and the} \qquad (1.19c)$$

$$\text{exitance } M = \frac{\partial\Phi_v}{\partial a} (\ell m\ m^{-2}) \qquad (1.19d)$$

Since the perfectly diffuse surface emits in 2π steradians according to the cosine law,

$$M = \int_0^{\pi/2} (L\cos\theta).2\pi\sin\theta d\theta = \pi L \qquad (1.20)$$

Example 1.3. Diffuse reflective surface and luminous power.

On a clear day, outside but not in direct sunlight, the illuminance E is about 10^4 $\ell m\ m^{-2}$. Assume a display device, emittive type with white spectrum, 5 % electro-optic conversion efficiency, A_4 screen with diffuse reflection coefficient $\rho = 0.3$. To make the electro-optical exitance equal to the reflected exitance $M_{rad} = M_{refl} = 0.3\ E = 3000\ \ell m\ m^{-2}$.

Over the display area $(0.06\ m^2)$ the power equals 180 ℓm. For a white spectrum $K = 200\ \ell m\ W^{-1}$, so that the electrical input to produce the desired electro-optical output (all dels "on") at the given efficiency equals 18 W. In dense text, the proportion of "on" to "off" dels is between 10 and 20 %; emittive type displays with negative contrast use only a few watts.

For any spectral distribution other than white, K is smaller. For instance, for the incandescent display with "on" filaments at 1900 K, the same active display area requires 200 times more power!

Example 1.4. Specular and diffuse reflection.

A luminous surface with area a_1 and luminance L_1 irradiates a second surface with area a_2, located at distance r measured along an interconnecting line. The angles of inclination between that line and the normals to the surface are ψ for a_1 and ϕ for a_2 respectively, see figure 1.16b. The solid angle ω subtending the apparent area a_2 from any point of surface a_1 is:

$\omega = r^{-2}.a_2.\cos\phi$.

The intensity of the emitted radiation (expression 1.19b) is $I_1 = L_1 a_1 \cos\psi$. For small areas a_1, ω can be considered constant, not depending on ψ. Thus the flux Φ_1 received by a_2 can be determined: $\Phi_1 = I_1 . \omega = L_1 a_1 a_2 r^{-2} \cos\psi\cos\phi$.

For specular reflection this flux is reflected about the normal to a_2 in the direction $- \phi$. Is the surface a_2 partly specular, partly diffuse, the specular reflected component $\Phi_2(s) = \rho(s).\Phi_1$, with $\rho(s)$ the specular reflection coefficient. The intensity $I_2(s)$ of the specular part of the reflected light is given by $I_2(s) = \Phi_2(s).\omega^{-1} = \rho(s).I_1$, since the solid angle of flux $\Phi_2(s)$ is the same as the angle ω before reflection. That being the case, $\Phi_2(s)$ can be considered as the fraction $\rho(s)$ passing through surface a_2. The diffuse surface a_1 is observed through a_2 and the observed luminance $L_2(s) = I_2(s).(a_1.\cos\psi)^{-1}$. Consequently, $L_2(s) = \rho(s).L_1(s)$ or, when taking colour dependence into account:

$$L_2(s) = \int \rho(s,\lambda).L_1(\lambda).d\lambda \tag{1.21a}$$

For diffuse reflection at surface a_2, the impinging flux is reradiated in all directions according to the cosine law. The irradiance E_2 of surface a_2 is $E_2 = \Phi_1.a_2^{-1} = L_1.r^{-2}.a_1.\cos\psi\cos\phi$. Assume a diffuse reflection coefficient $\rho(d)$, then the exitance $M_2(d) = \rho(d).E_2$ and the luminance $L_2(d)$ of surface a_2 becomes $L_2(d) = \rho(d).E_2.\pi^{-1}$ or

$$L_2(d) = a_1.(\pi r^2)^{-1}\cos\psi\cos\phi.\int\rho(d,\lambda).L_1(\lambda).d\lambda \tag{1.21b}$$

Example 1.5. The ratio of specular and diffuse reflections.
To illustrate the importance of expressions (1.21 a and b), consider a practical case where a_1 is the apparent area of a lamp with diffusing shade at uniform luminance L_1, a_2 is the surface of a flat CRT screen with specular reflection coefficient $\rho(s) = 0.04$ and diffuse reflection coefficient $\rho(d) = 0.25$. The specularly reflected lamp shade is seen at the luminance $L_2(s)$, whereas the part of the screen surrounding the reflected image obtains the luminance $L_2(d)$. The ratio $L_2(s)/L_2(d) = C_r$ is the contrast ratio of the reflected image.

$$L_2(s)/L_2(d) = \frac{\pi.\int\rho(s,\lambda).L_1(\lambda).d\lambda}{\mu\cos\psi\cos\phi.\int\rho(d,\lambda).L_1(\lambda).d\lambda} \tag{1.21c}$$

where $\mu = a_1.r^{-2}$, the solid angle subtending area a_1 from any point of surface a_2. With the screen passive, i.e. non-emitting, apparent shade area of 0.04 m^2 at distance r = 2 m, cosψcosϕ = 0.5 and ρ(s,λ) and ρ(d,λ) having the same spectral distribution characteristics, the so obtained contrast ratio C$_r$ = 40! Of course, this disturbingly high contrast ratio calculated here occurs only when the lamp is the single source of light illuminating the dead screen. With other light sources in the room, and with a luminance background produced by an active screen, the contrast ratio drops rapidly, depending on the luminances involved. For instance, if under the same conditions the shade luminance L$_1$= 200 cd m^{-2} and the background luminance L$_b$ of the active screen L$_b$ = 10 cd m^{-1}, the contrast ratio of shade against screen drops to 1.8!

In the expression (1.21c), the solid angle μ is in the denominator. Thus, reflection annoyance increases with the ratio of luminance over area of the reflected object: small objects produce difficult to suppress gleams in the screen. Also objects which are small in one dimension, like fluorescent tubes. If their presence cannot be avoided, anti-glare filters placed in front of the screen can improve the situation appreciably.

Example 1.6. Anti-glare filtering.
Anti-glare filters have been developed operating on different principles: neutral density filters, scattering layers, polarising filters, colour filters, micro-mesh filters and so on [1.10,1.11]. In this example only the first two are analysed.

The neutral density filter simply attenuates the light passing through; the transmission coefficient τ typically is τ = 0.3. Light from external sources passes twice through the filter when reflected by the display, the display light only once, thus improving the contrast ratio roughly by τ^{-1} = 3.3. However, such filters have two surfaces; if untreated themselves the problem is merely displaced from the display surface to the filter surface. Moreover the distance from the light emitting surface of the display to the front reflecting surface of the filter is increased, which usually tends to worsen the side-effects of anti-glare measures.

The better solution is to treat the surface of the display device. Such measures are particularly necessary for e.g. overlay touch panels. Early designs just increased the ratio of ρ(d)/ρ(s) by providing texture to the front surface, or an extra diffusing layer on the display front surface. This anti-glare filter, however, also scatters the display light, blurring the image (example 1.2).
To calculate the PSF causing the blurring effect of a perfectly diffusing

layer, see figure 1.17, assume a point source s at a distance ℓ from the
filter surface. The solid angle ψ subtending area da $= \omega.x.dx$ equals
$\psi = r^{-2}.\omega.x.dx.\cos\phi$. With source intensity I, the flux $d\Phi$ received by area
da is $d\Phi = I.r^{-2}.\omega.x.\cos\phi.dx$, the irradiance $E = d\Phi/da = I.r^{-2}\cos\phi$.

With a diffusion coefficient σ, the
luminance L at da becomes
$L = \sigma.I(\pi r^2)^{-1}.\cos\phi$.
Substituting $\cos\phi = \ell.r^{-1}$ and $r^2 = x^2 + \ell^2$
yields expression (1.13):

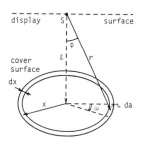

$$L = \sigma I \frac{\ell}{\pi(x^2+\ell^2)^{3/2}} \qquad (1.13)$$

This expression only covers the diffusion
characteristic of the filter, not other
properties like (colour dependent)

Fig. 1.17 Calculation of the   contrast enhancement. The point spread
PSF of a diffusing layer.   function is flat topped for $x \ll \ell$, thus
                                      broadening the actual dimensions of the
source. To minimise blur, ℓ should be small with respect to image dimensions
(lines widths, character strokes, dots). The perfect diffuser is not an
ideal add-on anti-glare filter for high resolution display devices.

1.4.2. Contrast definitions, contrast enhancement filters

The ratio of two luminances L_2 and L_1 as used in example 1.5, is the contrast

ratio: $C_r = L_2/L_1$ \hfill (1.22a)

widely used in engineering. Where the normalised difference of two luminance
levels is important, as in exponential functions, logarithmic detectors
(vision), visual contrast:

$C_v = (L_2-L_1)/L_1 = C_r-1$ \hfill (1.22b)

is preferred. Thirdly, if one wishes to describe a displayed image in terms of
deviations from a mean luminance level across the image, the modulation depth

$MD = (L_2-L_1)/(2+L_1) = (C_r-1)/(C_r+1) = C_v/(C_v+2)$ \hfill (1.22c)

is used. These definitions are well suited to describe bi-level achromatic or
monochromatic images, and experiments in threshold detection situations, where

only two levels are involved. When dealing with colour displays, the notion of colour contrast must be included; as colour is a psychological sensation it is dealt with under ergonomics (human factors) in chapter I.2. When considering achromatic or monochromatic displays with multiple grey levels, the statistics of the image become involved: the probability of occurence of each grey level (grey level histogram). On may, for instance, choose the first moment or average luminance L_a in combination with the second moment or variance σ^2.

In comparison with the bi-level definitions, L_a relates to $(L_2+L_1)/2$ and σ to $(L_2-L_1)/2$. Thus it seems appropriate to define:

$$MD = \sigma/L_a; \quad C_r - (L_a + \sigma)/(L_a - \sigma); \quad C_v = 2\sigma/(L_a - \sigma) \qquad (1.22d)$$

For $\sigma \ll L_a$, $C_r \simeq 2$ MD.

In table 1.1 below a comparison is made between C_r, C_v and MD. Also included is CS: contrast sensitivity, i.e. the ability to discern a just noticeable difference (JND). CS is the inverse of the threshold modulation depth necessary to produce positive recognition 50 % of the time: $CS = |MD|_{th}^{-1}$.

Table 1.1

C_r	C_v	MD	CS
0.1	−0.9	−0.818	1.22
0.2	−0.8	−0.666	1.5
0.5	−0.5	−0.333	3.0
1.0	0	0	∞
2.0	1.0	+0.333	3.0
5.0	4.0	+0.666	1.5
10.0	9.0	+0.818	1.22

Example 1.5, section 1.4.1, describes how the contrast ratio of specular over diffuse reflection at the face of a non-operating display, in an otherwise dark room, can be reduced by adding another source of light with uniform spatial distribution, in this case the background luminance of the display itself. The other source of light can also be a relatively large window situated at an angle such that the specular reflection is not visible to the observer of the display. The inverse process, contrast enhancement for the

desired image with respect to its background, is also possible by the choice of a combination of optical filters in front of the image forming layer of the display, at the inside or outside of the display cover glass. As was shown in example 1.5 the luminances associated with specular reflections are much higher (because of geometries) than those obtained from diffuse reflections. They should be our first concern, either by directing specular reflections away the observer, or by special anti-reflective coatings or filters. Typical reflection data are shown in table 1.2.

Table 1.2

| | reflection in % | |
	specular	diffuse
flat glass	4	0.2
flat plastic	2	
etched glass		40 to 60
λ/4 coated glass	0.1 to 0.25	0.02

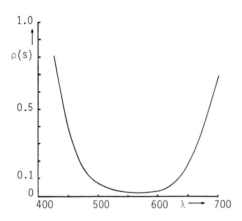

Fig. 1.18 Front surface reflectance of double sided quarter wave coated glass (data OCLI).

Because it does not induce blur as etched glass does (example 1.6) the high efficiency antireflection (HEA) coating based on quart wavelength layers (see figure 1.18 [1.12]) is to be preferred for treatment of the outside of the display front cover, regardless of combinations with other filters which are necessary e.g. to suppress internal specular reflections of the display or to increase contrast by reducing diffuse reflections. Of course, such anti-reflection filters can also be used to suppress internal reflections in the faceplate which otherwise would induce "halo" effects [1.13].

Such filters can be based on methods like:

- polarisation: specular reflection changes the direction of polarisation of light. This fact can be used to attenuate the strength of the reflected image by a polarisation selective filter (usually a circular polariser);
- geometry dependence: light reflected off the internal parts of the display travels twice through the filter, light emitted by the display only once.

Thus a neutral density (ND) filter with transmission coefficient $\tau < 1$ can increase contrast.

Also filters constructed of woven nylon fibers and fused fiber plates which, besides a transmittance < 1 also exhibit directional properties, belong to this category;

- colour dependence: the spectral distribution of $\tau(\lambda)$ is matched to the spectral distribution of the light emitted by the display, favouring the latter over reflections of environmental light providing that this has a very different spectral distribution. Thus luminance contrast is increased more than with the ND filter, however at the expense of colour contrast since the hue of the background now appears the same as that of the display foreground [1.12].

Example 1.7. Contrast enhancement, anti-glare.

Assume a display with foreground luminance $L_f(\lambda)$ and background luminance $L_b(\lambda)$: e.g. bi-level or bi-colour. The reflection coefficients of the display materials are given by a diffuse part $\rho_1(d,\lambda)$ and a specular part $\rho_1(s,\lambda)$. The display is provided with a front cover plate which consists of a colour filter with directionally dependent (θ is viewing angle) transmittance $\tau_1(\lambda,\theta)$; coated with $\lambda/4$ layers at the outside to suppress direct reflections. The anti-reflection layers have transmittance $\tau_2(\lambda)$, specular reflectance $\rho_2(s,\lambda)$ and diffuse reflectance $\rho_2(d,\lambda)$.

Assume further that this display is situated in a workplace where the observer of the display cannot avoid to see the reflection of a lit sign with luminance $L_e(\lambda)$. Nearby is a window which produces diffuse illuminance $E(\lambda)$ at the front of the display.

The block diagram of this optical situation is depicted in figure 1.19, which serves to calculate the contributing luminances as shown. Quantitative effects which are negligeable (such as $\rho_2(d,\lambda)$) have been omitted. It follows that

$L_1(\theta) = \int \tau_1(\lambda,\theta).\tau_2(\lambda).L_f(\lambda).d\lambda$ (transmitted part of L_f).

$L_2(\theta) = \int \tau_1(\lambda,\theta).\tau_2(\lambda).L_b(\lambda).d\lambda$ (transmitted part of L_b).

$L_3(\theta) = \int \{\tau_1^2(\lambda.\theta).\tau_2^2(\lambda).\rho_1(s,\lambda)+\rho_2(s,\lambda)\}.L_e(\lambda).d\lambda$ (specular reflections).

$L_4(\theta) = \pi^{-1} \int \tau_1^2(\lambda,\theta).\tau_2^2(\lambda).\rho_1(d,\lambda).E(\lambda).d\lambda$ (diffuse reflections).

Now three contrasts must be considered:

$G_{v1} = \{(L_1+L_4)-(L_2+L_4)\}/(L_2+L_4)$, the foreground (L_1+L_4) against the background next to the reflection of the sign (L_2+L_4); similarly

$C_{v2} = \{(L_1+L_3+L_4)-(L_2+L_3+L_4)\}/(L_2+L_3+L_4)$, the contrast of the displayed image against the background augmented by both the window and the lit sign:

$C_{v3} = \{(L_2+L_3+L_4)-(L_2+L_4)\}/(L_2+L_4)$, the contrast of the image of the lit sign against the diffuse background. The general expressions for C_{v1}, C_{v2} and C_{v3} are worked out below, their numerical value obtained by using the appropriate parameters τ_1, τ_2 and ρ_2 and variables L_f, L_b, L_e and E. This will be done for three optical situations: no filter (C_v), ND filter (C_v^*) and colour filter (C_v^{**})

$$C_{v1}(\theta) = \frac{\int \tau_1(\lambda,\theta).\tau_2(\lambda).\{L_f(\lambda)-L_b(\lambda)\}.d\lambda}{\int \tau_1(\lambda,\theta)\tau_2(\lambda).\{L_b(\lambda)+\pi^{-1}\tau_1(\lambda,\theta).\tau_2(\lambda).\rho_1(d,\lambda).E(\lambda)\}.d\lambda}$$

$$C_{v2}(\theta) = \frac{\int \tau(\lambda,\theta).\tau(\lambda).\{L_f(\lambda)-L_b(\lambda)\}.d\lambda}{\int [\rho_2(s,\lambda).L_e(\lambda)+\tau_1(\lambda,\theta).\tau_2(\lambda).[L_b(\lambda)+\tau_1(\lambda,\theta).\tau_2(\lambda) \times}$$
$$\times \{\pi^{-1}\rho_1(d,\lambda).E(\lambda)+\rho_1(s,\lambda).L_e(\lambda)\}]]d\lambda$$

$$C_{v3}(\theta) = \frac{\int \{\tau_1^2(\lambda,\theta).\tau_2^2(\lambda).\rho_1(s,d)+\rho_2(s,d)\}.L_e(\lambda).d\lambda}{\int \tau_1(\lambda,\theta).\tau_2(\lambda).\{L_b(\lambda)+\pi^{-1}\tau_1(\lambda,\theta).\tau_2(\lambda).\rho_1(d,\lambda).E(\lambda)\}.d\lambda}$$

Assume now that the luminances $L_f = \int L_f(\lambda).d\lambda = 100$ cd m^{-2}, $L_b = \int L_b(\lambda)d\lambda = 10$ cd m^{-2} and $L_e = \int L_e(\lambda).d\lambda = 500$ cd m^{-2}; the illuminance E at the display the face by diffuse light from the window: $E = \int E(\lambda).d\lambda = 1000$ lux.

The typical parameters of a glass display face are:

$\rho_1(s) = \int \rho_1(s,\lambda).d\lambda = 0.04$, $\rho_1(d) = \int \rho_1(d,\lambda).d\lambda = 0.25$.

For the situation without filter $(\tau_1 = 1, \tau_2 = 1, \rho_2 = 0)$ and $\rho_1(d)$, $\rho_1(s)$ independent of λ, the three contrasts are;

$$C_{v1} = \frac{\int L_f(\lambda).d\lambda - \int L_b(\lambda).d\lambda}{\int L_b(\lambda).d\lambda + \pi^{-1} \int \rho_1(d.\lambda).E(\lambda).d\lambda} = \frac{100-10}{10+0.25.\pi^{-1}.10^3} = \frac{90}{10+79.5} = 1.0$$

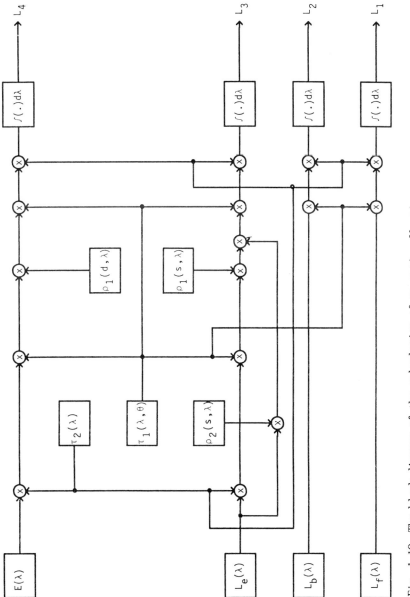

Fig. 1.19 The block diagram of the calculation of an anti-reflection, contrast enhancement filter (display cover).

$$C_{v2} = \frac{90}{89.5 + \rho_1(s) . \int L_e(\lambda).d\lambda} = \frac{90}{89.5 + 20} = 0.82$$

$$C_{v3} = \frac{20}{89.5} = 0.22$$

Conclusion: the contrasts C_{v1} and C_{v2} are sufficient to read the display, although not comfortably (see section 1.4.3 and I.2.4); but the image of the lit sign is a nuisance which, in prolonged use, can become intolerable. To improve the situation, a neutral density filter, anti-reflection coated, is placed in front of the display face.

Assume $\tau_1 = 0.3$, $\tau_2 = 0.95$, $\rho_2(s) = 0.002$. Then:

$$C_{v1}^* = \frac{\tau_1.\tau_2.90}{\tau_1.\tau_2.10 + (\tau_1.\tau_2)^2.79.5} = \frac{90}{10 + (0.285).(79.5)} = 2.76$$

$$C_{v2}^* = \frac{0.285.90}{1 + 2.85 + (0.285)^2.(20 + 79.5)} = 2.15$$

$$C_{v3}^* = \frac{\{(0.285)^2.0.04 + 0.002\}.500}{2.85 + (0.285)^2.(79.5)} = 0.28.$$

Conclusion: the image of the lit sign is still visible, the reading contrasts are very comfortable. From the optimisation of C_{v3}^*, with $\tau_1(\theta).\tau_2$ as variable, one obtains $\tau_1(\theta)|_{\theta=0} = 0.5$. The resulting C_{v3}^* now becomes $C_{v3}^* = 0.24$; still clearly visible! Another constraint is that the denominator of C_{v3}^* (the observed diffuse background) should match the surround luminance within a factor of 3. With the obtained $\tau_1 = 0.3$, $\tau_2 = 0.95$, $\rho_1(d) = 0.25$ and $E = 1000$, one finds for office systems

$$10 \le 0.285 \, L_b + (0.285)^2.\pi^{-1}.250 \le 90,$$

since 30 cd m^{-2} is a good estimate for the surround. Taking the lower interval from 10 to 30 cd m^2, it follows that $12.5 \le L_b \le 30$ cd m^{-2}. Choose e.g. $L_b = 25$ cd m^{-2}. The numbers so obtained for the contrasts are:

$C_{v1} = 1.58$; $C_{v2}^{*'} = 1.47$; $C_{v3}^{*'} = 0.19$.

Comparison of C_{v3} and $C_{v3}^{*'}$ tells us that the comparatively low figure without filter is due to wash-out by the background caused by the window;

the latter contribution is much attenuated by the low τ_1 of the ND filter. Thus one may conclude that the figures for C_{v3} obtained above are typical except when other ways are applied to ameliorate the desired contrast and fight the glare; e.g. the use of a colour filter. Of course, the colours used must not have too much contrast to the direct environment (surround) of the display. Providing that this constraint is fulfilled, the extra attenuation of specular reflection is easily obtained.

Assume e.g. that the emissive display emits light in three wavelength intervals, each 20 nm wide. Thus the total effective interval equals 60 nm out of the entire luminous interval of about 300 nm.

From expression (1.18) it follows that, for $\Phi(\lambda)$ constant, the interval limitation to 60 nm means an attenuation of 5 times. Assume that $E(\lambda)$ and $L_e(\lambda)$ both are white (i.e. constant in the luminous interval). Then, if the colour filter possesses three windows with wave intervals to match the emission intervals of the display (where $\tau_1 = 0.285$ while $\tau_1 = 0$ elsewhere), the display light is transmitted without attenuation while the average transmittance $\bar{\tau}_1$ for $\{E(\lambda) + L_e(\lambda)\}$ equals 0.2.

Applying these figures, one obtains ($L_f = 100$ cd m^{-2}, $L_b = 10$ cd m^{-2})

$C_{v1}^{**} = 6.2$; $C_{v2}^{**} = 4.7$; $C_{v3}^{**} = 0.13$.

Considering the large contrasts C_{v1}^{**} and C_{v2}^{**} there is still room to improve C_{v3}^{**}, by raising L_b. The average background with $L_b = 10$ cd m^{-2} turned out to become 12 cd m^{-2}, so L_b can be increased to $L_b' = 20$ cd m^{-2}, yielding $C_{v1}^{**\prime} = 3.3$; $C_{v3}^{**\prime} = 0.09$.

With these contrasts, reading is very comfortable while the image of the lit sign is barely visible.

1.4.3. A relation between threshold contrast and spatial resolution

The eye has detection thresholds, measured by just noticeable differences (JND) in luminance. The corresponding contrast sensitivity CS depends on:

1) the area a of the region with foreground luminance L_f (the target);

2) its contrast with the local background luminance L_b at the surrounding area;

3) on the spectral distribution $L(\lambda)$ of the light; and

4) on the average luminance.

For target areas subtended by angles of 1° or larger, the CS remains constant at about 300 to 500 depending on L_b. For decreasing target areas the CS decreases also: [1.14] and expression (1.24), see also section I.2.4.

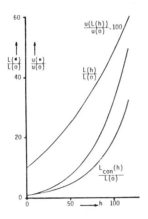

Fig. 1.20 Spot luminance
L(∗) vs the number of
increments h for steps ξ.

1) L_{con}(h): ξ assumed

constant = 0.03

2) L(h)}: ξ varies

according to (1.24)

3) u{L(h)}: driving

voltage for L(h), γ = 2.2.

The law of Weber: the threshold level ΔL/L is constant, applies to larger target sizes and under photopic conditions. The perceived sensation resembles a logarithmic function of the stimulus. Translated to display engineering this law states that, if the contrast between adjacent regions exceeds the JND, grey level contouring (see section 1.4.4) is perceived: the smooth gradient is broken–up in (a number of) grey steps, as shown in figure 1.24.

The driving signal of the display should be conditioned such that the displayed luminance increases in steps $\Delta L = \xi \cdot L$ where ξ is the threshold C_v associated with CS: $\xi = C_{vth} = \frac{2}{CS-1} = 2.CS^{-1}$ for the case of large area targets and comfortable $L_b = L(0)$. The first step then is L(1) = L(0)+ξL(0) = L(0)(1+ξ), the second L(2) = L(1)+ξL(1) = L(0)(1+ξ)², the h-th step:

$$L(h) = L(0).(1+\xi)^h = L_b.(1+\xi)^h \tag{1.23}$$

In figure 1.20 the relationship (1.23) between h and L_{con}(k)/L(0) is shown for constant ξ = 0.03. But ξ is not constant!

The relation between ξ, target area a and L_b was determined [1.14] for round, homogeneously illuminated targets and backgrounds. Figure 1.21 shows the results graphically for threshold contrast defined as 50 % probability of detection of single discs (dels). Of course this relation is only half the story, because the resolving distance between two adjacent dots and/or lines is equally important [1.15,1.16]. For small single dots and at photopic levels the relation between dot size (d), threshold contrast C_{vth} and L_b can be empirically modelled:

$$C_{vth} = \xi = \frac{L_f - L_b}{L_b} = 0.2 \ [1+300L^{-0.7}\{(d+0.4)^{-2}+0.014\}]^{0.5}\{(d+0.4)^{-2}+0.014\} \tag{1.24}$$

with d the linear angle of the dot in minutes and L in cd m^{-2}. Relation (1.24) is within 10 % valid for office conditions: $3 < d < 120$; $10 < L < 3000$.

The threshold contrast C_{vth} is a useful measure to determine the number of bits required to display continuous images (see section 1.4.4). For text and graphics the 50 % detection probability is insufficient. Even when the contrast is raised (by a factor of 3) to satisfy a 99.9 % detection requirement, the image still looks ghostly. For comfortable viewing a further factor of 5 is necessary [1.15]. So instead of e.g. $\xi \simeq 0.028$, one obtains a minimum contrast increment of $C_v = 0.42$ or $C_r = \sqrt{2}$. Furthermore, if the surround luminance is much brighter (e.g. outside and near windows) still more contrast is desirable: it has been suggested that display contrast should increase by 2 % of the ratio surround over background luminances.

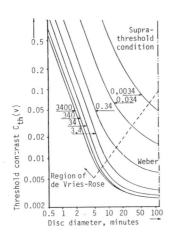

Fig. 1.21 Contrast thresholds (required modulation depth MD) versus visual angle of round target (del) at 7 background luminances [1.14].

At high background brightness the influence of L_b on ξ vanishes rapidly, leaving as the main effect that, for the equal brightness sensation of dots of different size, the product of contrast C_v and object sized must be constant.

This law provides the possibility of anti-aliasing illusion by contrast modulation, see figure 1.22 [1.6,1.7].

Fig. 1.22 Aparent centre of luminance shift method to decrease aliasing (jaggedness).

Assume a driving characteristic $L(u) = L(0).u^{\gamma}$ where u is the display input.

Then: $C_v = \dfrac{dL}{L} = \gamma \dfrac{du}{u}$.

Putting $\gamma \dfrac{du}{u} = \xi$ yields:

$$u(h) = (1+\frac{\xi}{\gamma}).u(h-1) \qquad (1.25)$$

For a cathode ray tube γ typically is 2.2; spotsize d = 3'.

Table 1.3 shows how, to satisfy expressions (1.25) and (1.24), the L(h)/L(0) curve in figure 1.20 is accordingly modified with respect to $L_{con}(h)$. The drive voltage u as function of h is also given.

The contrast sensitivity for sinusoidal or bar gratings (instead of single discs) is measured by observing at what modulation depths the grating is just noticeable (threshold MD). Supra-threshold sensitivity is measured in matching experiments at higher contrasts than "just noticeable", comparing perceived contrast of a sample with the contrast ratio C_{gr} of reference samples of different spatial frequencies.

Table 1.3

Just noticeable contrast increment for d = 3',
drive voltage increment ξ/γ for γ = 2.2.

L_b(cd m^2)	ξ	ξ/γ
10	0.051	0.023
20	0.042	0.019
50	0.033	0.015
100	0.029	0.013
200	0.026	0.012
500	0.024	0.011
1000	0.023	0.010
2000	0.023	0.010

Curves of constant experienced (supra-threshold) contrast as function of the spatial frequency of the grating [1.17] are shown in figure 1.23. The underlying describing model is $C_{st}(\rho)$ = 0.14$\{C_{gr}-C_{th}(\rho)\}$ wherein C_{st} is perceived supra-threshold contrast, C_{gr} the actual grating contrast and C_{th} the threshold (JND) contrast; ρ is spatial frequency in cycles per degree. When again a 99.9 % probability of detection is equated with comfortable viewing conditions (C_{st} = 3 C_{th}, independent of spatial frequency) then one obtains $C_{gr}(\rho)$ = 22.5 $C_{th}(\rho)$, a figure 50 % larger than above quoted from [1.15]. However, they are of the same order of magnitude. Given the variability in the parameters of humans and also questioning whether 99.9 % probability of valid responses is a realistic assumption in relation to humans, the advice to realise 15 $C_{th}(\rho)$ < $C_{gr}(f)$ < 23 $C_{th}(\rho)$ seems reasonable.

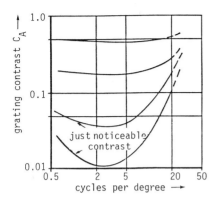

Fig. 1.23 Curves of constant perceived contrast against actual grating contrast at supra-threshold conditions [1.17].

The inverse graphical notation: contrast response for constant input contrast as function of spatial frequency, is the eye's socalled "Modulation Transfer Function". Socalled, because the transfer function is a notion defined in linear theory whereas the eye operates non-linear. This shows among others, by the fact that the minima in figure 1.23 shift to the left for lower local average luminance and vice versa (see also section I 2.4).

1.4.4. Contrast resolution in images with more than two grey levels

The question now arises to what metrical resolution a natural image must be digitised or, how much quantisation noise can be tolerated. If there is correlation between metrical and structural data, a continuous gradient is faithfully displayed when adjacent display elements (e.g. $(i+1)\Delta x, (j+1)\Delta y$; $i\Delta x; j\Delta y$) obtain contrast steps equal to the just noticeable difference (JND) in shades of grey. Under normal office working conditions ($\xi \simeq 0.03$), this difference can be modelled by equation (1.23): $L_{i+1,j} - L_{i,j} = 0.03\, L_{i,j}$. So the scale of grey levels is:

$$L(k) = (1.03)^k L(0) \qquad\qquad (1.26)$$

k is the rank of the selected grey level. Operations such as image filtering (for noise suppression, image enhancement) usually require linear magnitude representation, thus the display samples $F_d(m,n)$ also are a linear representation of required luminance if no function table is inserted. Assume that the magnitude is given by h increments q, in addition to a background F(0):

$$F_d = h.q + F(0) \qquad\qquad (1.27a)$$

or, with q = 0.03: $F_d = (1+0.03h)F(0) \qquad\qquad (1.27b)$

The human visual system responds roughly logarithmically (Weber's law) (the display actually should be driven according to expression (1.24)). The inequality now exists:

$(1.03)^{(k-1)}L(0)+0.015L(0) < (1+0.03h)F(0) < (1.03)^{(k+1)}L(0)-0.015L(0)$ or, with $F(0) = L(0)$:

$$0.985 < \frac{1+0.03h}{(1.03)^k} < 1.015 \tag{1.28}$$

This inequality only holds for h = k when both are small: for increasing h the generated luminance remains behind so that at certain magnitudes two successive linear levels remain within the boundaries of one exponential level (grey level contouring) as shown in figure 1.24 where a natural image is depicted in 9 shades of grey. Numerically this effect is demonstrated in table 1.4:

Table 1.4

h	0	1	2	3	4	5	6	7	8	9	10	11	12	13	14	15
k	0	1	2	3	4	5	6	6	7	8	9	10	10	11	12	13

wherein the perceived corresponding levels are calculated for JND:

$C_v(1) = \frac{L(k)-L(k-1)}{L(k-1)} = 0.03$ or $C_r(1) = \frac{L(k)}{L(k-1)} = 1.03$; with $C_v(1)$ the perceived contrast for 1 JND, $C_r(1)$ the corresponding contrast ratio.

Fig. 1.24 Image CART radiometrically underquantised; step size q equal to and 0.5 standard deviation of the image grey level histogram.

It is now in order to calculate the required luminance range L_f-L_b, foreground luminance L_f minus background luminance L_b, of a display. Since the display reflects environmental light L_e, the L_b cannot be zero. Suppose that $L_b = L_e = L(0)$ (expression 1.23), that the grey level representation in the display video input is linear (interval scale), and the maximum number h of grey levels is h = ma. Then (linear video) L(h) = 0.03hL(0) or

$$L_f(h) = L(h)+L_e = (1+0.03h)L_e \qquad (1.29a)$$

and

$$C_r(ma) = \frac{L_f(ma)}{L_e} = 1+0.03 \ ma \qquad (1.29b)$$

The number of perceived shades of grey is smaller (table 1.4); it can be calculated from (1.26) and (1.29)

$$k(max) = \frac{\log L_f(ma)-\log L_e}{\log C_r(1)} = 77.9 \ \log \ (1+0.03ma) \qquad (1.30)$$

The results of (1.29b) and (1.30) are tabulated below in table 1.5; these should be compared with the contrasts calculated in example 1.7.

Table 1.5

Addressed vs perceived shades of grey (achromatic light)

ma	$C_r(ma)$	k(max)
63	2.89	35.9
127	4.81	53.1
255	8.65	73.0

Thus, in order to make maximum use of 8 bits magnitude resolution, i.e. 73 shades of grey, the display must be able to generate, say 216 cd m^{-2} and the workplace must be kept in subdued light (25 cd m^{-2} reflection). But, this calculation is for the CRT too optimistic, since the effect of overlapping dels is not taken into account.

For adjacent spots on the face of the CRT the total luminance distribution is as depicted in figure 1.25. The resulting contrast ratio, given a background luminance of L_e, is decreased by the overlap since $\frac{L(h)+L_e}{L_s+L_e}$ is smaller than $\frac{L(h)+L_e}{L_e}$. In table 1.6 the maximum number of perceived contrast levels is given as function of the separation of neighbouring spots assuming that the spots have a Gaussian distribution; the separation is given in standard deviations σ. It is seen that the separation of 3σ yields a very smal number of separable shades of grey.

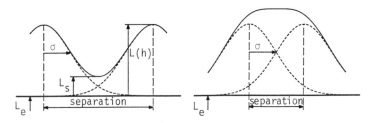

Fig. 1.25 Spot resolution decrease due to overlap.

For matrix displays with disjoint square dels the situation depicted in figure
1.25 is, without optically interpolating front cover (spatial low pass
filter), not valid; table 1.5 remains the correct reference.

Table 1.6

Addressed vs perceived shades of grey on a CRT screen,
for pixel separation in standard deviations:

ma	$C_r(ma)$	3σ	4σ	5σ	6σ
63	2.89	9.0	22.0	30.7	34.5
127	4.81	11.9	29.2	43.4	50.4
255	8.65	12.8	35.1	55.6	67.7

In the reproduction of metrical information, the data and the background can
be corrupted by several faults in the display itself:
- del-to-del non-uniformity in sensitivity causes mottled background,
  multiplicative noise in the data. This effect was conspicuous in early
  (experimental) designs of displays, has now been reduced to a few percent so
  that under most conditions it is not visible;
- gradually changing non-uniformity in sensitivity: shading. Shows mostly in
  the background, does not affect local contrast;
- non-uniform threshold level: the background is disturbed by additive noise.
  In some technologies it can be caused by a small but persistent memory
  effect due to prolonged presentation of a static picture;
- non-uniform reflection and/or transmission factors, which modulate light
  from external sources. This includes faults in display covers.

1.5 Addressing

1.5.1. General

In simple display systems every display element is connected to its own controller: direct addressing. Surprisingly, this is also true for modern alpha-numeric displays supporting more than 10^5 dels. Each display element is stimulated by (thin-film) transistors, mostly in a bi-level mode; but the appropriate thin-film transistors (TFT) in the array are matrix addressed by selecting corresponding columns and lines as described below.

In the following we assume that the image to be displayed is defined on an orthogonal lattice (section 1.3.1) with the nodes corresponding to the dels on the display face. We do not consider "stroke-written" displays wherein continuous vectors can be written in any desired direction; their use is limited to specific applications like when small symbols must be rotated without distortion or when very high speeds combined with high resolution must be realised.

Addressing is the process by which the desired luminance and chrominance is assigned to every display element. It follows that addressing consists of two separate functions: the del (m,n) located at (mΔx,nΔy) on the display surface is

(1) selected by the addressing signals (m,n) to enable it

(2) to respond to a simultaneously applied video stimulus.

Non selected dels are not affected by the video signal.

The response of selected dels can range from extremely fast (picoseconds) to fairly slow (tens of milliseconds), depending on the technology. Similarly during the period of non-selection the del can either remain in the assigned state (long intrinsic memory) or it can lose that state gradually (short intrinsic memory, decay). The decay time generally ranges from a few ms to hundreds of ms. Thus the chosen addressing method determines the local and average luminance level and the faithfulness of reproduction of moving parts in the image. Moreover, there are the all important aspects of cost of realisaton of the addressing method, and side effects like generation of heat, of electromagnetic interference (EMI) and of legibility of strongly vibrating displays.

Theoretically, the above definition of addressing allows random selection of dels; there need not be correlation between successive m and/or n. Also the selection process is not restricted to one-del-at-a-time. The display surface can be subdivided into S disjoint sets of dels, each with its own one-del-at-a-time addressing scheme. In e.g. cathode ray tube based displays, this method requires a multi-cannon CRT providing several beams simultaneously (section II.1.4.7).

The image is defined on an orthogonal lattice. This fact is used to advantage to realise low cost multiplexing methods: the addressing process is organised to incrementally follow both the lines and the columns of the lattice. We speak here of <u>lines</u> instead of rows, because the term "row" is reserved for horizontally orientated sets of <u>symbols</u>, not dels. For electro-optic materials with fast decay of the addressed states of the dels, a necessary intermediate mapping parameter is time: periodical rewriting of the whole image. The inverse of the rewrite period is the frame rate or frame frequency. The frame is the pattern consisting of N parallel lines (rasterscan). Spot velocity within the boundaries of the frame is constant in both x- and y-directions. The frame can be split into S disjoint fields.

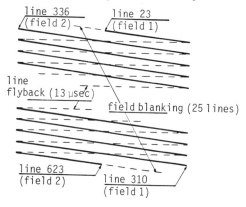

In TV S = 2 and the fields consist of either all the even or all the odd numbered lines; see figure 1.26 which depicts the TV-PAL standard. In this standard the fields are written one after the other, requiring a single spot CRT. Fields can also be written simultaneously when the number of beams equals the number of fields. The advantage of the latter method is that the address time per del is S times longer resulting in S times more luminance, while video bandwidth per beam is S times smaller.

Fig. 1.26 Rasterscan method with 2.1 interlace (TV-PAL system). The flyback times are not typical for high quality monitors.

Similar gains over one-del-at-a-time (resulting in every del addressed once per frame!) are realiseable in FPT, where the line and column electrodes (figure 1.27) all are simultaneously available. In the figure again there are 2 fields, but not interlaced; in each field many dels can be addressed at the same instant in time, provided they are located on the same line electrode: one-line-at-a-time addressing. The designated del(s) respond to the average or to the RMS field strength developed between the line and column electrodes, following a response law peculiar to the electro-optic material used. But contrary to the CRT response to a video stimulus: a monotonic and well reproducing luminance range of, say, two decades, most FPT materials have a rather small non linear response interval of width $F\Delta$ Vm^{-1}, separated from zero by a threshold fieldstrength F_{th} Vm^{-1}, see figure 1.28. In the simple

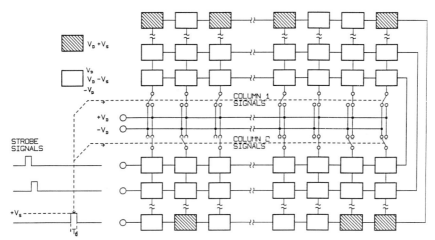

Fig. 1.27 Addressing scheme. Hatched dels are "on" (V_S+V_D); the others are "off" $(+V_D,-V_D,V_S-V_D)$. N lines need 0.5 N line drivers, 2 M column drivers.

addressing circuit of figure 1.27, the "off" dels still receive fieldstrengths proportional to $|V_D|$ with continually changing sign, depending on the displayed image on addressing progress. The "on" dels receive a stimulus proportional to (V_D+V_S).

The voltage sources V_D and V_S and the material parameters F_{th} and Δ may vary with e.g., temperature, pressure, time, age. Since the saturation parts of the

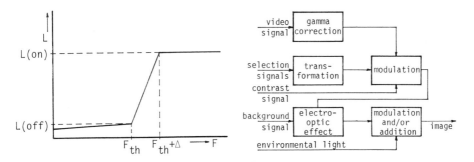

Fig. 1.28 Non-linear electro-optical response characteristic.

Fig. 1.29 Display system per del.

response characteristic (figure 1.28) most likely are not perfectly level, the results of such variations is a change in appearance of the display.

Of course, these effects are stronger in multi-grey level and colour displays. In displays with grey level response, the response law of figure 1.28 is not the desired display response law. The range $< F_{th} + \Delta$, $F_{th} >$ should be large and preferably the transition should be exponential to avoid bit loss (see section 1.4.4). In figure 1.29 the general layout of a VDU is given, per del. For one-del-at-a-time addressing (e.g. single beam CRT) the circuit also describes the system. For line-at-a-time addressing or multi-beam CRTs, it is necessary to provide as many modulators as there are columns or beams.

Another effect which needs mentioning here is the transformation of the alternating TV field lines to a spatio-temporal effect since the even and the odd fields are displaced by one line. Bars parallel to the display lines, with width of two or more display lines, will be written in spatially shifted positions for each field and consequently jump up and down at the frame rate. Thus temporal flicker is accompanied by spatial flicker, sometimes referred to as "twinkle". This dynamic spatial effect is far more annoying than the temporal modulation type flicker. Therefore, we will assume only non-interlaced addressing or, when interlaced, parallel addressing of fields. Let there be M del positions on a line, N on a column. In del-at-a-time addressing the frame period T_f equals MxN del periods T_d; if no other secondary operations are involved, like flyback. Under the most severe conditions, flicker is not perceived when the frame rate $f_f > 80$ Hz, or $T_f < 12.5$ ms. The next question is; how large must M and N be?

Example 1.8. Addressing/driving.

Assume that the displayed image should be of medium resolution quality (as compared to printed documents) and that the display surface must accommodate A_4 size (207×297 mm^2) without the upper/lower/left/right margins. It is known (section 2.4) that visibility of periodic gratings disappears for spatial frequencies > 60 periods per degree (ppd). Thus, the highest fundamental frequency of a periodic, non-harmonic perceived, phenomenon will be lower than 30 ppd. Consequently, the del diameter or diagonal required to realise this at a viewing distance of 342 mm will be 0.1 mm. For square dels, the display resolution then is 14 dels per mm. Ergonomic experiments [1.1,2.4] have shown that reading error scores settle to very low, satisfactory levels at 16 to 20 lines per symbol height; meaning that characters are 1.1 to 1.4 mm high which is typical for e.g. fine print. Thus the A_4 size (without margins about 160×250 mm^2) yields M = 14×160 = 2240 and N = 14×250 = 3500.

The del period becomes T_d = T_f/MxN → 0.0125/7,840,000 = 1.6 ns, or the selection rate is 625 Mdel s^{-1}. For cathode ray tubes this figure is even higher, because the flyback times must be subtracted from the frame period and the line writing time.

In the TV-PAL pattern, it is seen that allowance is made to lose 25 lines during each field flyback, a total of 3.2 ms referred to a frame time of 40 ms. The line period being set at 64 μs, one obtains the line writing time by subtracting the line flyback time. Making horizontal resolution on the screen equal to vertical resolution, the number of distinguishable points in a line must be 1.33x575 = 767 for an aspect ratio of 4:3. Taking the line flyback at 15 μs, the addressing time per point at PAL format is 64 ns which corresponds to a Nyquist frequency of 7.8 Mdel s^{-1}.

The del densities seemingly afforded by the addressing method usually are seldom realised in colour CRTs. E.g. the current shadow-mask pitch being 0.31 mm, the spot size should be about 0.9 mm. With 770 dels on a line, the screen diagonal becomes 1 meter; a size currently promoted for TV. In smaller tubes the spatial cut-off frequency of the MTF associated with spot standard deviation (see example 1.1) is too low in relation to the video bandwidth.

It stands to reason that, even with the PAL rates at a duty ratio (del period over frame period) of 16.2 10^{-7} and an average luminance of 50 cd m^{-2}, the peak-luminance-while-addressed must be prohibitively high. Fortunately the necessary energy can be supplied in the del period; it is converted into light during a much longer time (exponential decay, time constant being some ms) such that the duty ratio drops by about 10^4 and peak-luminance-while-addressed is easily achieved. Presently only the CRT can realise these sizes and speeds, although FPT performance is improving fast.

1.5.2. Line-at-a-time addressing

The line-at-a-time addressing method obtains better duty cycles than del-at-a-time addressing where the duty cycle is at best $(NxM)^{-1}$. For a bi-level display (alpha-numerics, no grey levels) the driving waveform can be (almost) rectangular with fast transitions through the interval Δ (figure 1.28); its duty cycle is N^{-1}. Given a technology which responds to the time average of the excitation and has fast rise and decay times (LED,EL), the resulting contrast is easily calculated. Assume that, for the case depicted in figure 1.27, the driving voltages (or currents, or fieldstrenghts) produce a peak "on" luminance L(on) and an "off" luminance L(off). The (spatial) average "on" luminance L(av,on) = N^{-1}{(N-1)L(off)+L(on)}; the average "off" luminance

$L(av,off) = L(off)$. Denoting the maximum obtainable contrast $C_v(m) = \{L(on)-L(off)\},/L(off)$, the average contrast

$$C_v(av) = N^{-1}C_v(m) \tag{1.31}$$

The technology in question must produce a peak "on" luminance N times larger than $L(av,on)$. This may be accomplished by hard overdriving; but that can be a serious constraint when $L(on)$ is restricted to a physically determined maximum which cannot be exceeded without damage to the device, e.g. shorter life or even destruction.

A second constraint on line-at-a-time addressing is that the peak current through the line driver equals the sum of all the "on" dels addressed by the columns; thus for a horizontal line displayed: M times the individual del current.

Similar considerations are valid for technologies which respond to the RMS value of the excitation, or when the response is characterised by a large decay time as is the case with the CRT (instantaneous storage of energy which leaks relatively slowly away).

The response of an RMS sensitive technology (LCD, incandescent) is determined by the average power in the driving signals (figure 1.27). The RMS component in a pulse signal of low duty cycle is rather small; it is difficult to achieve a sufficiently large $\hat{V}(on)/\hat{V}(off)$ ratio in the driving signals necessary to overcome the transition range Δ above the knee, in the response characteristic of the electro-optic material (figure 1.28). In the following it will be shown [1.18] that, depending on Δ and F_{th}, one obtains an optimum number $N(m)$ of lines that can be accommodated in the display, unless extra measures are taken in the form of introducing other non-linear devices with steeper and invariant transfer characteristic.

Let in figures 1.27 and 1.28 the "on" response be $L(on) = K_1F(on)$ and the "off" response $L(off) = K_2F(off)$, where $K_1 > K_2$ due to the non-linearity. It is desirable that $F(off) \leq F_{th}$, the threshold; and that $F(on) \geq F_{th} + \Delta$, the knee of the characteristic. In dynamic operation the threshold and the knee may shift compared to their static values: the dynamic threshold is denoted by $\hat{F}_{th}$, the knee by $\hat{F}_{th}+\hat{\Delta}$ where the caret $\hat{}$ signifies effective value (RMS). Thus

$$\hat{L}(on) = \hat{K}_1.\hat{F}(on) \; ; \; \hat{L}(off) = \hat{K}_2.\hat{F}(off) \tag{1.32a}$$

$$\hat{F}(on) = \alpha \; (\hat{F}_{th} + \hat{\Lambda}) \; ; \; \hat{F}_1(off) = \beta \; \hat{F}_{th} \qquad (1.32b)$$

with $\alpha \geq 1$, $\beta \leq 1$. We consider also the case of the "off" del in the non-strobed condition, due to e.g. a non-addressed field in the display, or a shorted electrode or faulty line driver. Assume that the driving signals $\hat{V}$ relate linearly to $\hat{F}$, transforming (1.32b) to:

$$\hat{V}(on) = \alpha K_3 (\hat{F}_{th} + \hat{\Lambda}) \; ; \; \hat{V}_1(off) = \beta K_3 \hat{F}_{th} \; ; \; \hat{V}_2(off) \neq \beta \; K_3 \hat{F}_{th} \qquad (1.32c)$$

K_3 is a device parameter. For the signals of figure 1.27 the power expressions are

$$\hat{V}(on)^2 = N^{-1}\{(V_S + V_D)^2 + (N-1)V_D^2\} \quad \text{(strobed)} \qquad (1.33a)$$

$$\hat{V}_1(off)^2 = N^{-1}\{(V_S - V_D)^2 + (N-1)V_D^2\} \quad \text{(strobed)} \qquad (1.33b)$$

$$\hat{V}_2(off)^2 = V_D^2, \; V_S \approx 0 \qquad \text{(non-strobed)} \qquad (1.33c)$$

Following [1.18] the optimum number $N(m)$ of multiplexed lines and the resulting contrast are obtained:

$$V_D = 0.5 \; \{\hat{V}(on)^2 + \hat{V}_1(off)^2\}^{0.5} \; ; \; V_S = 0.5 \; N^{0.5}\{\hat{V}(on)^2 + \hat{V}_1(off)^2\}^{0.5} \qquad (1.34a)$$

so that $N(m) = \left[\dfrac{\hat{V}(on)^2 + \hat{V}_1(off)^2}{\hat{V}(on)^2 - \hat{V}_1(off)^2} \right]^2 \qquad (1.34b)$

The ratio $R_1 = \hat{V}(on)/\hat{V}_1(off)$ yields the contrast in the driving signals for the strobed condition, and also the luminance contrast ratio using (1.32):

$$R_1 = \left[\dfrac{N^{\frac{1}{2}} + 1}{N^{\frac{1}{2}} - 1} \right]^{\frac{1}{2}} \; ; \; C_{r1} = \dfrac{K_1}{K_2} R_1 \qquad (1.35)$$

The ratio R_1 is often used in LCD literature, see also section II.3.4.1.

These expressions should be related to the device parameters $\hat{F}_{th} + \hat{\Lambda}$ and $\hat{F}_{th}$. Denote [1.18] the ratio $\hat{\Lambda}/\hat{F}_{th} = P$, an (inverse) measure for the steepness of the characteristic of figure 1.28. Then, see figure 1.30:

$$N(m) = \left\{ \frac{\alpha^2(1+P)^2+\beta^2}{\alpha^2(1+P)^2-\beta^2} \right\}^2 \quad ; \quad R_1 = \frac{\alpha}{\beta}(1+P) \quad ; \quad C_{r1} = \frac{\alpha}{\beta} \cdot \frac{\hat{K}_1}{\hat{K}_2}(1+P) \qquad (1.36)$$

Since in the ideal device P approaches zero, the contrast ratio is then mainly determined by α/β and $\hat{K}_1/\hat{K}_2$; however for large N the ratio R_1 approaches 1 leaving very little room for shift of $\hat{F}_{th}$ due to e.g. viewing angle, temperature and age. To maintain constant image contrast, temperature compensation is required and a manual control to correct for long term shift.

In the non-strobed condition $\hat{V}_2(off)^2 = V_D^2$.

For large N both α and β approach 1 and the contrast between strobed "off" dels and non-strobed "off" dels $C_{r2} = \hat{V}_1(off)/\hat{V}_2(off)$ approaches $2^{\frac{1}{2}}$, quite visible.

In figure 1.30 the relation between N and P, and between C_{r1} and P, are

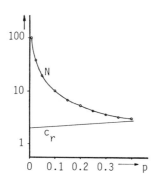

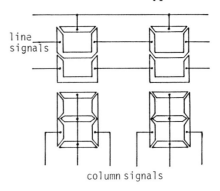

Fig. 1.30 N: Number of multiplexed lines; C_r: contrast obtained; as function of technology quality parameter P (see text).

Fig. 1.31 Segment multiplexing of numerals.

depicted. At lower multiplexing numbers the ratio α/β increases and the luminance contrast C_{r1} with it; but the gain is moderate. E.g. for N = 3 (alpha-numeric characters, figure 1.31) the ratio α/β = 1.93, the contrast C_{r1} = 1.93 $\hat{K}_1(1+P)/\hat{K}_2$. For large N, the resulting small difference between α and β (or $\hat{V}(on)/\hat{V}(off)$) obtained at high N becomes impractical, because of imperfect temperature compensation of $\hat{F}_{th}$ drift, or its sensitivity to viewing angle in twisted nematic LC. The performance can be improved by the addition of non-linear elements with better (switching,P) characteristics.

1.5.3. Active matrix addressing

Cascading a non-linear device (NLD) with every display element not only improves the switching characteristics but also the contrast gain by applying the method of slowly leaking energy. See figure 1.32 [1.19]. The capacitance in parallel with the del (intrinsic, or specifically made for the purpose) is

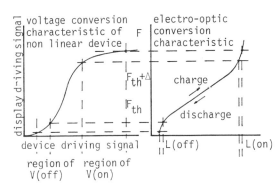

Fig. 1.32 Improvement of switching characteristics by cascading a non-linear element (see text).

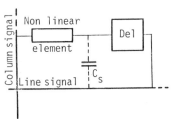

Fig. 1.33 Active matrix addressing: a non-linear device is in series with every del.

quickly charged by the driving current when the del is addressed "on" during the strobe period; in the non-addressed period the impedance of the NLD increases greatly so that the capacitance across the del discharges slowly. Since the information about the del being "on" addressed or not is carried by the driving voltage, the NLD must exhibit a non-linear voltage/current characteristic, like a diode, a varistor and the like. The circuit is depicted in figure 1.33.

The voltage across the del is developed by the large current i_1 through

$$V_1(\text{del}) = C^{-1} \int_o^{T_d} i_1 dt \quad \text{where the interval } 0\text{-}T_d \text{ is the addressing period.}$$

Discharge between strobe pulses follows $V_0(\text{del}) = V_1(\text{del}) - C^{-1} \int_{T_d}^{T_f} i_0 dt$. By making the ratio $i_1/i_0 > T_f/T_d$, the "on" addressed del remains "on" more than

37 % of the frame time T_f, being only partly discharged when the new strobe
pulse arrives. In direct drive, with time average responding technologies, the
"on" time is $T_f.N^{-1}$, with active matrix addressing more than $0.37\ T_f$, so that
the luminance "on" gain is considerable for large N. In RMS responding
technologies the calculation of the gain is more complicated (expressions 1.32
and so on) but again the gain is considerable.
Another advantage of active matrix addressing is that the non-linearity in the
series device makes the combination less sensitive to variations of the
threshold in the characteristic of the del (viewing angle, temperature, age).

One step further is the application of cascaded transistors. These have the
same advantages as the above non-linear devices but provide also for
separation (figure 1.29) of addressing power and display power because the
transistor has power gain. The basic lay-out is given in figure 1.34. In a)
the column address electrode still carries the display power; when grey levels
are desired the column drivers also are the video modulators. In b) the
address and video functions are separated; T_2 switches the display connection.
It is obvious that in the simple circuit of b), where the power input is
common to all display elements, line-at-a-time addressing is only possible for

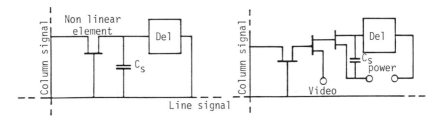

Fig. 1.34 Active matrix addressing examples. The transistors act as switches
driven by line and column voltages. In a) the address electrodes carry display
power b) address and video functions are separated.

bi-level (on-off) drive, for multi-grey level images del-at-a-time addressing
is required. The leaky memory must have a sufficiently large time constant.
A display requires an active matrix consisting of as many switching circuits
as the number of dels, e.g. for a 625x400 del display the complexity of the
active matrix also is 0.25 million switching circuits. Such amounts can only
be made (cost, size, integration) by photolithic techniques and the active
(semi-conductor) material must be applied by deposition techniques leading to
the concept of the thin-film transistor (TFT) matrix [1.19,1.20].

REFERENCES

[1.1]   Shurtleff, D.A., Legibility research, Proceedings of the Society for Information Display, <u>15</u>, no. 2, (1974), pp. 41-51.

[1.2]   Roufs, J.A.J. and Bouma, H., Towards linking perception research and image quality, Proceedings of the SID, <u>21</u>, no. 3, (1980), pp. 247-270.

[1.3]   Pratt, W.K., Digital image processing. Wiley Interscience, John Wiley and Sons, ISBN 0-471-01888-0, New York (1978).

[1.4]   Infante, C., On the resolution of rasterscanned CRT displays, Proceedings of the SID, <u>26</u>, no. 1, (1985), pp. 23-36.

[1.5]   Barten, P.G.J., Resolution of data display tubes, Proceedings of the SID, <u>25</u>, no. 1, (1984), pp. 35 12.

[1.6]   Oakly, D., Dejagging raster graphics by pixel phasing, digest of the International Symposium of the Society for Information Display, vol. XVII, (1986), pp. 344-347.

[1.7]   Ketcham, R.L., A high speed algorithm for generating anti-aliased lines, Proceedings of the SID, <u>26</u>, no. 4, (1985), pp. 329-336.

[1.8]   Snyder, H.L., Image quality: measures and visual performance. In: Flat panel displays and CRTs, Ed. L.E. Tannas, Van Nostrand Reinhold Company, ISBN 0-442-28250-8, New York (1985), pp. 70-90.

[1.9]   Riggs, L.A., Light as a stimulus for vision. In: Vision and visual perception, Ed. C.H. Graham, John Wiley and Sons, ISBN 0-471-32170-2, New York (1965), pp. 1-38.

[1.10] Jacobson, K.D., Optical performance of contrast enhancement filters, Proceedings of the SID, <u>24</u>, no. 1, (1983), pp. 43-48.

[1.11] Snyder, H.L. and Beaton, R.J., The display quality of glare filters for CRT terminals, Society for Information Display Symposium Digest, (1984), pp. 298-301.

[1.12] Christiansen, P., Design considerations for sunlight viewable displays, Proceedings of the SID, <u>24</u>, no. 1, (1983), pp. 29-41.

[1.13] Rancourt, J.D., Anti-halo coatings for CRT faceplates, Proceedings of the SID, <u>25</u>, no. 1, (1984), pp. 43-48.

[1.14] Blackwell, H.R., Contrast thresholds of the human eye, JOSA, <u>36</u>, no. 11, (1946), pp. 624-643.

[1.15] Carel, W.L., Pictorial displays for flight, Office of Naval Research. Available NTIS-AD 637669, (1965).

[1.16] Umbach, F.W., The perception resolution of two dots on a CRT for positive and negative contrast, Conference Record International Display Research Conference of the SID, San Diego (1985), pp. 123-126.

[1.17] Cannon, Jr. M.W., Contrast sensation: a linear function of stimulus contrast, Vision Research, <u>19</u>, (1979), pp. 1045-1052.

[1.18] Alt, P.M. and Pleshko, P., Scanning limitations in liquid crystal displays. IEEE Transactions Electron Devices, vol. ED-21, (1974), pp. 146-155.

[1.19] Howard, W.E., Active matrix techniques for displays. In: Seminar lecture notes, vol. II, Society for Information Display, San Diego, May (1986).

[1.20] Tannas, L.E., Flat panel display design issues. In: Flat panel displays and CRTs, Ed. L.E. Tannas, Van Nostrand Reinhold Company, ISBN 0-442-28250-8, New York (1985), pp. 91-137.

DISPLAY ENGINEERING: D. Bosman (Editor)
© Elsevier Science Publishers B.V. (North-Holland), 1989

I.2: AN ENGINEERING VIEW ON THE VISUAL SYSTEM-TECHNOLOGY INTERFACE

D. BOSMAN
University of Twente, The Netherlands

2.1. Introduction

The human-machine interface (HMI) will always be, in terms of hardware, a well defined collection of data channels. In the human to machine direction the inputs to the machine can be through selection of location coded devices (keys in a keyboard, areas in a touchpanel, push buttons, switches in an instrument panel); through spatial/angular displacements (mouse, slide wire, joystick, rotary controls) and through sounds, e.g. voice.
The machine to human direction is dominated by the visual and auditory channels where data transport is most dense, although small percentages (mainly feedback data) are through tactile receptors for contact measurands like forces, moments, form; and through proprioceptors which determine/ remember internal states like angular relations between fingers, part of the arms, body; head and spine, etcetera. This chapter is only concerned with the visual channel component (VSTI) of the HMI, in particular with the human factors.

At the instrumented side of the VSTI it has become possible to choose one out of several electro-optical conversion technologies. The cathode ray tube (CRT), the vacuum fluorescent device (VFD), the liquid crystal display (LCD), the electroluminescent panel (EL), the plasma or gasdischarge panel (PDP) and several other technologies of (as yet) minor importance. In chapters II.1-II.7 these technologies are discussed in general, including their optical properties which, in combination with and modified by the device parameters of particular designs, determine specific viewing characteristics. Technology properties involved are e.g. luminance and its spectral distribution; element/ electrode reflectances; limiting sizes, distances; possible forms of display elements and their particular point spread functions; background radiance/ reflectance; attainable contrast in a wide range of luminances (e.g. dimmability). Specific device parameters can modify the technology optical properties in ways which are in many cases open to analysis methods as discussed in chapter I.1 and, in turn, can also be modified by image-processing methods (display conditioning).

At the human side of the interface is the versatile. adaptive, operator who cannot be designed but who can be selected, trained, taught, experienced and sometimes be gifted. The VSTI should, in all its contributing factors and constraints, perfectly support the visual part of the operator's task. That

means it should updata/change the state of knowing of his apprehending mind
which needs and expects certain new data necessary to carry out the various
subtasks. The latter are strongly coupled to specific applications, examples
of which are given in part III: Applications. Many anatomical, physical and
psychophysical aspects of vision are independent of the information content in
the image and as such common to most display observation processes [2.1,2.2,
2.3]. They determine the quality of the perception of the image features where
the emphasis of our interest is on local features rather than global
properties. For instance, usually we are more interested in the effect of
local operators like the point spread function (PSF) [2.4], on e.g. the
distorsion of a font, or the blurring of character detail, rather than in its
Fourier transform (OTF) which predicts the effects of the PSF at a global
level like contrast reduction of periodic phenomena such as gratings (see
section I.1.3.2). Sometimes, however, one may prefer the transform domain
based on the suggestion that a similar transform takes place in the brain
[2.5].

Even when restricting ourselves to local effects described by physical and
psychophysical data, the relationships so obtained are not always conclusive
to use in an engineering model. For instance, many of the data published in
literature need conversion and/or formula fitting before being applicable; or
one has to choose one, the most appropriate, out of a number of suggested
relationships. Also, the validity of the data usually is restricted to short
duration experiments or operations; where display use can be a prolonged task.
But using the known facts still yields better VDUs and display panels than
when these facts are ignored. Thus in this chapter we propose to analyse what
display requirements there are (2.2) for legibility of symbols under various
circumstances and what the limits of visibility are for e.g. undesired
artefacts. With this in mind the three main aspects of the visual system are
addressed:
2.3 characteristics of the eye imaging system,
2.4 threshold and supra-threshold sensitivities,
2.5 colour in displays,
followed by some recommendations.

2.2. Legibility of symbols, robustness

In everyday use, the qualification "legible" for handwriting and print means:
"clear, easily read". Normally the contrasts involved are very high except for
some artwork; the types of symbols font traditional and continuous. That is
not the case with many matrix displays, see figure 2.1 for current CRT and EL

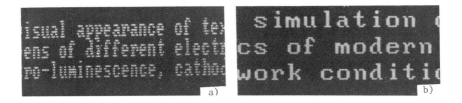

a)   b)

Fig. 2.1 Text on matrix organised images; a) cathode ray tube,
b) electroluminescence.

displays.

Furthermore the workplace geometries may differ from traditional reading
situations. Legibility can be defined for instance as a percentage of
incorrect readings in a paced reading task, or as a function of the time
required for self-paced reading of a representative text, or as the average
fixation time needed to read (nonsense) syllables at a selected error rate,
and variations thereof. It turns out that legibility is a function of many
parameters as shown in figure 2.2.

The symbol height determines the subtense under which the symbol is seen at a
given viewing distance. Legibility is different for given symbol subtense
under varying conditions of luminance, contrast, blur, resolution and
technology. For instance, for raster scan and dot-matrix displays, symbol
subtense must be larger for coarsely quantised symbols (6 lines/symbol → 36
minutes of arc, 12 lines/symbol → 15 minutes of arc) [1.1,2.6]. Symbol
quantisation is another important parameter. For the restricted range of
symbols such as the decimal numbers, the seven bar display is satisfactory.
For the range of alpha-numeric symbols the lower limit for symbol quantisation
is the 5x7 matrix. This yields just acceptable error performance under
favourable reading conditions and at maximum visual resolution.

Also involved is the parameter in figure 2.2 which is not objectively
measurable, namely "font" (see also chapter III.1). Several designs made
specifically for computer generated characters have been proposed [1.1,1.2]
and ergonomically validated [2.6,2.7]. For matrix displays with an orthogonal
grid with repertoire of only non-rotatable alpha-numerics the modified
"Huddleston" font presently appears to be optimal for the 5x7 del character
block (figure 2.3a). Newer proposals have been forwarded [2.7] using larger
arrays of non-square dels (figure 2.3b).

When the symbols should remain legible under rotation, like in the attitude
director indicator in aircraft, the size of the character block should measure
at least 11x14 dels to avoid excessive distortion of the symbols due to the

fact that the dels remain locked to the display grid. This effect is less
severe in cursive (stroke-written) symbols as only the endpoints of the
strokes written by the symbol generator are locked to the grid so that higher
effective resolution is obtained. The technique, like pixel phasing ([2.4],
section 1.3.3), is basically restricted to non-discrete technologies, e.g. the
CRT.

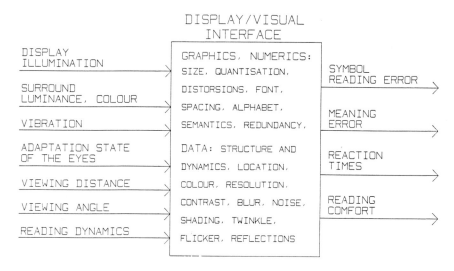

Fig. 2.2 Legibility performance factors.

Other factors which determine the legibility of symbols are:
- stroke width-to-height ratio or relative active area: the proportion of a
  symbol which actually emits or reflects light (important in dot-matrix
  representation);
- symbol width-to-height ratio, legibility is hardly affected by
  foreshortening at viewing angles up to 30°;
- symbol spacing, aids the recognisability of words, 25 to 200 % of average
  width seems acceptable with 50 % a good middle value;
- size of the alphabet, number of characters with maximally different features
  including sub- and superscripts.

Percent active area is important for displays with a uniform luminance
technology, e.g. reflective type displays, electroluminescence (EL). The
larger the area the more light output per dot, which benefits the legibility.
However, at constant luminous flux (e.g. LEDs) the luminance varies inversely
proportional with area, so that contrast increases with smaller active area.
Provided the resolution requirements of the eye are met, for LED type dot-
matrix displays small active area per dot or del is proven to be the more

Fig. 2.3 Samples of a) the 5x7 Huddleston (H), the Kinney (K) and modified
(M) fonts, b) proposed improvements [2.7].

economical way to increase legibility.

When a direct view display is vibrated with respect to the eyes, three effects
are possible which may cause deterioration of the legibility.

Firstly, the commonly experienced phenomenon with continuously (not
intermittently) energised displays is, that the contrast of edges
perpendicular to the sense of vibration become blurred. The reading error
depends on the frequency of vibration; at low frequencies the legibility
remains acceptable for the larger sizes of symbols [2.6].

Secondly, the effect associated with the periodically refreshed (flickering) type of display, with a frame rate higher than the critical fusion frequency and with moderate to long persistence (e.g. LCD, CRT). For large excursions in retinal angle (such as may occur in vehicles) multiple complete images are seen superimposed.

Thirdly, in matrix displays with short persistence dels (LED, DC plasma and EL) which are e.g. horizontally scanned and strongly vibrated vertically, the scanned lines appear to break-up to the extent that text and numbers become incomprehensible.

Collimated displays are much more tolerant to vibration.

Condensation, dust etc. may cause blur, but the display legibility must remain "sufficient". Blur degrades resolution and causes the MTF (section I.1.3.2) to fall off stronger at the higher spatial frequencies. Consequently, the symbol identification must not rely on high resolution. Assuming that symbol subtense is large enough compared to the PSF of the eye, actual practice has shown that 10 dels/symbol height (14 for TV) [2.6] are desirable (16 to 20 with requirement of subscript and superscript).

Under circumstances that errors have developed in the display (missing dels, missing lines, columns [2.1,2.8,2.9], the absence of confusion (robustness) is important. In figure 2.4 [2.9] an example is shown of the weakness of e.g. the numeral "8" which is easily changed into "6" and "9", or even "3" and "5", with very little dels missing; provided the losses are at the sensitive locations. In some technologies parasitic "on" dels also are probable, giving rise to other types of confusion. Sometimes very few incorrect dels can be responsible for a large change in meaning, as shown by the following examples:

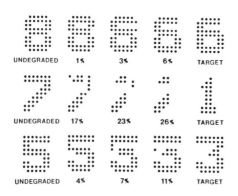

UNDEGRADED   1%   3%   6%   TARGET

UNDEGRADED   17%   23%   26%   TARGET

UNDEGRADED   4%   7%   11%   TARGET

Fig. 2.4 Lack of robustness against change of meaning due to a few missing dels [2.9].

– the car (cat) follows the truck (track), and
– he put his boot (foot,loot) on the table.

When the requirements of legibility and robustness have been met, the next goal is to improve acceptability [2.1,2.2], a notion which is even more difficult to make operational, but is a major commercial factor. At this stage of developments, the most important aspect yet is: proven absence of reading discomfort and of visual fatigue under conditions of intense, prolonged use. Research efforts into this area have not yet produced convincing results.

The key to display equipment design lies in understanding the physics of optics, materials and of electronics, in combination with signal (image) processing and computer science. It is impossible to model the operation of the human factors in similar exact terms, but for the visual channel this "scientific goal" has been pursued and sometimes with remarkable success (provided the cognitive system is not involved). In the next section this approach is followed: the eye is considered as a (non-linear) image-processor acting on luminous energy distributions.

2.3. Characteristics of the eye imaging system

2.3.1. General description

For display engineering purposes it is not necessary to consider the anatomy of the eye in detail. Like a camera it consists of an optical system, a light sensitive target and an image conditioning processor. Its operation is autonomous and self-adaptive. The optical system is made-up [2.10,2.23] of the cornea (strength 40 diopters), a variable size (2 to 8 mm) pupil and the crystalline lens with adjustable strength of 20 to 30 diopters. The focussing range of the young healthy eye is from 0.1 m to infinity. At rest the focus plane is a few meters away for wavelengths around 500 nm (green/blue green); but at infinity for 675 nm (red). The pupil area changes a factor 16 at most, insufficient for the 10 decades of luminance to which the eye can adapt given sufficient time. Clearly its function rather is to adapt to fast changing light conditions.

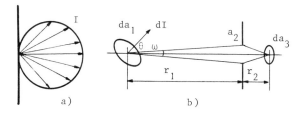

a)   b)

Fig. 2.5 a) Intensity distribution of ideal diffusing surface,
b) Image formation at the retina by the eye optics.

The scene is imaged onto the retina; it is interesting to note that only the luminance of the object and the size of the pupil determine the retinal illumination. Assume a flat, perfectly diffusing surface of area da_1 and luminance L,   see figure 2.5a which depicts the intensity distribution. It is mapped by the eye optical system with pupil area a_2, onto the "plane" of   the retina,   producing   the image of da_1 which obtains area da_3 (figure 2.5b). The lens system is considered (in simplified form) to be located in the   plane   of the pupil.   The   distance of da_1 to this plane is: r_1, and from the retina to this plane: r_2. From geometrical optics it follows   that   the   solid   angle ψ subtended   by   the   apparent area $da_1.\cos\theta$ is determined by $\psi = da_1.\cos\theta.r_1^{-2}$; which angle also equals $da_3.r_2^{-2}$ provided that da_1 is   sufficiently   large   to disregard smearing by diffraction and scatter:

$$da_1.\cos\theta = da_3.(r_1/r_2)^2 \qquad\qquad\qquad\qquad (2.1)$$

The   flux Φ in   solid angle $\omega = a_2.r_1^{-2}$ equals $d\Phi = I.\omega = Lda_1.a_2.r_1^{-2}.\cos\theta$, (compare [2.4] expression 1.19d) or, when $da_1.\cos\theta$ is replaced   by   its   image da_3: $d\Phi$ = $Lda_3.a_2.r_2^{-2}$.   The   same flux is projected onto da_3, be it that a transmission loss occurs represented by the transmission factor τ (= < 0.5). Consequently the retinal illumination $E = \tau d\Phi/da_3$ becomes:

$$E = La_2.\tau.r_2^{-2} \ [\text{lux}] \qquad\qquad\qquad\qquad (2.2)$$

The   eye   parameters τ and r_2 vary only slightly among individuals. Therefore the product $L.a_2$ is a   measure   of   retinal   illumination,   its   unit   is   the Troland; L in cd m^{-2}, a_2 in mm^2.
From   (2.2)   follows that the distance r_1 from the object to the eye, and also its surface size and inclination with the visual axis, do   <u>not</u>   influence   the illumination of the retina; <u>only its luminance and the area of the pupil</u>!

In   the   retina   millions   of receptors dissect the projected image into small elements. The receptors transform the photons captured in a certain (variable) integration time into a pulse train   through   the   intermediary   of   an   opto-chemical   reaction.   Retinal   post-processing   modifies pulse repetition frequencies such   that   the   signals   to   the   brain   emphasise   high   spatial frequencies.   The   intersection   of   the visual axis with the retina defines a region [2.10] of 90 mrad or 5   degrees   retinal   angle,   called   fovea,   where receptor density is most dense (about 200,000 mm^{-2}).

Example   2.1.   At   a viewing distance of 500 mm the area of sharp viewing is a fuzzily bounded region of 45 mm, which fits a 20   character   word   of   typed text,   or   the   business   end   of a pointer of an indicating instrument: see figure 2.6c. The total   picture   is   obtained   by   mentally   combining   such partial   images   while   these   are   obtained   in   a   succession of so called fixations (see figure 2.9).

There are two kinds of receptors: the rods which cover the luminance range of 10^{-5} cd m^{-2} to 3 cd m^{-2} (scotopic view, at the lower luminances completely achromatic), and the cones which cover the luminance range of 3.10^{-3} cd m^{-2} to 10^{5} cd m^{-2} and are colour sensitive (photopic view). The intermediate range where the receptor responses overlap is called mesopic view.

The density of the cones is maximum at the centre region of the fovea: the foveola, which comprises about 25,000 cones, no rods (see figure 2.6). The diameter of the light sensitive part of the cone, the outer segment, is about 1 μm, the pitch about 2.5 μm. Outside the foveola the proportion of rods over

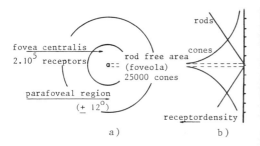

Fig. 2.6 a) Regions at the retina;
b) Density distributions of cones
and rods, [2.10]; c) Foveal viewing
at 500 mm distance.

cones increases rapidly, but the surface diameter of the receptors increases also, leading to decreasing receptor density with retinal angle.

The light quanta (photons) penetrating the outer segments of the receptors may, or do not, hit a molecule of the opto-chemical substance. When hit, the molecule reacts by separation into two parts which, incidentally, also involves a change in optical density: bleaching. The conversion rate is only a few percent. This reaction in turn increases the permeability of a cell membrane, which has consequences for the electrical state in the receptor. Through a complicated process an electric charge builds-up across the membrane of its output neuron. The charge, when it exceeds a threshold, releases an action potential that travels along a nerve fiber to the receiving neuron in the retinal system network. After suitable processing in the network the brain receives by means of 10^{6} nerve fibers an electrically coded replica of the image projected onto the retina, coded in the form of pulse repetition frequencies with rates between 5 and 40 pulses per second (pps) for "stationary" images and ultimately ranging up to 120 pps to follow dynamic changes. Given the enormous luminous range from low scotopic to high photopic, considerable adaptation of "the eye" must take place, and from a systems point of view mostly in the receptors themselves, since in front end stages saturation must be avoided. Fresh opto-chemicals must be made continuously at the rate given by the current adaptation level; when the eye is suddenly subjected to a much lower luminous level for which the compensation by the dilation of the pupil (figure 2.7) is insufficient, it takes a long time [2.11] (receptor volume over molecule fabrication rate) to fill the storage element so as to reach the required concentration associated with the new luminous level (figure 2.8). Under "normal" luminance conditions, it seems that both the cones and the rods tend to adapt to a sensitivity where the pupil size is most comfortable with respect to optical resolving power. But then the diameter is sufficiently small to cause diffraction effects; the Airy disc diameter (figure 2.13c) produced by a hole of 3 mm diameter which is 17 mm away from the target plane (in this case r_2 from pupil to retina) equals

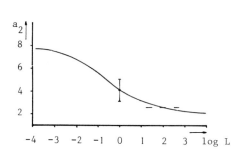

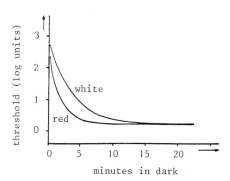

Fig. 2.7 Typical pupil diameter as function of luminance [2.10].

Fig. 2.8 Typical dark adaptation curve, [2.11].

Fig. 2.9 Saccades and fixations, [2.13].

7.5 μm for red, 5.8 μm for green and 4 μm for blue light. The foveola is thought to support a fixation spot, with the help of which image details are formed by means of eye and head movements. The major eye movements involved are saccades [2.1, 2.12], sudden jumps in viewing direction, shown in figure 2.9. At first the image part in the foveal region is taken in and provisionally interpreted (100 to 140 ms) before the decision is made whether to change the direction of viewing. By this sequential process much larger areas of the image than subdued by the fovea are seen sharp; the image is visually sampled by the saccadic movements of the eyeball. The fixation density is closely correlated with image features which evoke curiosity, or with difficulty of interpretation or seeing.

Example 2.2. The number of fixations in the text lines of figure 2.9 is larger for finer print; the angular span of the saccade tends to match the receptor density to image resolution. The duration of a fixation also varies with luminance, in a coordinated effort to maintain constant photon catch, depending on the adaptation state of the eye. See also example 2.1.

A saccade may span up to 20°, which takes about 70 ms, a small saccade ($\leq$ 1°) 20 ms [2.12]. For angular excursions larger than 20°, more than one saccade is required. It is impossible to keep the eyeball steady for prolonged time; as fixations last more than 140 ms, the period between saccades ranges from 160 ms to e.g. 1 s, depending on the situation. The accuracy of refixations is better than 2° to 3°, depending on the angle of redirection. Easy pointing angles of the eye without head movement is about + 25° (up), $-$ 35° (down), and 15° sideways [2.14]; as depicted in figure 2.10. In the area bounded by these limits the optimum viewing characteristics (foveal) can be utilised for primary displays. Display sampling frequency and sequential readings of non-correlated words or numbers done without re-orientation of the head, can attain reading frequencies ranging between 0.5 to 2 per second.

The ability to resolve detail in a still image is the static visual acuity, (SVA). It is measured under comfortable luminance and contrast conditions. Targets shown are e.g. Landolt rings, round fat rings with a gap equal to the ring thickness; in experiments the subjects must identify its orientation. Landolt rings have the advantage over e.g. Snellen letters (figure 2.11) that their MSD ([2.4], section 1.3.2) does not change with varying orientation, only their phase spectral distribution. In figure 2.12 [2.15] the SVA is given both as function of retinal angle and of retinal illumination. SVA decreases with eccentricity (larger receptor surface area, lower receptor density), but its dependence on retinal illumination needs further discussion (see example 2.5). The facts: the threshold of resolving power is about 0.4 minute for a gap or two short parallel lines; vernier acuity, the alignment error of two slightly displaced lines is even better. When predetermined information is available, like a long line should be straight, the healthy young eye can even detect a local deviation of about 2 arc seconds!

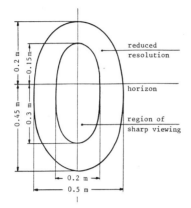

Fig. 2.10 Viewing area covered with
eye movements only, at viewing
distance of 0.35 m.

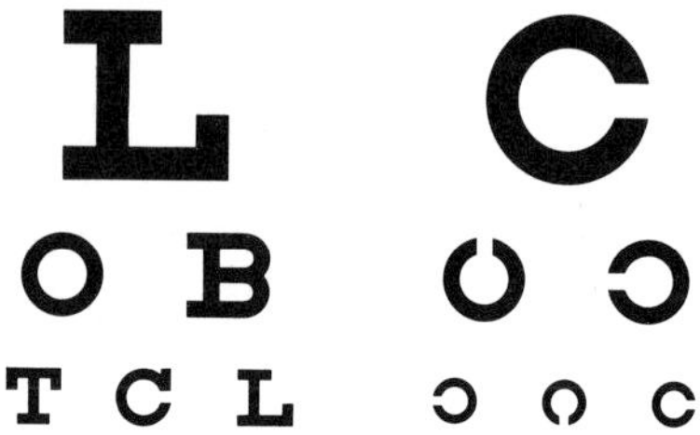

Fig. 2.11 Snellen letters and
Landolt rings.

Example 2.3.   The latter high sensitivity has severe consequences for e.g.
  CRT display design. At 400 mm viewing distance, the interline repeatability
  should be 4 to 10 microns at the screen; equivalent to a few nanoseconds
  depending on screen size. This is consistent with other findings [2.16]
  about the allowable standard deviation along the edge of a sharp, clean
  line.

SVA is impaired by bright light or reflections located within 1 rad (57°) of
the vision axis. The dynamic visual acuity (DVA), where the eye must follow a
feature in a moving image, is lower. For instance, at 1 rad/s angular velocity
the DVA is 60 % of SVA, at 2.5 rad/s only 20 %.

2.3.2. The eye as an image processing system

The optical system of the eye is far from perfect, reason to keep the iris
generally stopped at a small diameter to limit the magnitude of abberations
even though diffraction smears the image. At 2.4 mm pupil diameter the line
spread function (LSF, convolution of a PSF with a line) as depicted in figure
2.13a [2.3,2.17] is optimal in terms of minimum total smear: the centre part
(down to 50 % response, plus and minus 1 minute wide) is determined by
diffraction and the remaining tails which resemble back-to-back exponential
behaviour, are mainly due to diffusion by the neural tissue in front of the
receptors scattering the light ([2.4], example 1.2, section 1.3.2). Both the
diffraction and the diffusion low pass filter the image. A good model is

obtained as follows. The diffraction PSF of the pupil with area a_2 mm^2, projected onto the retina, is a Bessel squared function with effective area of $\frac{180}{a_2}$ μm^2. This spatial light distribution penetrates the neural tissue undisturbed by a fraction a; the remaining fraction 1-a is scattered in all directions after which 50 % (= $\frac{1-a}{2}$) adds to the undisturbed fraction. In mathematical form: the pupil PSF contributes at the retina the fraction a of the PSF(pupil); the neural tissue diffuses $\frac{1-a}{2}$ x PSF(pupil) * PSF(tissue) where * stands for convolution. The diffusion part dominates as to the luminous flux transmitted and, moreover, spreads it over a wide area (1440 + $\frac{180}{a_2}$) μm^2. The cross section of the PSF of the reflected light as measured by fundus photographs matches figure 2.13b. The magnitude part of its Fourier transform, the MTF ([2.4], section 1.3.2) vanishes at about 50 to 60 cycles per degree (cpd); that spatial frequency limit matches approximately the spatial Nyquist frequency of the receptor mosaic at the densely populated foveola, which is about 120 sampling points per (linear) degree.

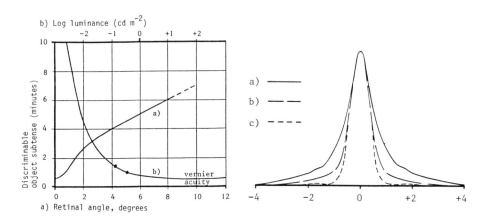

Fig. 2.12 Static visual acuity at very high contrast:
a) as function of the retinal angle,
b) as function of object luminance [2.15].

Fig. 2.13 Response of the eye optics [2.17]:
a) the line spread function,
b) the point spread function as calculated from a)
c) Airy disc PSF.

This matching optical presampling filter prevents the perception of aliasing ([2.4], section 1.3.3), only at the foveola, the point of regard [2.18]. The spatial sampling frequency decreases with retinal angle (figure 2.12a) while receptor size increases but nobody complains about Moiré perception in off-axis view, presumably because the non-regular (random) lattice makes it less conspicuous [2.18].

The spatial frequency response of the eye optics thus defines the upper limit of resolving power, measured as the lower threshold of seeing sinusoidal modulated gratings of varying frequency and contrast. The sensitivity so found tends to be larger than when measured with short bars where the lenght/width ratio is smaller than 5. The static visual acuity is closely connected to grating sensitivity. Both are dependent on the (average) retinal illumination, which is a fundamental property addressed below. It is shown that, due to the small receptor surface area combined with the short integration time (between 50 and 200 ms) and at luminous levels lower than the photopic range, the amount of photons contributing to the luminous sensation is very low e.g. in one receptor 300 per s at 1 cd m^{-2}; reason for pooling receptor outputs into receptive fields at the mesopic and scotopic ranges because the number of photons received by a receptor per integration time becomes too low to make effective use of.

Example 2.4. The receptor density is maximal in the foveola; their pitch is about 2.5 μm. Without corrugations etc. the surface area a_3 of the mosaic elements equals 5 μm^2. The amount of energy that impringes on it during the integration time $t(i)$, contributing to the photon counting process in the receptor, is determined by the illuminance of the retina (expression 2.2), decreased by the fraction of light that is reflected, absorbed and scattered away before entering the receptors. These factors lumped together yield a wavelength dependent spectral luminous efficacy curve $K.V(\lambda)$ (section I.1.4.1) with K defined as $K = \int \Phi_v(\lambda).d\lambda \,/\, \int \Phi_e(\lambda).d\lambda$; $K = 673$ ℓm W^{-1} for the photopic range, $V(\lambda) \leq 1$ and maximum for green light ($\lambda = 555$ nm). Using expression (2.2) with τ accounted for by K, the monochromatic radiation energy Q required to produce in one receptor the sensation of L lumen when La$_2$ Troland enters the pupil, is:

$$Q(L) = \{K.V(\lambda)\}^{-1}(La_2).r_2^{-2}.a_3.t(i) = 5.14 \frac{La_2}{V(\lambda)} a_3.t(i).10^{-6} \text{ Joule} \qquad (2.3a)$$

Given a receptor area of e.g. 5.10^{-12} m^2 and integration time $t(i)$ of 0.05 s, the radiation energy Q(L) becomes

$$Q(L) = 128 \frac{La_2}{V(\lambda)}.10^{-20} \text{ [J]} \qquad (2.3b)$$

For green light ($\lambda = 555$ nm) the spectral luminous efficiency $V(\lambda)$ equals unity; the energy q per photon: $q = 36.10^{-20}$ [J] so that the average number of photons $\bar{N}(L)$ associated with Q(L) is:

(green light): $\overline{N}(La_2) = 3.6(La_2)$ (2.4a)

A similar calculation can be made for (displays radiating or reflecting) white light. The average sensitivity of the eye over the entire luminous interval is obtained by substituting for $V(\lambda)$: $\int V(\lambda)d\lambda = 0.3$; and using average photon energy, yielding

(white light): $\overline{N}(La_2) = 11 \ (La_2)$ (2.4b)

For light with other spectral distributions (incandescent, fluorescent etc.), $\int V(\lambda)d\lambda$ becomes smaller and consequently more photons are required to produce the same luminous sensation. That fact may be the reason why older operators prefer green screen VDUs over e.g. white; apart from the extra colour contrast they also experience more photometric stimulation.

In steady state regime the pulse rate of the signals to the brain varies from 5 to 45 pulses per second for a luminance input range of 2 decades; the compression is brought about by a non-linear brightness function $B(La_2)$ which may be thought of as consisting of the product of two separable stages:
- the eye optic point spread function (figure 2.13) which produces a photon
  flux density distribution $\dot{n}(r)$ at the retina; the photon flux density at the
  receptor is thus determined by the PSF magnitude gain g at the location of
  the receptor;
- a function f converting the photon count $N(La_2)$ of a receptor into the optic
  nerve signal $B(La_2)$.
  Because the brain is more interested in local contrasts $C(v) = \Delta L/L$, than
  in spatially averaged intensities, the differential gain $b(La_2)$ is more
  important than $B(La_2)$:

$$b(La_2) = \frac{dB}{d(La_2)} = \frac{\partial B(La_2)}{\partial N(La_2)} \cdot \frac{\partial N(La_2)}{\partial (La_2)}$$ (2.5)

To obtain an engineering model which approximates the processes of visual imaging, one may proceed as follows.

A spot or disc of diameter d minutes and uniform luminance L over its area and against a dark background, would, without diffraction and scatter, produce at the retina an image with area $a_3 = \pi(3438)^{-2}(d/2)^2 r_2^2 = 19.2 \ d^2 \ \mu m^2$ and illuminance (expression 2.2) $E = (La_2)r_2^{-2} \tau = 0.0035\tau(La_2)$; a_2 in mm^2. The equivalent base area of the PSF(eye) is about $1440 + \dfrac{180}{a_2} \ \mu m^2$. In the convolution process of image and PSF both areas add, so that an average PSF gain $\bar{g}$ results

$$\bar{g} = \frac{19.2d^2}{19.2d^2+1440+180a_2^{-1}} \tag{2.6a}$$

Further there are lens abberations which smear the spot response. A first approximation is to increase d by a certain amount, which is consistent with the eye resolution, e.g. replace in (2.6a) d by (d+0.4):

$$\bar{g} \approx 0.014\{(1+0.125a_2^{-1})(d+0.4)^{-2}+0.014\}^{-1} \tag{2.6b}$$

For large discs (a background) $\bar{g} \approx 1$ as is the case for expressions (2.4); for a point source of d = 0.6: $\bar{g} \approx 0.013$. Thus, in general, the photon flux density $\dot{n}(r)$ at the retina is reduced; for a disc of size d and luminance L_2 against a large background of luminance L_1 the retinal contrast C(g) becomes

$$C(g) = \frac{\bar{g}_1(L_2-L_1)}{\bar{g}_2L_1} \approx \bar{g}_1 \frac{\Delta L}{L} = \bar{g}_1 C(v) \tag{2.6c}$$

Using the same reasoning which obtained expressions (2.4), using the relation $\dot{n}(r) = E/K.V(\lambda)$, one obtains

$$\dot{n}(r) = \bar{g} \frac{La_2}{r_2^2} \cdot \frac{1}{KV(\lambda)} \cdot \frac{1}{q} = \bar{g}K_0La_2; \quad K_0 = \frac{1}{K.V(\lambda)qr_2^2} \tag{2.7a}$$

For La_2 in Troland and white light, $K_0 = 44.10^{12}$; for slightly dimmed incandescent light as used in some ergonomic experiments [2.21], $K_0 = 50.10^{14}$. The number of photons received by a receptor surface of area a_3, during the integration time t(i), can now be determined:

$$N(La_2) = \bar{g} K_0a_3t(i).(La_2) \tag{2.7b}$$

and the factor $\frac{\partial N(La_2)}{\partial(La_2)}$ in expression (2.5) becomes

$$\frac{\partial N(La_2)}{\partial(La_2)} = \bar{g} K_0a_3t(i) \tag{2.7c}$$

The factor $\frac{\partial B(La_2)}{\partial N(La_2)}$ is more complicated, although not difficult, to calculate because both quantum noise and neural noise are involved causing a threshold of seeing as depicted in figure I.1.21 [2.21].

The quantum noise of the stochastic variable $N(La_2)$ has a standard deviation $\sigma\{N(La_2)\}$; the neural noise $\sigma(r)$. When the two sources are independent, the total noise variance $\sigma(t)^2$ obtained equals the sum of the variances

$$\sigma(t) = [\sigma\{N(La_2)\}^2 + \sigma(r)^2]^{0.5} \tag{2.8a}$$

To be perceived, a variation $\Delta N(La_2)$ in the image photon catch must exceed $\sigma(t)$ by a certain factor which, from detection theory [2.19], is known to be 0.86 for 50 % probability of detection:

$$\Delta N(La_2) \geq 0.86 \, \sigma(t) \tag{2.8b}$$

This is true for any area of the detector surface; for one receptor and also for combinations of receptors which cooperate (pooling) to counter adverse effects of quantum noise at low luminances. Under photopic conditions pooling is hardly necessary.

A simple model of the quantum noise is the following [2.20]. The number of photons hitting any particular area obeys the Poisson distribution, which has the important feature that the mean of the stochastic variable equals its variance; $\sigma\{N(La_2)\}^2$ can be replaced by $N(La_2)$ and expression (2.8a) turns into

$$\sigma(t) = [N(La_2) + \sigma(r)^2]^{0.5} \tag{2.8c}$$

so that the minimum required variation $\Delta N(La_2)$ becomes

$$\Delta N(La_2) \geq 0.86 \, [N(La_2) + \sigma(r)^2]^{0.5} \tag{2.8d}$$

Implicit in (2.8d) is the assumption that exceedance of the right hand term by $\Delta N(La_2)$ is equivalent to exceedance of one just noticeable difference (JND) in brightness. By definition, $\Delta B = 1$ JND is one increment on the brightness scale; therefore $\Delta B/\Delta N(La_2) = \{\Delta N(La_2)\}^{-1}$. There is no evidence that $B(La_2)$ is a piecewise defined function of L; the differential gain $\partial B(La_2)$ exists for any L. Since only the JND is observable, one can only assume that the differential gain is about equal to the difference response obtained from 1 JND:

$$\frac{\partial B(La_2)}{\partial N(La_2)} \approx \{\Delta N(La_2)\}^{-1}; \text{ so that (2.5) becomes}$$

$$b(La_2) = \frac{\bar{g}K_0a_3t(i)}{0.86[N(La_2)+\sigma(r)^2]^{0.5}}$$ (2.9a)

From ergonomic experiments [2.21] it is known that at high luminances and large d, where the quantum noise contribution becomes small compared to the neural noise, the threshold contrast becomes constant: $(\Delta L/L)_{th}$ = 0.0026. Consequently, our model assumptions lead to the following expression for the photon contrast at the retina:

$$\frac{\Delta N(La_2)}{N(La_2)} \geq 0.0026 \ [111,000\{N(La_2)\}^{-1} + 1]^{0.5} \ \text{or}$$ (2.9b)

$$\bar{g}.C_{th}(v) \geq 0.0026 \ [\frac{90}{\bar{g}La_2} + 1]^{0.5}$$ (2.9c)

using expressions (2.6c) and (2.7b) with $a_3 = 5.10^{-12} \ m^2$, $t(i) = 0.05$ and $K_0 = 50.10^{14}$ which is appropriate for comparison with measured data [2.21].

In figure 2.14 the block diagram associated with (2.9c) is depicted which,

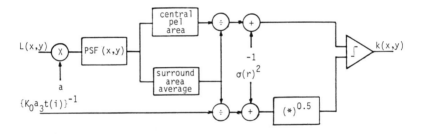

Fig. 2.14   Threshold contrast exceedance structure suitable for image-processing.

being based on point spreading, should also enable to estimate threshold data for object forms other than discs. It is often argued that the optic eye PSF is of no importance because it is largely compensated by a (deconvolving) cortical PSF. Expression (2.9c) shows that the non-linearity incurred by the threshold forbids the interchanging of smearing before detection and the for subsequent engancement. That is particularly true for more complicated objects such as two parallel lines, the smearing of which is analysed in

Example 2.5.   Consider the size of symbols and the resolving power needed.
   Assume that the luminance is sufficiently high so that no extra receptor pooling is required. The PSF/LSF of the eye optics produce fuzzy images at the retina. For ease of computation the LSF cross section after sampling is

approximated by the discrete receptor areas (figure 2.15a) resulting in a
symmetrical staircase with areas 0.15, 0.4, 1.0, 0.4, and 0.15. Simple
discrete convolution of the staircase LSF with the uniform receptor PSF
would again produce a staircase result, but the eyeball vibrates
rotationally (micro tremor) in a random manner with standard deviation of
between 1 and 2 receptor diameters at a centre frequency of about 70 Hz.
This phenomenon makes the convolution result again continuous as depicted by
the full line in figure 2.15b. The receptors take samples of the areas under
this curve, between the dashed lines, with width equal to the diameter of
the receptors (marked underneath the abcissa). Their outputs are
proportional to these areas. In figure 2.15c the result is depicted of such
a convolution with two parallel lines of width equal to the receptor
diameter; each partial contribution given by the dashed arrows. The sums of

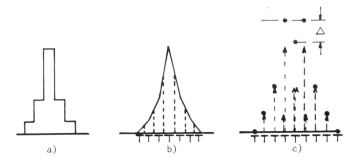

Fig. 2.15 On the explanation of line resolving power.

the arrows are the receptor outputs, shown by the fat dots. The response
character is a function of the distance d between the lines. For d about two
times the receptor diameter and larger, the contrast between the dip and the
peaks (figure 2.15c) should exceed the threshold contrast at the receptor
level (expression 2.9c); becoming progressively more discernible. For
smaller d contrast reversal takes place an the two thin lines are seen as
one.

It follows that for d < 1.5 receptor diameters no amount of luminance and
contrast can produce the sensation of two lines; it is a geometrical limit. In
practice this limit of d is slightly lower due to the hexagonal packing, but
the principle is confirmed by experiment.
The required amount of contrast can also be calculated by applying detection
theory. A relative check is easily made: increasing d from 1' to 1.5' reduces
the required luminance at fixed contrast by roughly a factor 2, evidenced by
the dots in figure 2.12. Of course, the location of these dots provides
information on the range of luminance where maximum SVA may be expected.

When reliable detection by each receptor can no longer be assured because the accumulated average energy N(L) becomes too low, the retina can counteract by pooling outputs of locally related receptors into socalled "receptive fields". [2.23]. Positive pooling reinforces the output due to spatially correlated luminance contributions; the uncorrelated (random) contributions tend to cancel, i.e. average out. The total variance of the receptive field is given by the sum of the variances of the contributing receptor outputs involved; leading to an SNR(C) improvement like in the following PSF example.

Example 2.6. In a hexagonal packing the first ring consists of six, the second ring of twelve receptors. Assume (see figure 2.16) that the first ring receptors each contribute 30 % of the central receptor output signal, those of the second ring 5 %. When combined linearly the total output, due to spatially 100 % correlated local features, equals 3.4 times that of the central receptor. If the image noise is spatially uncorrelated, the noise gain of this linear combination (pool) can be found by summing the 19 variances of the weighed noise signals and taking the square root. At the central receptor the weight of the variance equals unity; for each of the first ring neighbours

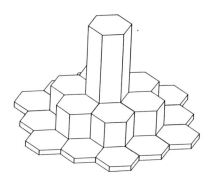

Fig. 2.16 PSF properties, discrete approximated.

the transfer is obtained by multiplying the (central) variance discrete approximated by the squared amplitude transfer, i.e. 6 times $\{0.3\}^2$, and so on.

Then the standard deviation σ(out) in the PSF output signal equals $(1+6 \times 0.09+12 \times 0.0025)^{0.5}$ = 1.25 times the input standard deviation σ(in). This simple spatial low pass filter thus achieves a SNR(C) improvement of 3.4/1.25 = 2.7 times (at the expense of some resolution).

It has been suggested [2.24] that the resulting PSF base area is inversely proportional to luminance in that regime, with the consequence that the low pass spatial cut-off frequency of this retinal action is proportional to the square root of local average luminance: compare figure 2.12b for $\{\log L\} < -1$. The receptive fields need not be circular [2.23] but, it seems, can take on forms with preferred directions e.g. along contrast features in the image. Then the "MTF" becomes image dependent, preserving both the local SNR(C) and the resolution across a contrast gradient. For instance, the contrast sensitivity CS ([2.4], section 1.4.2) for targets of comparable sizes in all directions (circles, squares etc.) is lower than the CS for oblong objects with size ratio > 5 (e.g. gratings).

It turns out that, besides all-positive pooling, there are also (image dependent) neuronal interactions with an enhancing character. In vertebrate eyes the operation is believed to be linear; the associated (cortical) PSF can be regarded as having negative sidelobes (figure 2.17a). Convolved with an image, such oscillatory PSFs show a gain for the region of spatial frequencies with periods similar to the period of the waviness in the PSF. With only one sidelobe as depicted in figure 2.17a the region of gain is not sharply defined, i.e. so broad, that the MTF (figure 2.17b) seems to differentiate in the lower region of the spatial frequency spectrum. Consequently the response to edges produces overshoots as shown in figure 2.17c.

The process can functionally be explained by assuming that each receptor output is decreased through a joint action of direct neighbours, by a factor which is proportional to a weighed combination of the outputs of these neighbours [2.23]:

$$f(i,0) = f(ni,0) - k \sum_{m=1}^{n} a_m . f(i,m) \qquad (2.10)$$

In this expression $f(i,m)$ is the (decreased) output frequency of receptors in the m-th ring around the inhibited cental (m = 0) receptor; $f(ni,m)$ is the uninhibited response of the receptors. The coefficients a_m are the weight

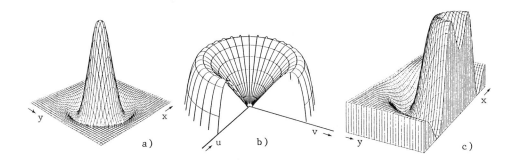

a)   b)   c)

Fig. 2.17 a) The apparent PSF of the eye with enhancement of high spatial frequencies [2.25]; b) The MTF associated with the PSF of a); c) disc response with lateral inhibition, notice overshoots [2.25].

factors; these are very small ($a_1 \simeq 5$ %, $a_2 \simeq 1.5$ %, etc.) due to the fact
that so many neighbours partake in the process. Operation (2.10) can only
decrease signal magnitude, nowhere in the spectrum is actual gain: a byproduct
of the linear enhancement process is the loss of signal to noise ratio. Figure
2.18 is an example of this operation with noise, SNR(C) $\simeq$ 3; the peak and

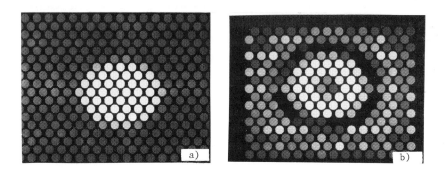

Fig. 2.18   Intensity plot of a) luminance distribution of disc target
on background (1 pixel gradient; b) response of eye with post retinal
enhancement process, note noisy character.

troughs (overshoots) are clearly visible in the otherwise noisy picture. The
eye can only afford this process when the SNR(C) is sufficiently high, i.e. in
the photopic range. Therefore, at low luminance levels the peak of the MTF
graph disappears.

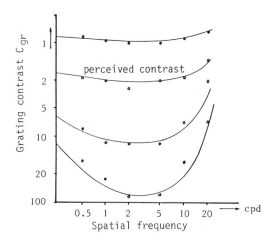

Fig. 2.19   MTF curves for supra-threshold
conditions, curves of constant perceived
contrast [2.27].

At supra-threshold levels
the left hand part of the
MTF graph also flattens with
respect to the peak as shown
in figure 2.19 [I.1.17,
2.27]; now not because of a
shortage of photons. This
suggests that as the SNR(C)
increases, not only the
areas of receptive fields
decrease, but also the depth
of the negative sidelobes,
i.e. the magnitude of the
lateral weight factors in
(2.10). The curves in figure
2.19 are modelled by [1.17]:

$$C_{st}(\rho) = 0.14 \{C_{gr} - C_{th}(\rho)\} \tag{2.11}$$

wherein $C_{st}(\rho)$ is perceived contrast at supra-threshold, C_{gr} is the actual, physically verifiable contrast of the grating used in the measurements and C_{th} the perceived contrast at absolute threshold (just noticeable difference); ρ is the spatial frequency in cpd.

Supra-threshold sensitivity can be determined in e.g. matching experiments, comparing perceived contrast of a testgrating with that of a reference of differing frequency. Expression (2.11) is verified in the interval $1 \leqslant \rho \leqslant 12$ cpd. At very low (scotopic) luminances the amount of available photons is so low, that statistical improvement processes such as described above become difficult or impossible; the eye is really in a state of photon counting and post retinal processes must interprete the sparsely available information. The average display engineer is insufficiently aware about this regime: it takes specialists in night vision.

2.4. Threshold and supra-threshold sensitivities

2.4.1. Threshold data

The perception of luminance variations (in time and distance) is restricted by lower bounds, the thresholds. The threshold is defined as the just noticeable difference (JND) at 50 % probability of detection. Obviously in display viewing the JND is not acceptable; detection probability should be 99 % or better.

There are several kinds of threshold:
- contrast as function of size of the object or of spatial separation of two or more objects,
- temporal variation of stimulus and/or background, also as function of size,
- spatio-temporal variations, as experienced in e.g. movements in the image,
- colour contrast.

Example 2.7. Calculation of luminance contrast threshold.
Expression (2.9c) in combination with (2.6) calculates the contrast threshold as function of object luminance and size, at a given pupil size and corresponding eye PSF:

$$C_{th}(v) = \frac{\Delta L_{th}}{L} = 0.180[1+6600 \frac{(1+0.125a_2^{-1})(d+0.4)^{-2}+0.014}{La_2}]^{0.5} \; x$$

$$x \; \{(1+0.25a_2^{-1})(d+0.4)^{-2}+0.014\} \qquad (2.12)$$

The coefficients have been slightly modified to improve the fit with the Blackwell's data [2.21].
In the following table the calculated threshold data (M) are compared with the

Table 2.1

L[cd m^{-2}]	3430		343		34.3		3.43	
a$_2$[mm^2]	2.27		3.14		5.7		11.3	
d[^{1}]	M	B	M	B	M	B	M	B
3.6	0.0155	0.0154	0.0179	0.0190	0.0272	0.0270	0.0528	0.0476
9.7	0.0045	0.0046	0.0047	0.0047	0.0059	0.0057	0.0097	0.0089
18.2	0.0031	0.0035	0.0032	0.0035	0.0038	0.0037	0.0060	0.0053
55.2	0.0026	0.0028	0.0027	0.0028	0.0031	0.0029	0.0048	0.0037
121	0.0025	0.0027	0.0026	0.0027	0.0031	0.0030	0.0047	0.0034
360	0.0025	0.0027	0.0026	0.0027	0.0031	0.0028	0.0046	0.0033

measured data (B). The agreement is very good: except for low luminances because the more sensitive rods come into play, and for small disc size because the average gain $\bar{g}$ is insufficient to describe the PSF shape effect.

2.4.2. Brightness scaling

A model for threshold response is not sufficient to determine visual responses except for display artefacts such as visibility of separation lines, faulty dels etcetera. To obtain insight in the displayed image as such, we need a model of supra-threshold response in terms of how many (threshold) levels have been exceeded. Returning to (2.9a) one obtains [2.22]

$$\frac{dB(La_2)}{d(La_2)} = \frac{K_0a_3t(i)}{0.86[\{N(La_2)\}+\sigma(r)^2]^{0.5}} \qquad (2.13a)$$

Remembering that under non-quantum limited conditions $\sigma(r)/N(La_2) = 0.003$, and using again the data $K_0 = 50.10^{14}$, $a_3 = 5.10^{-12}$ and $t(i) = 0.05$, one finds (incandescent light)

$$\frac{dB(La_2)}{d(La_2)} = \frac{1}{0.0026} \frac{g}{(gLa_2)^{0.5}(gLa_2+90)^{0.5}} \tag{2.13b}$$

and the brightness B is obtained by integration:

$$B(La_2)= \frac{1}{0.0026} \int_{g_1L_1a_2}^{g_2L_2a_2} \frac{du}{u^{0.5}(u+90)^{0.5}} = 770 \, \ell n \, \frac{(g_2L_2a_2)^{0.5}+(g_2L_2a_2+90)^{0.5}}{(g_1L_1a_2)^{0.5}+(g_1L_1a_2+90)^{0.5}} \tag{2.14}$$

The integration boundaries are the product of the amount of light through the pupil (Troland) and the eye PSF (average) gain $\bar{g}$ for the object ($\bar{g}_2$) and the background ($\bar{g}_1$).

The result $B(La_2)$ must be rounded to the nearest integer to obtain the number k of JNDs associated with the object stimulus. The structure of the image processing software which realises (2.14) is shown in figure 2.20.

In the interpretation of this result two things need attention. Firstly the model assumption of an energy sensitive detector with threshold yields a logarithmic function instead of a power function [2.28] with exponent 0.33 as suggested in literature. Without discussing the relative merits of the choice, one may observe that both functions are very much alike (within 10 % over a range of 70, compare $\ell n \, x$ with $(x-1)^{0.33}$).

Secondly the subtraction of the results of the two paths is known in image processing as "enhancement by unsharp masking": the operation of expression (2.10) is performed with f(ni,o) the response of the receptor channel at the

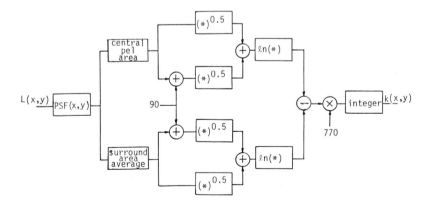

Fig. 2.20 Brightness calculation $k(x,y) = f\{L(x,y)\}$. Note enhancement by "unsharp masking" technique: weighted subtraction of central and surround responses.

centre of the eye PSF and k Σ a$_m$.f(i,m) the response of the channel to the
surround area average. When the weight factors of both channels are equal, and
the image is uniform, the output equals zero: only transitions (e.g. edges)
produce a change in k(x,y).

In figures 2.21, 2.22 and 2.23 three results relevant to display design and

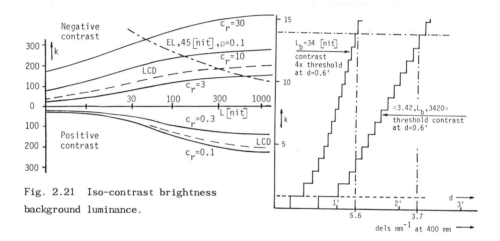

Fig. 2.21 Iso-contrast brightness
background luminance.

Fig. 2.22 Brightness k vs disc size d.

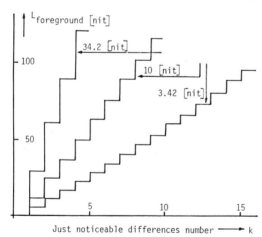

Fig. 2.23 Required luminance vs
brightness k for point source d = 0.6′.

evaluation are presented, for
circularly symmetric processing.
In line with evidence that
contrast features can be extracted
with directional preference,
neither the PSF at the beginning
of figure 2.20, nor subsequent
PSFs acting on k(x,y), need be
circular symmetric.

Additional look-up tables (LUTs)
may convert the computed k's into
probabilities of seeing certain
line segments, to implement the
notion of legibility. And, of
course, the diagram may be
extended to cope with temporal

effects (different delays/filters in the two branches) and colour detection
effects. These complicate the issue very much; the greatest benefit is
obtained from simulating early static visual processes in the image-processor.

Put together, the major eye characteristics discussed in this section lead to an example of the sequence of operations on a character displayed on a CRT face (except for the retinal sampling which is too fine to be reproduced here) as shown in figure 2.24.

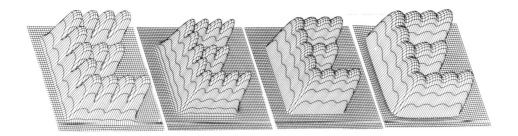

Fig. 2.24 Sequence of operations in the process of visual acquisition:
a) letter E as formed on a CRT screen by typical PSF;
b) the image as projected onto the retina, smeared by eye optics; (geometrical distorsion by curved retina omitted);
c) response of the receptors (log operation);
d) image after enhancement by lateral inhibition.

2.4.3. Flicker thresholds

The eye is sensitive to temporal changes [2.29] in brightness, in order to see movement in images. Brightness, not retinal luminance (in Trolands) is the variable: a thin line of uniform width and varying luminance seems to vary in width; a small blinking source seems to vary in size. At high temporal frequencies (> 80 Hz) the perception of flicker is gone, one sees an average intensity. The frequency at which the sensation of flicker just disappears, is the critical fusion frequency (CFF).

In display engineering the sensitivity for temporal high frequency variations is a nuisance; it forces the display designer to choose a frame rate higher than the CFF even when the image refresh rate is much lower. Flicker sensitivity can also be used to advantage to draw the observers' attention: with blinking cursors, highlighting text, and in analogue indications (e.g. bargraphs). Such temporal modulation generally occurs at the frequency where flicker sensitivity is maximum (see figure 2.25).
Models of temporal sensitivity are [2.30]

– the Talbot-Plateau law which postulates linearity in perception of intermittent stimuli L:

$$\bar{B}(L) = B(\bar{L}); \quad \bar{L} = \frac{1}{t} \int_0^t L \ dt \tag{2.15}$$

- the Ferry-Porter law

$$CFF = c_1 \ell n \ La_2 + c_2; \ La_2 > 10 \tag{2.16}$$

with magnitude La_2 in Troland

- the Granit-Harper law which gives the relation between CFF and the size of the flickering source

$$CFF = c_3 \ \ell n \ d + c_4; \ d < 7^o \tag{2.17}$$

where d is the size of a square or round stimulus in degrees. The expression is claimed to be valid over 7 decades range, be it that the parameters c_3 and c_4 must be adjusted.

In figures 2.26 and 2.27 the measured data [2.30] are given.

The PSF gain $\bar{g}$ (expression 2.6b) is not operative because the area of the receptors involved is much larger that the 1500 μm^2 or so of the PSF(eye).

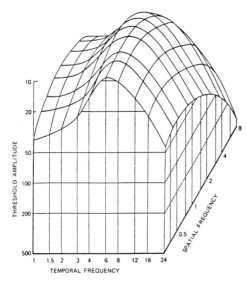

Fig. 2.25 Spatio-temporal amplitude threshold surface, for circular (Bessel, J_0) gratings; 16° degree field, 10^3 Td, natural pupils [2.31].

That this must be so can be understood by considering that the t(i) of a receptor is long compared to the period of the CFF. To obtain a sufficient sampling rate (at least one every 6 ms), groups of at least 10 receptors must cooperate to establish the temporal modulation at every spatial location; otherwise, aliasing would cause temporal low frequency interference. For positive detection even at threshold, the brain needs corroboration by adjoining groups of receptors; the higher the flicker frequency, the more receptors are involved and the lower the spatial frequency response (resolution) see figure

2.25. The combined effects of expressions (2.16) and (2.17) are described by

$$CF = 8 \ \ell n \ La_2 + 4 \ \ell nd + 5 \tag{2.18}$$

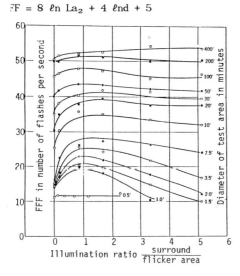

Fig. 2.26 Foveal fusion frequency and illumination of surrounding field [2.32].

d in degrees, La_2 in Troland CFF in Hz; and which is fairly useful for $La_2 > 100$, $1 < d < 30$. Of course, the coefficients depend on the spectral distribution of L (refer to 2.7a) and in this case the source is incandescent. For other sources e.g. phosphors, EL, LED the coefficients change. Flicker tends to be more annoying and fatiguing in peripheral view where the sensitivity is also higher than in front: i.e. in displays to the side of the screen that is watched.

In expression (2.18) surround luminance is not accounted for; the peripheral average luminance influences the CFF a little but even for the push button display the effect is small, see figure 2.26.

In displays without intrinsic memory the consequence of raster scan and line-at-time addressing is that each del is fully turned on for a short period. It is argued [2.33] that the fundamental component in a train of short pulses equals twice the average. This is also approximately true for electro-optic exponential decays (persistence) with time constant which is short compared to the del addressing period. The modulation depth MD of the fundamental frequency then is given by the ratio of the amplitude of the fundamental component over total background luminance: i.e. the sum of the average del luminance plus the reflected luminance. In figure 2.27 the CFF is given for pulse excitation of phosphors with different decay times. The duty cycle was 2 %, the (sharp edged) spot size 4 mm. With a P1 phosphor, 4 kV accele ration, and pulse rate of 30 s^{-1}, the average spot luminance is 17.1 cd m^{-2} (5 fL). Average del luminance is proportional to pulse rate. It is clear that the CFF of the display eye combination is higher for short persistence phosphors. Because EL and LED electro-optic conversions have very short turn-on and decay times the addressing rate of these types of display must also be high; in environments where vibration exists even up to 400 Hz.

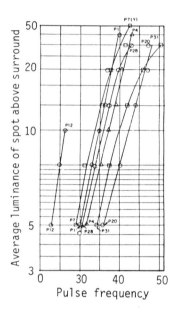

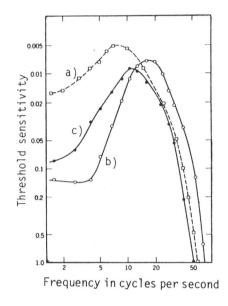

Fig. 2.27 Fusion pulse rate vs
phosphor type [2.33].

   average of 3 observers; 5/32
   inch spot, 3 f.L. surrounds.
   4 KV acceleration for P1, P7
   10 KV acceleration for P4
   P12, P20, P28, P31
   2 % duty cycle.

Fig. 2.28 Modulation depth [2.34] for
flicker   fusion   as   function   of
frequency. Harmonic modulated white
light with retinal illumination of 100
Td. Testfields: a) 2^O, light surround;
b) edgeless; c) 4°, dark surround.

At   very low temporal frequencies f_t <
10 Hz the CFF curve peaks between 7
and 14 Hz, depending on circumstances.

For large, edgeless fields [2.34], the very low frequency threshold MD   equals
MD ≈ 0.13 while at the peak f_t ≈ 13 Hz,   MD ≈ 0.0065: a factor of 20
difference! But for a sharply defined 2^O flickering field the max gain   occurs
at f_t ≈ 9 Hz, MD ≈ 0.008 compared with MD ≈ 0.013 at f_t ≈ 1 Hz.
The   surface of figure 2.25 shifts vertically for other spectral distributions
of the luminous source, giving tremendous variations in attention value at low
f_t, but little variation in CFF.
In the image-processing model of figure 2.20 the low frequency behaviour up to
about 10 Hz to 15 Hz can be simulated (not: structurally modelled) by
inserting different group delays in the two branches with the consequence that
if the delay difference Δt = $(2f)^{-1}$, the unsharp masking process favours
temporal   frequencies f_t = $(2\Delta t)^{-1}$. The data for edgeless fields (figure 2.28)

provide information about the parameters involved. The model can be used to estimate the perceived flicker of more complicated structural forms.

2.5. Colour in displays

2.5.1. The value of colour in displays

The use of colour in display technology [2.35,2.36,2.37,2.38] is mainly for coding purposes, to improve perception or attention value by adding colour contrast to luminance contrast. The concept of just noticeable differences (JND) is more appropriate to use of colour in displays than the colour sensation proper. One cannot very well describe the mechanisms involved, without the psychological concepts of perceived colour [2.10,2.23].

Achromatic light is defined as not possessing a hue. Hue is the attribute of a colour perception denoted by red, yellow, green and so on: the degree of saturation being the "difference" between the perceived colour and the (almost) achromatic stimulus most resembling it. Brightness sensation is caused by the perceived luminance of a luminous source, independent of its colour. Chromaticness of a colour is determined by the attributes hue and saturation.

Perceived colour cannot be measured objectively. For instance, the perceived colour depends on the eye's adaptation state which, among others, is influenced by the dominant colour preceding the colour stimulus and by the prevailing colour of the surround. Even under ideal conditions a specific colour sensation Q can be induced by radiant fluxes with different spectral distributions (metameric colours). With the eye adapted to other conditions, these metameric colours are perceived with differing hues and saturations. Similar effects are possible with isomeric (identical spectral distributions) colours. An example of simultaneous colour contrast induction is the perceived colour (hue) of e.g. a green line which extends into both brown and blue backgrounds; in order to maintain the same hue along the line the objective colour (as measured with a spectra radiometer) must be changed at the intersection with the two backgrounds. Another example is that certain colours change hue with luminance and with ambient (surround) illumination [2.37], [2.38,2.39]; the detection thresholds are also colour dependent, and may vary considerably from person to person.

The colour vision properties of the eye are poor to absent for very dim, low light level scenery (scotopic vision); also at the other extreme, high level (surround, veiling) luminance. Most experiments with eye response to display colour were carried out at comfortable luminances; especially in the last two decades much attention is given to the extrema. At very low light levels, the general threshold (luminance) varies from 0.003 Td to 0.008 Td depending on wavelength; the specific threshold (hue) coincides with the general threshold for red (0.003 Td) but a colourless interval exists for green to blue (threshold 0.0023 Td). These figures apply to the sensitivity of a dark adapted eye and field size of 1.5°. With white light added the thresholds increase considerably. In the range of 200–1000 Td the ratio of white light stimulus over threshold chromatic stimulus is about 45 for red, 5 to 10 for yellow–green and 100 to 200 for blue [2.39]. Apparently red and blue are more saturated colours than yellow–green. However, blue should not be used for small symbols since in the fovea the blue cones are not well represented like green and red and thus resolution is poor. Yellow–green needs high intensity for maximum saturation. At high to very high luminance, all colours wash–out [2.35,2.40]. Observing these facts, the number of just noticeable luminance levels [2.10,2.35,2.36,2.37,2.41] vary strongly with luminance, wavelength and saturation. Similarly the just noticeable difference in (dominant) wavelength is far from constant; for positive identification of a colour under a wide range of conditions specified for a display, the number of useable colours is small (e.g. 5 to 7, [2.35,2.41]) and limited to hues wide apart. The use of hue is also restricted by accommodation difficulties of the average healthy eye and by the existence of certain colour anomalies. Yet under favourable conditions the wavelength JND is only a few nm which places severe requirements on the tolerances in the driving signals of a RGB display.

Degree of saturation is a very poor code. In addition, it was found [2.37] that very saturated colours, used in low to normal surround luminance conditions, induce fatigue and must be avoided. Because brightness coding also affords low resolution, colour coding must mainly be limited to small selections and combinations of hues; form coding remaining the most versatile (size, shape, orientation, alpha–numerics).

2.5.2. Colour characteristics of displays

The results of numerous colour coding experiments are not always comparable nor directly applicable. Some were obtained in trials with coloured reflecting surfaces, or with coloured light projected onto a diffusely reflecting surface, others with light emitting symbols. However, they make use of the same psycho–physical concepts [2.10,2.42] which are objectively measurable. In the following these concepts and their interrelations are briefly introduced, and defined where applicable to the field of displays.

Monochromatic colour: an electromagnetic wave $\vec{Q}(\lambda)$, its radiated energy located at wavelength λ and confined to a small band $\Delta\lambda$ around λ. The total luminosity interval is bounded by $380 < \lambda < 750$ (in nm). The colour of radiance $E.\vec{E}$ ($\vec{E}$ in power per sr per m^2; E a scalar) can be matched by a suitable mixture of three monochromatic base colours, $\vec{R}$, $\vec{G}$ and $\vec{B}$: its representation in R, G, B three dimensional orthogonal colour space given by (figure 2.29).

$$E.\vec{E} = R.\vec{R} + G.\vec{G} + B.\vec{B} \qquad (2.19a)$$

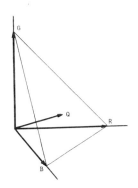

The equality sign means the matching by experiment of hue, saturation and brightness of the left and right hand stimuli. The wavelength of the primary colours $\vec{R}$, $\vec{G}$ and $\vec{B}$ have been standardised in 1931 by the Commission Internationale d'Eclairage (CIE, ICI) at respectively 700 nm (red), 546.1 nm (green) and 435,8 nm (blue); their strength can be determined according to constraints imposed by a given application e.g. whether it is desired to remain in the luminous power domain as in

Fig. 2.29 R,G,B colour space.

the following. The scalars R, G and B are the tristimulus values of the colour $\vec{Q}$ in the trichromatic system; E = R+G+B (preservation of flux). The colour parameters independent of E are the coordinates of the intersection of $\vec{Q}$ with the unit plane R+G+B = 1:

$$r = \frac{R}{R+G+B} , \ g = \frac{G}{R+G+B} , \ b = 1-r-g \qquad (2.19b)$$

Because the sum equals unity, two coordinates suffice to characterise the colour completely in terms of hue and saturation (not strength) as in fig. 2.31 which depicts the locus of spectrally pure colours (curved line), closed by a straight line (purples). Desaturated colours are within this closed domain.

Both figures show that r and g are negative in parts of the luminous interval. This means that the corresponding primary stimuli must be added to the source to be matched, because negative light does not exist. When this is not possible such as in displays, only a small portion of fully saturated colours can be realised, the remaining colours (tints) being non-saturated: i.e. they can be regarded an additive mixture of a saturated colour (dominant wavelength) and white (the achromatic point).

The property of additive mixing (first law of Grassman) implies that the addition of two stimuli of differing colour and strength produces (second law of Grassman) another colour with coordinates:

$$E_t \cdot \vec{E}_t = E_1 \cdot \vec{E}_1 + E_2 \cdot \vec{E}_2 = (R_1 + R_2)\vec{R} + (G_1 + G_2)\vec{G} + (B_1 + B_2)\vec{B}$$

$$\text{or } r_t = r_1 \frac{E_1}{E_1 + E_2} + r_2 \frac{E_2}{E_1 + E_2}; \ g_t = g_1 \frac{E_1}{E_1 + E_2} + g_2 \frac{E_2}{E_1 + E_2} \qquad (2.19c)$$

and r_t, g_t are on a straight line in the chromaticity diagram connecting (r_1, g_1) and $(r_2 + g_2)$. See also example 2.8.

One may place specific constraints to the matching experiments. For instance that the stimulus E to be matched is spectrally pure and of unity magnitude: $E(\lambda) = 1$. Because the perception of luminance is strongly dependent on wavelength, so are the resulting spectral tristimulus values, here denoted by $\bar{r}(\lambda), \bar{g}(\lambda)$ and $\bar{b}(\lambda)$ [2.10].

$$1 \cdot \vec{E}(\lambda) = \bar{r}(\lambda) \cdot \vec{R} + \bar{g}(\lambda) \cdot \vec{G} + \bar{b}(\lambda) \cdot \vec{B} \qquad (2.20a)$$

The magnitude of the stimulus being constant and equal to unity, these spectral tristimulus values are called the colour matching functions. They are

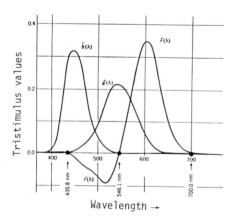

Fig. 2.30 Colour matching functions [2.10].

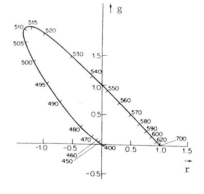

Fig. 2.31 Chromaticity coordinate diagram showing the locus of spectrally pure colours [2.10].

depicted in figure 2.30 [2.10]. The units of $\vec{R}$, $\vec{G}$ and $\vec{B}$ can be chosen such that $\int \bar{r}(\lambda)d\lambda = \int \bar{g}(\lambda)d\lambda = \int \bar{b}(\lambda)d\lambda$ with the consequence that the achromatic point in the chromaticity diagram r,g is at (1/3,1/3), see figure 2.31 [2.10]; the radiant power ratios of $\vec{R}$, $\vec{G}$ and $\vec{B}$ being about 72:1.4:1.0. The chromaticity coordinates describing the spectrum locus become

$$r(\lambda) = \frac{\bar{r}(\lambda)}{\bar{r}(\lambda)+\bar{g}(\lambda)+\bar{b}(\lambda)} \; ; \; g(\lambda) = \frac{\bar{g}(\lambda)}{\bar{r}(\lambda)+\bar{g}(\lambda)+\bar{b}(\lambda)} \; ; \; b(\lambda) = 1-r(\lambda)-g(\lambda) \quad (2.20b)$$

The end points of the spectrum locus determined by $\vec{R}$ and $\vec{B}$ are connected by the (straight) purple line (expression 2.19c).

The luminous experience associated with $1.\vec{E}(\lambda)$ is given by the luminous efficacy $K(\lambda)$ of the eye (figure I.1.15). The luminances $K(\lambda).1.\vec{E}(\lambda)$ of the stimulus and of the primaries $\vec{R}$, $\vec{G}$ and $\vec{B}$ must be weighed by the luminous efficacies at these wavelengths:

$$K(\lambda).\vec{E}(\lambda) = \bar{r}(\lambda).K(700).\vec{R} + \bar{g}(\lambda).K(546.1).\vec{G} + \bar{b}(\lambda).K(435.8).\vec{B}$$

or, with $V(\lambda) = K(\lambda)/673$ ($V(555) = 1.0$, $\vec{R}:\vec{G}:\vec{B} = 72:1.40:1$):

$$V(\lambda) \approx 0.22 \, \bar{r}(\lambda) + 0.98\bar{g}(\lambda) + 0.013\bar{b}(\lambda) \quad (2.21)$$

For the general case where $E_v(\lambda)$ varies with λ, the resulting luminance L_v becomes

$$L_v = 0.22 \int L_v(\lambda).\bar{r}(\lambda)d\lambda + 0.98 \int L_v(\lambda).\bar{g}(\lambda)d\lambda + 0.013 \int L_v\bar{b}(\lambda)d\lambda \quad (2.22a)$$

and the chromaticity coordinates (expression 2.20b):

$$r_v = \int r(\lambda)d\lambda, \; g_v = \int g(\lambda)d\lambda, \; b_v = \int b(\lambda)d\lambda \quad (2.22b)$$

Consequently, with the CIE primaries, one lumen of equal energy white is obtained by the additive mixture of 0.22 ℓm red, 0.98 ℓm green and 0.013 ℓm blue. For the NTSC phosphors, with different chromaticity coordinates and less saturation, the match against one lumen of white light from a blackbody at 6500 K yields tristimulus values of 0.290 ℓm red, 0.606 ℓm green and 0.105 ℓm blue.

Likewise, from expression (2.22a) it follows that

$$R_V = \int E_V(\lambda).\bar{r}(\lambda)d\lambda, \quad G_V = \int E_V(\lambda).\bar{g}(\lambda)d\lambda, \quad B_V = \int E_V(\lambda).\bar{b}(\lambda)d\lambda \qquad (2.22c)$$

It would be convenient if the tristimulus values associated with the stimulus could be transformed such that only one is proportional to luminance. This leads to the choice of non-existing primaries which, in the r,g diagram, lie on the line obtained by putting $L_V = 0$ in expression (2.22a), known as the alychne: g+0.207r + 0.013 = 0. By maximising the area occupied by real colours in the chromaticity diagram associated with the new primaries, two zero luminance primaries X and Z on the alychne have been defined; the result also being that the new chromaticity coordinates remain positive. The linear transformation from r,g,b space to x,y,z space is given by [2.10]:

$$x = \frac{0.49000r+0.31000g+0.20000b}{0.66697r+1.13240g+1.20063b} \qquad (2.23a)$$

$$y = \frac{0.17697r+0.81240g+0.01063b}{0.66697r+1.13240g+1.20063b} \qquad (2.23b)$$

$$z = \frac{0.01000g+0.99000b}{0.66697r+1.13240g+1.200063b} \qquad (2.23c)$$

Note that the sum of the coefficients of the numerator equal unity and of the denumerator 3. It follows that also in the x,y chromaticity diagram the coordinates of the achromatic point are (1/3,1/3). Because it is a linear transformation, linear combinations of two stimuli again have coordinates on a straight line.

The primaries so obtained have chromaticity coordinates

R (700 nm)	x = 0.73467	y = 0.26533	z = 0.00000
G (546.1 nm)	x = 0.27376	y = 0.71741	z = 0.00883
B (435.8)	x = 0.16658	y = 0.00886	z = 0.82456

see figure 2.32. The tristimulus values X,Y,Z are obtained:

$$X = \frac{x}{y} L_V, \quad Y = L_V, \quad Z = \frac{z}{y} L_V$$

with L_V the luminance of the stimulus Φ_V.

Analogous the colour matching functions $\bar{x}(\lambda)$, $\bar{y}(\lambda)$ and $\bar{z}(\lambda)$ are

$$\bar{x}(\lambda) = \frac{x(\lambda)}{y(\lambda)} V(\lambda), \quad \bar{y}(\lambda) = V(\lambda), \quad \bar{z}(\lambda) = \frac{z(\lambda)}{y(\lambda)} V(\lambda) \qquad (2.25)$$

which are shown in figure 2.33. Returning to figures 2.29 and 2.32, a straight line connecting any two points P, R within the triangle determines the hue are

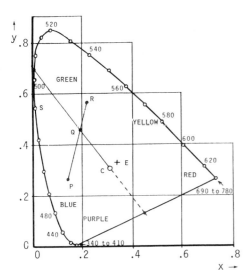

Fig. 2.32 CEI, x, y chromaticity diagram.

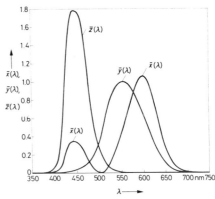

Fig. 2.33 Spectral tristimulus values $\bar{x}(\lambda)$, $\bar{y}(\lambda)$, $\bar{z}(\lambda)$ of monochromatic stimuli E of unit radiant power, with respect to the three fixed primary stimuli R, G, B [2.10].

purity of the colour Q obtained by the additive mixture of light sources with the chromaticity coordinates of the endpoints P and R. Standard CIE white sources A, B, C and D are close to the equal energy stimulus E. Within the colour triangle, the relative distance $\frac{QC}{SC}$ between any point Q to the achromatic point represents the purity of the colour. The intersection S with the spectral locus of the line, connecting the chromaticity point Q of a colour with that of C, determines the wavelength of the dominant colour of Q. Its intersection T with the spectral locus in the opposite direction determines the complementary colour. Colour measurements are covered in e.g. [2.10].

Example 2.8. Assume two light sources (e.g. phosphors of a CRT) with luminance spectral densities $L_1(\lambda)$ and $L_2(\lambda)$. Addition means that the tristimulus values are added: $X_t = X_1 + X_2$, $Y_t = Y_1 + Y_2$ and $Z_t = Z_1 + Z_2$. From the definition of the chromaticity coordinates x, y and z it follows that $X = \frac{x}{y}.Y$ and $Z = \frac{z}{y}.Y$. Using the fact that Y is proportional to the

amount   of light one obtains $X_t = \frac{x_1}{y_1} L_1 + \frac{x_2}{y_2} L_2$, $Y_t = L_1 + L_2$ and $Z_t = \frac{z_1}{y_1} L_1$
$+ \frac{z_2}{y_2} L_2$; the new chromaticity coordinates become

$$x_t = \frac{\dfrac{x_1}{y_1} L_1 + \dfrac{x_2}{y_2} L_2}{\dfrac{L_1}{y_1} + \dfrac{L_2}{y_2}} , \quad y_t = \frac{L_1 + L_2}{\dfrac{L_1}{y_1} + \dfrac{L_2}{y_2}} , \quad z_t = \frac{\dfrac{z_1}{y_1} L_1 + \dfrac{z_2}{y_2} L_2}{\dfrac{L_1}{y_1} + \dfrac{L_2}{y_2}} \qquad (2.24)$$

Simple arithmetic shows that the x_t, y_t shift along the line connecting the
coordinates (x_1, y_1) and (x_2, y_2) with weight factor $y_1 L_2/(y_2 L_1 + y_1 L_2)$. This is
also true when one of the sources is achromatic: adding more achromatic light
shifts the chromaticity coordinates towards E, desaturing the colour.

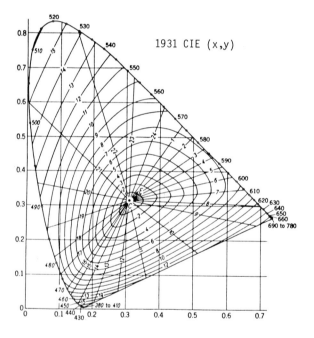

In figure 2.34 the
loci of DIN hue and
saturation values are
shown, including the
area covered by a CRT.
It is clear that the
relatively small area
covered by the CRT
affords little freedom
in choosing colour
domains which are both
sufficiently saturated
and separated to ensure
positive identification
of the colours under
all circumstances [2.41,
2.45].

Fig. 2.34 Loci of constant hue (straight lines
through C) and constant saturation (closed
curves around C) [2.43].

To determine the
required separation, it
must be quantifiable in
terms of a number of
JNDs, of both the
luminance and the colour contrast. Of the several chromaticity transformations
that exist, a uniform colour scale (UCS 1976) is noteworthy: it provides
measures for colour difference, chroma and saturation [2.10].
For constant L, iso-hue lines are straight, iso-saturation loci are circles.
The colour difference CD can be computed from $CD = [(\Delta L^*)^2 + (\Delta u^*)^2 +$
$(\Delta v^*)^2]^{0.5}$ with u^* and v^* the chromaticity coordinates, the quantity L^*

correlates with "lightness". Lightness is the ratio of the brightness of an area in proportion to that of a similar area perceived as white: relative brightness.

$$L^* = 116 \ [\frac{Y}{Y_n}]^{0.33} - 16; \ u* = 13 \ L^* \ (u'-u'_n); \ v* = 13 \ L^* \ (v'-v'_n) \qquad (2.25a)$$

$$u' = \frac{4X}{X+15Y+3z} \ , \ v' = \frac{9Y}{X+15Y+3z} \qquad (2.25b)$$

$$u'_n = \frac{4X_n}{X_n+15Y_n+3Z_n} \ , \ v'_n = \frac{9Y_n}{X_n+15Y_n+3Z_n} \qquad (2.25c)$$

The tristimulus values X_n, Y_n and Z_n are references associated with a nominally white stimulus [2.10]. The exponent 0.33 in the lightness expression reminds of Stevens' cube root law [2.28]. The correlates for chroma C_{uv}^* and saturation S_{uv}^* are: $C_{uv}^* = [(u^*)^2 + (v^*)^2]^{0.5}; \ S_{uv}^* = \frac{C_{uv}^*}{L^*}$ \qquad (2.26)

Under office conditions CD should be chosen between 0.03 and 0.06.

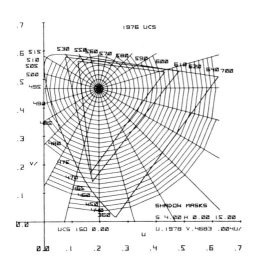

Fig. 2.35 Shadow-mask CRT boundaries plotted on iso-hue, iso-saturation contours in 1976 UCS diagram [2.41].

Figure 2.35 gives an example of this chromaticity diagram for constant L^*, with iso-hue and iso-saturation loci.
A practical indicator of the combined contrast due to differences in luminance and chromaticity is the Discrimination Index (DI) [2.44]. Although not generally accepted in the vision community, it is being used by display designers because it provides reasonable estimates. The luminance difference EL is defined as the logarithmic response to contrast ratio: $EL = \log L_1/L_2$. Based on the assumption that the JND in visual contrast on the average equals about 5 %, the

threshold in EL becomes 0.021. Also assuming that for comfortable discernibility a contrast ratio of 1.4 is required ([2.4], section 1.4.3) about 7 EL threshold intervals are necessary: $\log 1.4/\log 1.05 \approx 7$. Taking this as a reference, the luminance Index IDL of a display contrast C_r is defined as:

$$IDL = \log C_r/\log 1.4 = 6.7 \log C_r \qquad\qquad (2.26a)$$

The chromatic distance EC (in the 1960 CIE-UCS diagram) is defined as $EC = (\Delta u^2 + \Delta v^2)^{0.5}$.

The threshold chrominance distance is estimated [2.46] at 0.00384. Again the assumption is made that 7 intervals are necessary for comfortable discernibility and that number: $7 \times 0.00384 = 0.027$ is taken as the reference. Thus a Chrominance Index IDC is obtained:

$$IDC = (\Delta u^2 + \Delta v^2)^{0.5}/0.027 = 37(\Delta u^2 + \Delta v^2)^{0.5} \qquad\qquad (2.26b)$$

and the discrimination index becomes

$$DI = (IDL^2 + IDC^2)^{0.5} = 37 \{0.18(\log C_r)^2 + \Delta u^2 + \Delta v^2\} \qquad\qquad (2.26c)$$

An example of the use of the discrimination index is in reference [1.12] and in chapter III.3. Obviously the discrimination index is no more than an order of magnitude indicator because both the luminance JND and the chrominance JND vary considerably with surround/background luminance and hue, size of the area displayed, influence of adjacent colours (induction) [2.37].

2.6. Recommendations

From the foregoing it can be inferred that it is difficult to satisfy every observer of a colour display, especially when a large palette is desired. In that case detail knowledge of colour generation (the physics) and of colour perception (psychophysics) is required.

For less demanding applications selfconstraint pays off. A number of generally accepted recommendations for displays in instrumentation are:

- spatial resolution: display element pitch < 1.7 minutes of arc (5 per mm at viewing distance of 400 mm),
- display size: A_4, 250×170 mm^2 useable area,

- negative contrast of emissive display with screen illuminance E lux and display reflectance $\leq$ 0.03: display minimum (background) <u>emission</u> 0.01 E < L_b < 0.04 E, neutral or very desaturated colour. Contrast ratio foreground luminance L_f over screen illuminance:
  i)   bilevel display: 0.04 E < L_f < 0.2 E,
  ii) 4 grey levels display: 0.04 E < L_f < 1.0 E, grey level contrast steps of 1.4,
  iii) continuous display: 0.04 E < L_f < 1.0 E ([2.4] section 1.4.3).
       Positive contrast of emissive display with screen illuminance E lux and display reflectance $\leq$ 0.03: display minimum background emission 0.02 E < L_b < 0.04 E, neutral or very desaturated colour. Contrast ratio of (darker) foreground luminance L_f over screen illuminance in bi-level display: 0.01 E < L_f < 0.025 E.
       It is not recommended to use more than two luminance levels in a positive contrast emissive display.

- Number of colours addressable: six plus white, the chromaticity coordinates of the dominant wavelengths of preferred colours [2.41,2.45]

$\approx \lambda$ [nm]	x	y	colour
610	0.657	0.330	red
585	0.536	0.451	amber
550	0.287	0.684	green
480	0.094	0.122	cyan
460	0.139	0.037	blue
–	0.340	0.081	magenta

- Generally full saturation must be avoided (except for small patches and thin lines) because of colour dependent accommodation which may cause visual fatigue: desaturated colours require less refocusing. Also red through green needs little refocussing, switching from red or green to blue does. Therefore, and also because the eye is "blueblind" at the central part of the fovea, small blue patches and print must be avoided. Similarly, edges between adjacent colours which differ only in the amount of blue are difficult to discern.

- The addition of white (e.g. ambient illumination) desaturates the colours. Since the chromatic distance of yellow to the achromatic point is the smallest, pure yellow appears less saturated, washes-out easily. Consequently, to cope with increases in ambient illumination, the colours as addressed must also have a lower bound on saturation.

- In instrument/office displays subtle colouring in the image is not adviseable, particularly at low luminance, because
  i) the large variability among individuals in their ability to distinguish small colour differences,
  ii) desaturation and hue shift effects for smaller patches/print and short duration stimuli.

REFERENCES

[2.1] Roufs, J.A.J. and Bouma, H., Towards linking perception research and image quality, Proceedings of the SID, 21, no. 3, (1980), pp. 247-270.

[2.2] Snyder, H.L., Image quality: measures and visual performance. In: Flat panel displays and CRTs, ed. L.E. Tannas Jr., pp. 70-90.

[2.3] Granrath, D.J., The role of human visual models in image-processing, Proceedings of the IEEE, 69, no. 5, (1981), pp. 552-561.

[2.4] Bosman, D., Image characteristisation and formation, Chapter 1 of this book.

[2.5] Blakemore, C. and Campbell, F.W., On the existence of neurones in the human visual system selectively sensitive to orientation and size of retinal images. J. Physiol. 203, (1969), pp. 237-260.

[2.6] Buckler, A.T., A review of the literature on the legibility of alpha-numerics on electronic displays, Techn. Memorandum 16-77, US Army Human Engineering Laboratory, Aberdeen Proving Ground, Md. (1977).

[2.7] Van Nes, F., A new teletext character set with enhanced legibility, Proceedings of the SID, 27, no. 3, (1986).

[2.8] Laycock, J., The effect of picture element failure on the legibility of a matrix display image, Displays, 6, no. 2, (1985), pp. 70-78.

[2.9] Uphaus, J.A. and Pastor, J.R., Investigating the correlation between reading errors and degraded numerics. Proceedings NAECON, 2, IEEE, New York (1982), pp. 734-738.

[2.10] Wyszecki, G. and Stiles, W.S., Colour Science. 2nd edition, John Wiley and Sons, ISBN 0-471-02106-7, New York (1982).

[2.11] Bartlett, N.R., Dark adaptation and light adaptation. In: Vision and Visual Perception, ed. C.H. Graham, John Wiley and Sons, ISBN 0-471-32170-2, New York (1965), pp. 185-207.

[2.12] Laycock, J., The measurement and analysis of eye movements. In: Search and the human observer, eds. J. Clare, M. Sinclair, Taylor and Francis, ISBN 0-85066-193-5, London (1979), pp. 163.

[2.13] Den Buurman, R. Eye movements and visual observation, (in Dutch), De Ingenieur, 91, no. 32/33, (1979), pp. 551-554.

[2.14] Dreyfuss, H., The measure of man, human factors on design, Whitney, Library of Design, New York (1966).

[2.15] Riggs, L.A., Visual acuity. In: Vision and Visual Perception, ed. C.H. Graham, John Wiley and Sons, ISBN 0-471-32170-2, New York (1965), pp. 321-349.

[2.16] Hamerly, J.R. and Dvorak, C.A., Detection and discrimination of blur in edges and lines. JOSA, 71, no. 4, (1981), pp. 448-452.

[2.17] Gubisch, R.W., Optical performance of the human eye, JOSA, 57, (1967), pp. 407-415.

[2.18] Yellott Jr. J.I., Wandell, B.A. and Cornsweet, T.N. The beginnings of visual perception: the retinal image and its encoding. In: Handbook of Physiology. Section I: the nervous system, Volume III: sensory processes, Part I. American Physiological Society, Brathesda, Maryland. Distributed by The Williams & Wilkins Company, Baltimore, ISBN 0-683-0-1108 1, Maryland 21202, (1984).

[2.19] Tanner, W.P. and Swets, J.A., A decision-making theory of visual detection, Psychological Review, 61, no. 6, (1954), pp. 401-409.

[2.20] Rose, A., The sensitivity performance of the human eye on an absolute scale. JOSA, 38, (1948), pp. 196-208.

[2.21] Blackwell, H.R., Contrast thresholds of the human eye, JOSA, 36, no. 11, (1946), pp. 624-643.

[2.22] Grind van de, W.A. et al., The concepts of scaling and refractioness in psychophysical theories of vision, Kybernetik, 8, (1971), pp. 105-122.

[2.23] Cornsweet, T.N., Visual perception, Academic Press, Library of congress catalogue card no: 71 107570, New York (1970).

[2.24] Cornsweet, T.N. and Yellot Jr., J.I., Intensity dependent spatial summation, JOSA-A, 2, no. 10, (1985), pp. 1769-1786.

[2.25] Blommaerts, F.X.X. and Roufs, J.A.J., The foveal point spread function as a determinant for detail vision, Vision Research, vol. 21, (1981), pp. 1223-1233.

[2.26] Campbell, F.W. and Robson, J.G., Application of Fourier analysis to the visibility of gratings, Journal of Physiology, 197, London (1968), pp. 551-556.

[2.27] Georgeson, M.A., Sullevan, G.D., Contrast constancy: deblurring in human vision by spatial frequency channels, J. Physiol. 252, (1975), pp. 627-656.

[2.28] Stevens, S.S., The surprising simplicity of sensorymetrics, Am. Psychologist, 17, (1962), pp. 29-39.

[2.29] De Lange, H., Research into the dynamic nature of the human fovea-cortex systems with intermittent and modulated light, JOSA, 48, (1958), pp. 777-784.

[2.30] Brown, J.L., Flicker and intermittent stimulation. In: Vision and Visual Perception, ed. C.H. Graham, John Wiley and Sons, ISBN 0-471-32170-2, New York (1965), pp. 251-320.

[2.31] Kelly, D.H., Adaptation effects on spatio-temporal sine wave thresholds, Vision Research, 12, (1972), pp. 89-101.

[2.32] Berger, C., Illumination of surrounding field and flicker fusion frequency with foveal images of different sizes, Acta Physsiol. Scand., 28, (1954), pp. 161-170.

[2.33] Turnage, R.E., The perception of flicker in cathode ray tube displays, Information Display, 3, (1966), pp. 38-52.

[2.34] Kelly, D.H., Effect of sharp edges in a flickering field, JOSA, 49, (1959), pp. 730-732.

[2.35] Krebs, M.J. et al, Colour display design guide, ONR Contract N00014-77-C-0349. Honeywell Systems and Research Center. Available DDC-AD A 066630, Springfield, C., Virginia 22161, USA (1978).

[2.36] Teichner, W.H. et al, Colour research for visual displays contract, N00014-76-C-0306. Available NTIS-AD A 043609, Springfield, C., Virginia 22161, USA (1977).

[2.37] Walraven, J., Perceptual problems in display imagery, SID Symposium Digest, Orlando (1985).

[2.38] Snyder, H.L., The visual system: capabilities and limitations. In: Flat panel displays and CRTs, L.E. Tannas, ed. Van Nostrand Reinhold Company, ISBN 0-442-28250-8, New York (1985).

[2.39] Purdy, D., Mcl., Spectral hue as a function of intensity, The American Journal of Psychology, 43, (1931), pp. 541-559.

[2.40] Tyte, R.N. et al, Legibility of a light emitting diode dot array in high illuminance, Proceedings of the SID, 21, no. 1, (1980), pp. 21-29.

[2.41] Laycock, J., Selected colours for use on colour cathode ray tubes, Displays, 5, no. 1, (1984), pp. 3-14.

[2.42] RCA Electro-optics Handbook, Technical Series EOH-11, Lancaster, PA, USA (1978).

[2.43] Richter, M., Grenzmuster Tafeln auf der Grundlage des Farben systems der DIN Farben karte, Farbe, 3, (1954), p. 157.

[2.44] Galves, J.P. and Brun, J., Colour and brightness requirements for cockpit displays: proposal to evaluate their characteristics, AGARD, AvP Conference Proceedings CP-167, Paris (1975), pp. 6-1 to 6-8.

[2.45] Corte de, W., High contrast sets of colours for colour CRTs under various conditions of illumination, Displays, 6, no. 2, (1985), pp. 95-100.

[2.46] Jones, A.H., Optimum colour analysis characteristics and matrices for colour television cameras with three receptors, Journal SMPTE, 77, (1968), pp. 108-115.

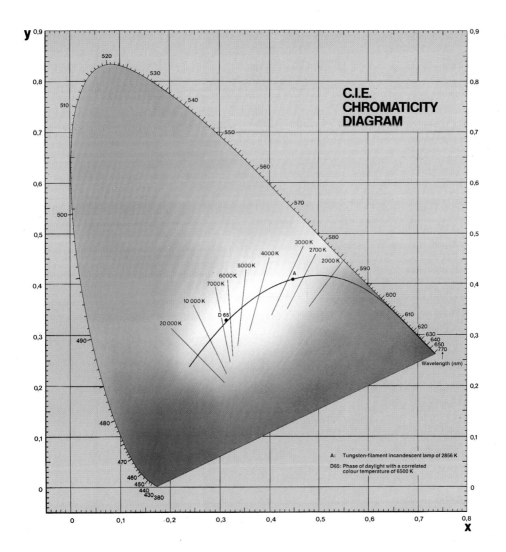

y

0,9
0,8
0,7
0,6
0,5
0,4
0,3
0,2
0,1
0

520
530
540
550
560
570
580
590
600
610
620
630
640
650
770

Wavelength (nm)

510
500
490
480
470
460
450
440
430 380

3000 K
2700 K
4000 K
2000 K
5000 K
6000 K
7000 K
A
10 000 K
D 65
20 000 K

**C.I.E.
CHROMATICITY
DIAGRAM**

A: Tungsten-filament incandescent lamp of 2856 K

D65: Phase of daylight with a correlated
        colour temperature of 6500 K

0   0,1   -0,2   0,3   0,4   0,5   0,6   0,7   0,8

X

Colour Triangle, CIE 1931 x,y System.

PART II

TECHNOLOGIES

DISPLAY ENGINEERING: D. Bosman (Editor)
© Elsevier Science Publishers B.V. (North-Holland), 1989

II.1: CATHODE RAY TUBES

G.H. HUNT
Royal Aircraft Establishment, United Kingdom

1.1. Historical survey

The cathode ray tube (CRT) is one of the earliest electro-optical devices, and its name derives from experiments in which a greenish glow was produced on the glass surface of an evacuated tube when a high potential caused electrons to be stripped from the cathode and to strike the glass. Although 'electron beam tube' would be a much more accurate description the earlier name is so well established that it is unlikely ever to be displaced.

The phenomenon that cathode rays (electron beams) are deflected by electric and magnetic fields led to the use of the tube as a laboratory measuring device. The successive developments to transform this simple device into a practical display include the incorporation of a flat phosphor display screen and the use of a heated filament cathode. By 1908 Campbell Swinton was suggesting the use of the CRT for transmitting and reproducing television images, and during the 1920s and 1930s the CRT was used in much of the radio and electronic research then under way.

By the beginning of World War II in 1939, two principal applications of cathode ray tubes had emerged. The first was in domestic television in which the CRT screen was continuously re-drawn with a raster pattern, the luminous intensity of which was controlled by modulation of the electron beam. The second application was for the radio navigation and radar systems then coming into use, in which the CRT was generally used to display transient waveforms for which conventional electrical instruments were inappropriate. This application included use on ships and in aircraft, as well as in ground-based systems.

The use of CRTs in television has continued to be the largest application for CRTs in terms of numbers of tubes manufactured. But from the 1950s onwards there has been a steady growth in its application to the data processing field in which its versatility as a display medium has been unmatched by any other form of display. Thus it is currently in use as the display device in almost all the many systems in which digital computers are used, e.g. in the commercial world, banking, process control, computer-aided design and electronic games, as well as for small personal computers.

Many of the past and current requirements for CRTs have been adequately satisfied by monochrome tubes. During the past 40 years there have been significant improvements in performance, and particularly in recent years the

computer graphics requirements have led to tubes of very high resolution [1.1,
1.2].   But   the   desire   for   multi-colour   CRTs,   particularly   for   domestic
television, resulted in the development by RCA of the shadow-mask tube which
first went into production in 1954. Although originally intended for TV, the
shadow-mask and its derivatives such as the Trinitron have been used for many
applications in the computer display field, in which the improvements in
legibility or cosmetic appearance justify the extra costs inherent in a colour
display. Specially bright and rugged colour tubes are now becoming available
for the particularly demanding needs of aircraft displays; but CRTs have not
found significant use in cars or other land vehicles because of the development
of cheaper alternatives such as VFDs.

1.2. Principles of operation

1.2.1. Monochrome tubes

The basic principle of the monochrome CRT in its simplest and most usual form
is shown in figure 1.1. The main structure consists of an evacuated envelope in
the form of a glass or metal bottle with a narrow neck flaring towards the
front display surface which, for most applications, is of rectangular shape.
Within the neck is an electron gun which projects a beam of electrons towards
the phosphor screen deposited on the front surface, the electrons being
accelerated by the large potential difference (typically in the range 10 kV to
30 kV) established between the gun and the anodes and screen. The beam is
deflected so that it can strike any part of the screen, and is modulated in
intensity so as to vary the perceived luminance of the screen.

As shown in figure 1.1, the CRT may be considered as incorporating four
electro-optic regions, each of which will be briefly described in turn. These
are:
(a) The electron beam formation region.
(b) The beam focussing region.
(c) The beam deflection region.
(d) The screen region.

a) The electron beam formation region, frequently referred to as the gun, is
   formed of a heated cathode together with electrodes to modulate the beam
   current, and electrodes to accelerate the electrons away from the cathode.
   To obtain the small size of electron beam spot at the screen which is
   increasingly required for high resoluton tubes, the diameter of the cathode
   emitting area must also be very small, and hence the current density must be

as large as possible. Oxide cathodes are generally used and consist of a thin layer of a mixture of barium-strontium oxides deposited on the top of a nickel cylinder internally heated by a filament to a temperature of about 1100 K. Maximum current density is about 0.3 A cm^{-2} giving an average cathode life of over 10.000 hours. If higher current densities are required it is necessary to use an alternative cathode material such as the barium-impregnated tungsten sponge which forms the basis of the new "dispenser" cathodes now coming into limited use in CRTs. Current densities of 1 A cm^{-2} have been obtained, and long lives are claimed although considerable care has to be taken to avoid poisoning from contaminants within the CRT envelope, a requirement which leads to somewhat more expensive tube processing. The remaining electrodes within the beam forming region may be arranged in a triode or tetrode form, there being a large variety of alternative electrode shapes. In some cases the beam forms a cross-over in front of the cathode, and in other cases the field is arranged to give a more uniform laminar flow beam pattern.

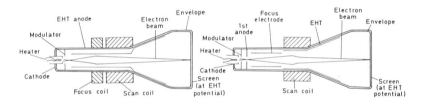

Fig. 1.1 Magnetic focus tube.   Electrostatic focus tube.

b) The focussing of the electron beam can be by either electrostatic or electromagnetic fields. Electromagnetic focus is produced by a uniform magnetic field parallel to the tube axis which causes the electrons to spiral towards that axis. Theoretically, it can give the best possible resolution but requires very careful and elaborate adjustment to obtain the theoretical optimum. Improvements in electrostatic focus design have, however, resulted in spot sizes which are very similar to those which can be achieved in practice using magnetic focus. In addition, electrostatic focus results in considerable weight saving and ease of replacement of tubes. Electrostatic focus is achieved either by means of an einzel lens (low voltage focus) or by a bi-potential lens (high voltage focus).

c) Beam deflection can also be achieved either electrostatically or magnetically. It is invariably necessary to deflect the beam in both directions (X and Y) relative to the tube axis (Z), the same method being used for both directions. The choice between the two techniques is a complex one, and for many applications either method is feasible. However if a combination of high screen luminance, high resolution and large scan angles is required, it is usually necessary to use magnetic deflection. Because of the inductance of magnetic deflection coils, they are particularly suitable for use in resonant circuits operating at constant frequency as in raster formats; but for stroke-written formats which require wide bandwidth deflection, magnetic coils consume considerable power. For high resolution tubes the deflection coil assembly frequently incorporates additional coils or permanent magnets for dynamic focussing to correct deflection defocussing, for correcting astigmation and for centring the display format.

d) The screen is formed of one or more layers of powdered material deposited on the tube faceplate; this material is normally described as a "phosphor" and the process by which light is emitted from it when excited by high-energy electrons is called cathodo-luminescence. Many crystalline materials have been used as phosphors and are commonly based on zinc sulphide although recently there has been much use of the rare earth activated oxides and oxysulphides. The main factors to be considered in the choice of phosphors are: luminous efficiency, decay time, colour and life. The luminous efficiencies are generally in the range 10 to 50 ℓm W^{-1}, with the higher values being for emissions in the green wavelengths where the eye has maximum sensitivity. These efficiencies are a factor of 10 or more greater than those for other types of electroluminescence, which is one of the principal reasons for the continued pre-eminence of CRTs for many applications. Phosphor decay times range from less than a microsecond to several seconds, the longer persistance phosphors allowing significant integration which can be valuable for reducing or eliminating flicker and noise in relatively stationary images, but causing smearing of dynamic images. Phosphors can be obtained for all colours in the visible spectrum, and for enough of the spectral triangle to allow reasonable reproduction of most colours. As for life, it is possible to de-sensitise phosphors by excessive beam current, but for most practical current levels phosphors are now available with lives of several thousand hours provided they are operated within the linear part of their operating characteristics.

Most phosphors used for CRT screens are listed in the JEDEC P number register operated by the US Electronic Industries Association who have also issued guidelines on phosphor measurement methods. More than 50 types have been registered to fulfill the different needs, including multi-colour phosphors of various types and the so-called cascade phosphors which incorporate a UV emission from one material which energises a second phosphorescent material.

From 1982, the P register has been replaced by the Worldwide Phosphor Type
Designation System (WTDS) [1.3] also operated by EIA, but it will probably be
some years before the familiar P numbers are completely superseded. It should
be noted that considerable variations in performance may be obtained from
different samples which are nominally the same, due to different proprietary
manufacturing techniques.

The phosphor screen, which is deposited on the faceplate, has a thin aluminium
backing to provide a conducting path for the electrons and to increase the
light output by reflecting to the front the light initially emitted to the
back. The whole screen is at a potential of 10 kV or more relative to the
cathode. For some applications a filter may be bonded to the front of the
screen to improve the visual characteristics, and possibly to absorb some of
the X-rays which are generated if the EHT voltage levels are very high.

The simple CRT described above produces a monochrome picture, the colour of
which is determined by the phosphor characteristics. The generation of a multi-
colour picture is a more difficult task and requires a significant complication
of the basic tube technique. Several alternative methods of producing colour
pictures have been devised, the most widely used and promising of which are
described below and in section 1.4.

1.2.2. Shadow-mask colour tubes

The type of colour CRT which has been most extensively used is the shadow-mask
and its derivatives, indeed none of the alternative types which are described
later has made any significant penetration in the commercial market. As
originally developed by RCA for commercial TV application and shown in figure
1.2a, it incorporates three separate guns, together with a shadow-mask and a
patterned screen. The phosphor screen is printed in the form of an array of
'triads', each triad consisting of a dot of each of the primary phosphor
colours, red, green and blue. The geometry of the guns, shadow-mask and screen
is arranged so that the beam from each gun can fall on only one coloured dot in
each triad, the other two colours being masked from the gun by the shadow-mask.
Thus the three guns can be independently fed with the required modulation
signals of the three primary colours. The three beams pass through the same
magnetic deflection coils with the same electron velocities and are therefore
deflected together, by approximately the same amount, and the electrostatic
focus elements are generally also common so that a single focus control is
sufficient.

This arrangement requires very exact alignment of the guns, mask and phosphor pattern, but the high yield and low cost in large-scale production are clear evidence that the alignment techniques are now well established. The most recent developments in shadow-mask CRTs for the domestic market have concentrated principally upon improving the visual contrast through the use of a black matrix screen to fill in the areas between the phosphor dots, and the reducton in length of tube by increasing beam deflection angles, and with these improvements this type of shadow-mask has continued to satisfy the requirements of luminance, contrast and resolution appropriate to the high volume TV market.

For other applications, particularly those concerned with high definition graphical displays such as those use in computer-aided design systems, as well as for other displays which are viewed from relatively close distance, an improvement in resolution at the screen is called for. To meet these requirements variants of the shadow-mask tube incorporating different geometries have been developed. Figure 1.2 shows the three possible arrangements, of which the delta gun has already been described. The in-line gun can be used with either a circular hole shadow-mask or with a slotted

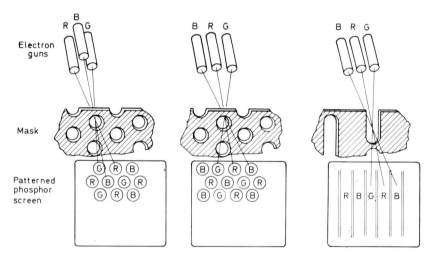

Fig. 1.2   Delta gun with   in-line gun with   in-line gun with
           circular hole mask   circular hole mask   slotted mask.

shadow-mask, the latter giving a higher transmission but with larger hole-pitch and hence more limited resolution. The in-line gun combined with circular holes and circular screen dots has been offered by several manufacturers as the best solution for very high resolution colour CRTs for character-graphic displays, as described in [1.4,1.5,1.6,1.7].

In order to achieve high resolution, the width of the electron beam and the diameter of screen dots must both be very small, and to avoid spurious colour fringing the beam diameter must be typically about three times the dot diameter. To achieve a fine-pitch screen requires a correspondingly fine-pitch shadow-mask and a method of assembling the complete tube to very high accuracy. To achieve the narrow electron beam requires a very small cathode size, a large aperture gun structure and good dynamic focus both at the centre and corners of the screen. In some cases (e.g. [1.7]) it has been found necessary to tilt the array of holes on the shadow-mask to match the distortion of the in-line trio of beams, in order to prevent the generation of colour impurities. For all these reasons it is clear that high resolution tubes are much more expensive to manufacture than the tubes used in domestic TV, but the performances which have been obtained are very impressive. A dot pitch of about 0.2 mm on a 20-inch tube (360x270 mm) giving a horizontal resolution of over 1500 dots is representative of the performance now available.

The difficulties described above are compounded if the CRT is required to give both high resolution and good colour purity in a high brightness environment. The first luminance problem arises because the luminous efficiency of red and blue phosphors is significantly less than for the high efficiency green phosphors. A second difficulty is that with monochrome tubes low luminance can to some extent be compensated by using narrow band coloured filters in front of the tube to minimise reflection, whereas with shadow-mask and other colour tubes it is not easy to produce filters with bands matching all three phosphor colours.

The most serious difficulty in designing high brightness, high resolution shadow-mask CRTs is associated with the considerable loss of electron beam current by absorbtion at the shadow-mask, so that typically no more than 20 % of the average beam current reaches the phosphor. Apart from the loss of luminance, this also creates the problem in high brightness tubes that the energy absorbed in the shadow-mask can cause thermal expansion and thereby cause colour distortion by upsetting the exact alignment between the shadow-mask and the phosphor screen. This has led to the development of pre-tensioned flat shadow-masks in combination with flat screens, using techniques described in [1.8].

A particularly demanding application for high brightness, high resolution tubes is in aircraft cockpits, where the environment is made more difficult by significant levels of vibration. It has taken many years of development for shadow-mask tubes to reach a standard adequate for aircraft use, but they are now widely deployed in civil aircraft and, as reported in [1.9], tubes to meet the exceptionally severe requirements of military aircraft are also now available.

1.2.3. Penetration screen and current-sensitive tubes

An alternative technique for colour CRTs is the Penetration Screen or 'Penetron' tube, which is of much simpler construction than the shadow-mask type. Its construction is generally similar to that of the monochrome tube shown in figure 1.1, but the multi-colour characteristics are obtained by the particular construction of the phosphor screen [1.10].

In one arrangement of the screen (figure 1.3b) it is formed of a combination of two types of phosphor material, one of which is generally chosen to emit green light and the other of which emits red light. The green phosphor particles are coated with an 'onion-skin' of barrier material and then further coated with a layer of red phosphor. It is also arranged that the EHT potential applied to the screen can be set at any value between upper and lower limits which are typically 17 kV and 10 kV. At the lower limit the beam current electrons are unable to penetrate the barrier layer surrounding the green phosphor particles, and only red light is emitted from the screen due to electron excitation of the red phosphor material. At the higher EHT limit the electrons penetrate the barrier and excite the green phosphor particles, so that both green and red emissions are produced. However because the efficiency of the green phosphor material is generally greater than that of the red, the resultant colour is a reasonably pure green.

For intermediate EHT values, a range of different colours is obtained resulting from the addition of red and green in different proportions. However the use of only these two primary colours necessarily restricts the range of colours which may be generated to a single line on the colour triangle, and in fact it is generally possible to produce only four distinct colours (red, orange, yellow and green), and in high ambient illumination possibly only three. Other spectral characteristics can be obtained by the use of different phosphors, and variations in temporal characteristics are also possible by choosing phosphors with different storage/decay properties.

The principal advantages of this type of tube as compared with the shadow-mask tube are the rugged mechanical construction and the high resolution. The absence of any fragile component such as the shadow-mask, or of the need for accurate geometrical registration, means that it can be used in any vibration environment for which a monochrome tube is suitable, e.g. an aircraft cockpit. Also, the absence of any pattern on the phosphor screen means that tube resolution is limited by beam width rather than by the size of the phosphor spots or lines, so that high resoluton is possible. Patterned contrast filters can also be used without forming Moiré fringing effects.

Although the luminance is not reduced by the same mechanisms as in the shadow-mask tube, it is still significantly below that of the monochrome CRT. This is principally for the red colour, and is due in part to the relative inefficiencies of red phosphors and also due to low excitation voltage. Moreover the green colour is also generated inefficiently, by comparison with monochrome CRTs, mainly due to energy loss in the barrier coating. An improvement in red luminance can be obtained by using a "multilayer screen" [1.11] instead of the onion-skin method. In this, as shown in figure 1.3a, a

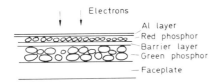
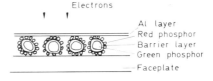

Fig. 1.3 a) Multilayer type.   b) Onion-skin type.

layer of green phosphor is deposited on the glass tube-face, then covered by a layer of barrier material, then with the red phosphor, and finally a thin-film of aluminium is deposited. This allows the red emission to be excited more efficiently and the loss of luminance due to internal absorption to be minimised.

As with the beam index tube, described later, mechanical simplicity is obtained at the cost of electrical complexity. It is necessary to switch high EHT voltages at fairly high speeds and, in addition, the deflection sensitivity and focus change with EHT so that it is necessary also to change the current conditions in these circuits. In practice it has been found impractical to alter all these circuit conditions in the 100 ns which would be necessary if colour variations were to be produced during the writing of a raster TV frame; therefore this type of tube is normally operated in a field sequential colour switching mode, which complicates the design and can lead to colour flicker problems. For cursively-written data the colour can be changed between groups of characters and symbols.

Although there has been some usage of penetration screen tubes, particularly in avionics applications, the limited colour capability has prevented any very wide penetration of the CRT marketplace.

A rather similar tube has been reported [1.12], which incorporates a phosphor screen the colour of which is a function of the electron beam current, rather

than voltage. The screen is formed of a mixture of two phosphors, both with
non-linear luminance/current density characteristics. Hence it is possible to
arrange that with a change in current density, the ratio of the two emissions
changes significantly, and with the two phosphors chosen to emit in two colours
(e.g. red and green), the resultant emitted colour will change with current.
The total luminance will also change, so that an independant luminance control
has to be provided by the variation of on-off ratio in a beam current chopping
circuit.

By comparison with the penetration tube, the complexity arising from this beam
control circuit is small compared with the difficulty of switching EHT screen
voltage. However there remain the limitations of a restricted colour range and
a   relatively low brightness, and these have prevented any significant
application thus far.

1.3. Physical characteristics

The overall shape of conventional cathode ray tubes, both monochrome and
shadow-mask, is generally that of a long-necked bottle, the neck of which
contains the electron gun and the base of which is the screen. The depth of the
tube is principally determined by the type of gun arrangement and by the
maximum deflection angle of the electron beam; for high precision tubes and for
tubes operating at high luminance levels which require large EHT voltages, a
deflection angle of 70° is normal, whereas many commercial and domestic CRTs
are 110° tubes. For a 70° tube, with a typical gun size, the overall tube
length is about 1.2 times the screen diagonal dimension (e.g. for an avionics
tube with a screen of 175 mm x 175 mm, the length is about 300 mm). The
screen's usable area is less than the overall area of the front glass surface,
a non-usable border of 10 mm being typical.
Although the tube-length is a major disadvantage since it determines the depth
of the complete packaged display, the bottle-shape of the tube and in
particular the narrow neck allow much of the associated electrical circuits and
deflection coils to be mounted into a complete package very little larger than
the overall tube dimensions. In some cases, when the display tube is being
incorporated into a total system with other optical components, or where it is
being fitted into some confined space, such as with an aircraft head-up
display, the geometry of the complete package can be designed to suit these
external contraints.
Direct view displays for control and monitoring purposes, range 50 mm diagonal
to 300 mm diagonal, are usually rectangular or square in shape. Head-up
displays in aircraft use circular screens up to 100 mm diameter. CRTs for

helmet-mounted displays have been developed [1.13], and with these it is
necessary to minimise both weight and size of the CRT; a typical tube has a
20 mm diameter useful screen and a length of 100 mm.

For use in computer terminals, screen sizes are typically 300 mm to 350 mm
diagonal. However when complex graphics are to be displayed, such as in CAD
systems, sizes up to 600 mm diagonal may be used [1.7]. Figure 1.4 shows some
representative modern monochrome tubes.

Fig. 1.4 Representative monochrome CRTs (Thomson-CSF, photo Mathieu).

The requirement for high EHT voltages is frequently quoted as a disadvantage of
CRT displays, but although such voltage supplies require special circuits,
these are well developed and cause little problems. Even for aircraft displays
operating at high EHT levels and at high altitudes, design techniques have
evolved so that 25 kV is now considered standard practice. Similarly, the use
of a large evacuated glass tube, which might be thought operationally
undesirable, has not in practice caused real problems, and although tubes have
been constructed using metal and ceramic envelopes this is now not generally
necessary. However for use in very severe applications such as military
aircraft, care has to be taken with the mounting arrangements; tubes are
sometimes manufactured with an integral metal mounting collar or frame. This
has the further advantage of providing simpler and more accurate assembly for
maintenance purposes, a concept which is often further extended by supplying
the tube already assembled with integral magnetic deflecton coils.

Because of the relatively large size of the tube, and the use of magnetic and
high voltage components, the weight of displays based on CRTs is generally
rather large. For many purposes this may not be significant but size and weight
are a major disadvantage when displays are required for small portable

computers, and in applications such as aircraft and helicopters the weight penalty could be significant since a typical airborne display unit with a screen size of 200 mm x 150 mm may weigh about 7.5 kg.

1.4. Novel types of CRT

The conventional CRT, in both its monochrome and multi-colour forms as described in section 1.2, has become established as an effective and versatile display device, well matched to a wide range of applications. It is therefore quite understandable that several attempts have been made to develop novel forms of tube which would, in some respects, improve on the conventional types; particularly by aiming the development at some identified part of the market. Thus, for example, novel tubes have been manufactured for miniature portable television sets, for TV sets intended for outdoor use, for low voltage digital addressing, and for high brightness applications. At the present time, none of these tubes has yet achieved any widespread usage but a description is given below of several of the more promising developments which in the medium term may find some applications. However it appears extremely unlikely that any of them will succeed in winning more than a very small place in the future CRT market. No description is given of the various forms of storage tube which have been developed, since it now appears that with the advent of solid-state digital memory devices, the use of storage tubes in the future will not be large.

1.4.1. Flat cathode ray tubes

The large bulk of conventional CRTs has always been a substantial disadvantage; and attemps were made by both Aiken [1.14] and Gabor [1.15] in the 1950s to make flat CRTs. In each case the gun was arranged to be parallel to the screen face, the electron beam then requiring one electrostatic deflection field to provide line scan before being finally deflected by a separate orthogonal field into the screen to give frame scan. The complexity of the electro-optic field design, and resultant poor resolution, coupled with the need to switch very high voltages at field rate, were the main difficulties which prevented these tubes from becoming successful.
More recent developments by Sinclair and Sony, aimed particularly at small portable TV receivers, follow similar arrangements but now have the advantage of being able to use LSI circuits to correct some of the deficiencies in electro-optic performance. The Sinclair tube, shown in figure 1.5 and described in [1.16], has dimensions of 100 mm x 50 mm x 20 mm, and the phosphor screen on the rear of the tube is viewed through a Fresnel lens incorporated into the

front glass screen. Both Sony and Sinclair brought these tubes into limited production in 1984.

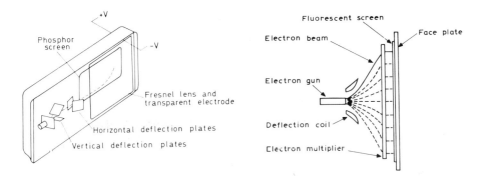

Fig. 1.5 Flat cathode ray tube.   Fig. 1.6 Channel multiplier tube.

Further development of the flat CRT is described in 1.4.2 below.

1.4.2. Channel multiplier tubes

The channel multiplier CRT is a recent development by the Philips company, which has been comprehensively described in a series of papers by Woodhead and his colleagues [1.17,1.18,1.19,1.20]. The principle of operation is shown in figure 1.6, from which it will be seen that the deflection region of the tube is separated from the screen region by the channel multiplier. This has the feature that the electron beam in the deflection region is a low voltage, low current beam with significant advantages in minimising the power requirements for the deflection coils, reducing the demands on the cathode/gun design, and generally improving the distortion and de-focussing problems.

A high voltage is provided between the multiplier and the screen anode, which provides the power necessary to give a high brightness display on the screen. The overall luminance characteristics are determined by the gain of the multiplier and the EHT voltage on the screen, and are almost completely decoupled from the characteristics of the deflection region, though the luminance of individual points on the screen remains a function of the modulator in the gun.

To achieve a high brightness, high definition display, the channel multiplier must have the following characteristics:

(a) High gain.

(b) Uniform gain.

(c) High maximum output current.

(d) High spatial resolution.

[1.18] describes in detail the design of the multiplier which was originally developed by Philips and which substantially meets the above requirements. The construction of the multiplier is shown in figure 1.7; it consists of a number of metal dynodes supported by insulators, the channel surface of the dynodes being coated by a good secondary electron emitter. An electron entering a channel starts a cascade of electrons which increases in intensity at subsequent collisions with the following dynodes. A gain in the region of 500 to 1000 is generated by seven dynodes each at 300 V potential relative to the previous one.

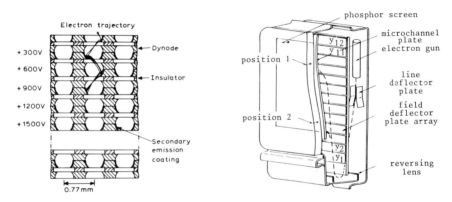

Fig. 1.7 Dynode type electron multiplier.   Fig. 1.8 Cut-away sectional view
of flat CRT (SID).

Performance figures for a 225 mm diagonal experimental tube are quoted in [1.19]. The electron gun forms a 1 mm spot at the centre of the tube, and the pitch of the channels in the multiplier assembly is 0.77 mm. Gain and uniformity are good although there are some problems due to electron backscattering at the input to the channel.

The feasibility of incorporating channel multipliers into CRTs having been established, further advances have been made as reported by Emberson et al in [1.21], who describe a tube which combines the principle of separation of deflection and acceleration fields with the concept of the flat CRT which was described in 1.4.1 above. The advantage of this combination is that the complex beam deflection region can be operated at low current and low voltage levels, thus reducing the extent of distortion and aberration which are generated in this region. Figure 1.8 is a sectional view of the flat CRT showing how the electron beam is initially deflected by electrostatic line deflection plates,

Fig. 1.9 Flat CRT based on folded electron beam and channel multiplier (Photograph courtesy Philips Nederland).

is then reversed at the base of the tube, and is finally deflected into the required vertical position by a voltage applied to one of the series of field deflection plates.

The other major advance provided by the tube described in [1.21] is the use of a channel multiplier of very much higher resolution than the dynode multiplier shown in figure 1.7. The channel plate is formed of a series of glass tubes fused together, the inner surfaces of the tubes giving an electron secondary emission ratio much greater than unity, so that after multiple impacts of electrons against the tube wall an overall current gain of several thousand is achieved with an applied voltage of 1.4 kV. The electrons are finally accelerated by a potential gain of 15 kV before reaching the phosphor screen.

The pitch of the tubes which form the channel plate is less than 50 μm, so that the overall tube resolution is determined by electron beam diameter rather than pitch dimension. A resolution of 720x540 pixels on a 300 mm diagonal CRT is claimed, see figure 1.9. Eight bit grey scale capability up to 600 cd m^{-2} at 50 Hz field rate has been demonstrated; the current linearity of the channel plate is not sustained at high current levels equivalent to 1000 to 1500 cd m^{-2}, causing grey scale distortion.

The possibilities for developing a multi-colour version of this flat tube appear rather limited. However, it is very well suited to the penetration phosphor technique described in 1.2.3 since the electron beam energy can be altered independently of the beam-deflection region of the tube.

1.4.3. The guided beam display

For very large screen displays, particularly television displays for use in conference rooms and similar venues, projection techniques are generally used. An alternative approach which has been taken by RCA is the development of a large-screen flat panel CRT, suitable for wall mounting. The aim is to produce a 75 cm x 100 cm flat tube with full-colour capability.

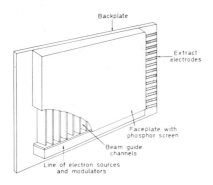

Fig. 1.10 Guided beam display.

The technique is described in [1.22], see figure 1.10. A horizontal line of discrete electron sources is used, each with a discrete channel which guides the electron beam vertically up the rear of the phosphor screen. Between the open channels and the screen is a series of electrodes which extract the beam onto the screen at the appropriate vertical point. This ladder of electrodes combined with the open channel forms a beam guide which has very high transmission efficiency and can give an output luminance up to 350 cd m^{-2}.

To obtain colour, a shadow-mask technique is used, each of the open channels carrying additional scan electrodes to enable one of the three colour dots at each location to be selected.

Although results of preliminary experiments were reported in [1.22] in 1982, development has been discontinued before it reached the production stage.

1.4.4. Digitally–addressed CRTs

The combination of digital addressing with the high luminous efficiency of cathodo–luminescent phosphors appears potentially attractive for computer and airborne use. The most extensive development programme of a tube using such techniques was that of the Northrop Corporation, under the name 'Digisplay', and has been described by Goede [1.23]. Instead of a single scanning electron beam, the Digisplay uses a large number of individual beams, each striking the phosphor layer at a fixed spot. Addressing is carried out by controlling the average intensity of the individual beams, using a series of aperture plates interposed between the large area cathode and the phosphor screen. Patterns of conducting material around the apertures can be digitally–addressed with 30 V potentials so that any individual electron beam can be switched on or off.

Several versions of the display are described by Goede of which the most useful for airborne use is a 512 character display using 7x5 fonts on a 135 mm x 95 mm active area, with an average spot luminance of 820 cd m^{-2}. A colour version was also made but the development programme was subsequently stopped and no other tubes of this type are currently available except for the Character Display Tubes from EEV (UK) which show only a single 7x5 font character.

An interesting variant on the digitally–addressed flat tube is being developed by Siemens, and is reported in [1.24]. This is in many ways similar to the 'Digisplay' tubes, but the large area cathode is replaced by a plasma cathode which generates a cloud of electrons which are accelerated through a matrix addressing array onto the screen. See also chapter II.4, section 4.4.7.

1.4.5. Light valve tubes

Instead of using a conventional cathodo–luminescent phosphor, the screen of a CRT may be formed of a material whose transmission can be a function of an electron beam incident upon it. Liquid crystal materials have been used for this purpose and, as reported in [1.25] they are conceptually very attractive as a way of producing "light valves" for projection TV. Cholesteric and nematic liquid crystals (chapter II.3) have both been used. The electro-optic axis, centred on the gun axis, has to be displaced from the centre of the tube to allow the optical projection system to shine through the rear of the screen, but the design of the tube is otherwise conventional. Good resolution and contrast are reported, although at TV rates the response time of the LC material causes some "smear" with dynamic displays.

1.4.6. Beam index colour tubes

An alternative to the shadow-mask tube which is similar in some respects but avoids the structural problems of the shadow-mask is the beam index tube [1.26]. This is shown in figure 1.11. The phosphor screen is arranged in a pattern of stripes similar to those of the 'Trinitron' CRT, but instead of using a mechanical mask to ensure that only the correct phosphor is energised by the beam, it uses electronic control. A single gun is used and the beam is scanned across the stripes in a raster mode (cursive writing is not generally possible), the beam current being modulated so that the appropriate levels of current are used to excite each of the three colours in turn.

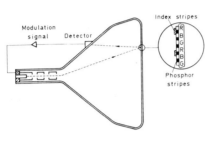

Fig. 1.11 Beam index CRT.

To achieve good colour purity the beam width must be very narrow, and the beam modulation must be exactly synchronised to the beam position. This synchronisation is achieved by putting onto the rear of the phosphor a fourth series of stripes in the form of a UV or X-ray emitting material or an interlaced conductive comb structure. In each case the emission is detected and its timing is used to synchronise the beam current modulator to the scan deflection waveform, through the fast acting control loop. The phase response and stability of this loop must be very good, and problems arise if the stability is affected by any crosstalk between the feedback signal from the index stripe and unwanted signals from the colour phosphor stripes. To minimise this effect, Ando et al [1.27] have used two index stripes for each phosphor triad, subtracting the signals from each and thereby cancelling the crosstalk. They have also simplified the structure of the tube, and thereby reduced manufacturing costs, by printing both the index stripes and the colour phosphor stripes in the same operation.

The advantage of considerable mechanical simplification inherent in the beam index tube is obtained at the cost of electronic complexity. In spite of many attempts to develop a viable tube of this type, the difficulty of making a control loop with the required phase stability at high frequency, together with colour purity and brightness problems, have so far prevented this type of tube from being produced in large scale, although a 120 mm x 120 mm high brightness tube for aircraft application has recently been marketed.

1.4.7. Multiple beam CRTs

With a conventional CRT of the type described in 2.1, the video bandwidth
necessary to provide a given level of resolution is simply calculated as the
number of lines multiplied by spatial bandwidth per screenwidth, multiplied by
the refresh rate. Alternatively for CRTs such as those used in high resolution
applications, the equivalent pixel rate applied to modulate the electron beam
is the product of number of pixels on the screen and refresh rate, to which a
further factor must be added to account for flyback time. Thus with a CRT
displaying 1500 lines, each of 2000 pixels, and refreshed at 60 Hz, the data
needs to be input at 240 megapixel rate. Faced with the implications of this
upon video-amplifier bandwidth, CRT designers have explored the concept of
multiple beam CRTs in which several beams are generated, scanned and focussed
as a group. Modulation of each beam is carried out independently so that the
bandwidth required is reduced in inverse proportion to the number of beams. A
further advantage is that the overall luminance at the screen is increased
compared with the luminance which would be achieved by single beam operation
with the same screen size and the same individual density.

The principal difficulty with this technique is to generate several beams and
to focus and scan than as a single group so that the resultant pattern on the
screen is visually uniform and accurately positioned. Published papers (e.g.
[1.28,1.29]) give good accounts of the problems and of the extent of the
corrections necessary to provide accurate spatial characteristics for 8 and 16
beam CRTs. Thus far there has been no significant application of multiple beam
CRTs, and it remains to be seen whether the trade-off between the complications
of very high video bandwidths versus that of complex tube design will
eventually justify this variant of conventional CRT design.

1.4.8. Filtering and similar techniques used with CRTs

For completeness it is convenient to describe here some techniques which have
been applied to modify the appearance of electronic displays, particularly
CRTs, and thereby to give improved or significantly altered optical
characteristics.
The first of these is the field-sequential colour filter, previously described
by Shanks [1.30]. A CRT display incorporating a white or multi-colour phosphor
is viewed through a filter screen made of a material such as a liquid crystal,
which is arranged so that using its birefringent characteristics it can be
switched to pass a different colour of light depending on the voltage applied
to it. Such a display/filter combination can be made very rugged and of
reasonable brightness. Most arrangements of the liquid crystal filter permit

the use of only two principal colours, and the disadvantages are then similar
to those of the penetration tube, a limited colour range based on the two
principal colours, and the possibility of colour flicker due to field-
sequential colour switching. Difficulties have also been found in producing
large liquid crystal panels of good cosmetic appearance. Nevertheless, [1.32],
work is continuing on this type of display, and by using a sequence of several
polarising filters and liquid crystals cells full-colour displays have been
demonstrated. An evaluation of the visual performance of LC-shutter technology
CRTs as compared with the shadow-mask type [1.31], concludes that the two
technologies do not differ significantly in visual fatigue or visual
performance that they produce.

An alternative application of LC-filter technology has been proposed by Hunt
[33]. The concept is shown in figure 1.12, the LC-shutter being divided into
horizontal segments, the segments being switched in synchronism with the
raster-scanned CRT so that they have high transmission when the CRT trace
behind them is energised, and low transmission when it is not. This results in

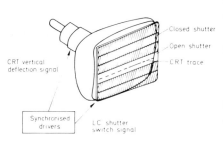

minimal attenuation of the emitted trace
and maximal attenuation of the ambient
light scattered from the CRT face.
Results of initial trials with a
filter/CRT combination operating on
these [1.34], show that the theoretical
performance predictions were largely
achieved by the experimental trials. The
technique appears to have promising

Fig. 1.12 Active LC-shutter filter.   potential for CRT applications in high
                                        luminance level ambient, e.g. in
aircraft or for monitors used out-of-doors.

In addition to the Shanks technique for obtaining multiple colours, another
method is to optically combine the luminous outputs from two or three CRT
display surfaces of different colours. The 'COMED' display [1.35] uses lenses
and mirrors to combine a monochrome CRT and an illuminated map and a similar
arrangement could be used with any display surfaces. The use of optical
combination implies limited viewing angles and hence is particularly
appropriate to vehicle drivers and pilots where reduced luminance can be made
acceptable by correct positioning of the exit pupil. However such a display has
not yet been used in e.g. aircraft, due presumably to the penalties of volume
and complexity.

1.5. Addressing/driving

The monochrome CRT described above has a uniform screen face which can be
energised instantaneously at one spot only, the position of that spot being
determined by analogue currents in the X and Y deflection coils (or in the case
of electrostatic deflection by potentials on the X and Y plates). Therefore to
produce a complete 'field' of information, the spot must be deflected according
to some pattern, which may cover all or part of the whole screen, and upon
completion of the pattern it must within a short period restart the process and
write the next pattern. Because of the response characteristics of the eye, the
fields are usually written at repetition frequencies of 50 Hz or more, so that
the screen appears to emit continuously and little or no flicker is perceived.
Two alternative types of scanning pattern are available, generally known as the
raster and cursive techniques.

1.5.1. Raster scanning

In the raster method (figure I.1.26) the spot always follows a fixed pattern of
parallel straight lines, spaced closely together, to cover the whole screen
area. The lines are usually horizontal and written from the top of the screen
to the bottom. The picture or symbolic message is then written by modulating
the beam current so that the emitted luminance can be varied independently at
any point on the screen. The modulation can either be simple 'on-off' to
produce the equivalent of a black-white screen picture, or can be continuously
varied by an analogue input signal to produce a 'shades of grey' picture.

Common standards for raster writing are essential, and these have evolved
largely from television systems. The two most common are the European standard
format of 625 lines at 50 fields per second, 25 frames per second, and the
American standard of 525 lines at 60 fields per second, 30 frames per second.
In both these, two fields make-up a frame, each field comprising half the total
lines which are interlaced with the other half, to form the whole frame. This
is to obtain the apparent repetition frequency of at least 50 Hz while
maintaining adequate resolution with a modulation signal bandwidth of no more
than 5 MHz; see also section 1.5.1 in chapter I.1.

For many purposes it is possible to use these TV standards, which give
considerable economic advantages because of the wide availability of circuit
components, recorders and other equipment matched to them. But in parallel with
the development of higher resolution CRTs has been the evolution of higher
bandwidth standards. A standard of 875 line, 30 frames per second, 2:1
interlace represents a modest increase, and beyond that is the EIA-RS-343

standard for a 1023 line raster, but several other line counts (e.g. 1029 and 1225) are in common use, and this number seems to be increasing. Most of these new standards are being developed for computer graphic purposes, but although the concept of high definition TV is being pursued in several countries, great difficulty is being experienced in agreeing new international standards for broadcast TV. For aircraft displays, a new NATO standard (STANAG 3350) is being prepared [1.36] which is generally similar to the established TV standards quoted above.

One of the principal advantages of raster operation is that the deflection currents follow a set pattern and the deflection circuits can have a relatively low bandwidth, 50-100 kHz being typical with a high Q circuit, and this results in lower energy losses in the deflection system than for cursive operation at typical writing speeds.

1.5.2. Cursive beam control

The cursive method of writing allows the spot to travel over the screen face in any direction. Generally this technique is used with a fixed beam current, so that provided the spot velocity is constant the luminance of the resultant lines is also uniform. The method is particularly useful for line writing and is only secondarily used for area shading, but the written lines can be either straight or curved and may include a repertoire of fixed characters. Because of the random nature of the writing pattern there are no accepted standards. Writing speeds depend upon the amount of data to be displayed, and are typically in the range 30-100 m s^{-1}. The need for uniform spot velocity with a randomly moving spot implies that resonant deflection circuits cannot be used and that broad band operation of these circuits is necessary with consequent power losses.
The use of multi-colour tubes introduces some complications. Shadow-mask tubes written in raster or cursive mode merely require three modulation signals to the three independent guns. Penetration tubes are usually driven cursively, and require alternate frames to be written in one of the two phosphor colours; and therefore the EHT must be switched between frames. In addition it should be noted that the standards for raster writing mentioned above are not appropriate for this type of tube because the use of sequential colour fields at frame frequencies of 25 to 30 frames per second introduces colour flicker. Beam velocity switching requires to rapidly switch the EHT (in about 1 ms). Considerable energy is dissipated because of the input capacity of the CRT. A further complication is deflection and focus circuits which can cope with the variations in tube sensitivity caused by the changes in EHT voltage.

1.6. <u>System interface</u>

For many applications of CRTs, the information to be displayed is of a pictorial nature, and derives directly from a sensor such as a TV camera, an infra-red camera or a radar equipment. In the simplest of cases the sensor will produce its output signal in a format identical to the raster format required by the CRT display system, so that no separate interface equipment is required. However, particularly if several sensors are fed into a common system with a single display it will be necessary to transform the sensor output from one format standard to another, and converted into an analogue form appropriate to the CRT driving circuits. Numerical information is presented on the CRT in the form of either numerics or a scale, pointer or vector. Special interface units known as waveform generators are used for this conversion and can be made for either raster or cursive operation.

Although early waveform generators were generally made up of dedicated hardware elements to perform specific roles of character generation, geometric transformation etc., it has recently become possible to use the high speed

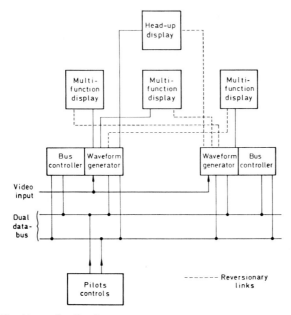

Fig. 1.13 Aircraft display system.

capabilities of modern digital processors to carry out all the functions in a general purpose computer in which all the required characters and symbols are defined by software. The availability of large (e.g. 64 K) random access memories at low cost is important in this type of design as it enables a store location to be assigned to each picture point in the display with the content of the store being updated at intervals from the computer. Modern techniques such as this have the advantage that formats can be modified by software and they have now become the standard form of waveform generator. The store is read out at fixed intervals and may be modified to improve the cosmetic appearance of lines or symbols before being converted to analogue form to be fed to the deflection and modulation amplifiers. When a combination of symbolic and pictorial information is to be presented, the appropriate signals can be mixed together either in the digital store or at the analogue stage.

Figure 1.13 shows in block diagram form the arrangement of a display system in a modern military tactical aircraft.

1.7. Visual characteristics

1.7.1. Resolution—monochrome tubes

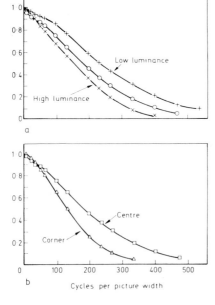

a

b

Cycles per picture width

Fig. 1.14 MTF curves for a monochrome CRT.

The phosphor screen is made up of a layer of fine granules of the phosphor material, typically of around 10 μm diameter and except in the case of the smallest tubes the size of these should not be significant in determining tube resolution. For monochrome tubes, therefore, the resolution of the image on the screen is principally determined by the diameter of the electron beam and spot-spread within the phosphor, and by the resolution characteristics inherent in the beam writing pattern. In practice, for raster displays, these two factors tend to be approximately equal; clearly there is no merit in making either factor grossly different from the other. Therefore for raster tubes operated at the TV line standards described in section 1.5 above, a resolution at the centre of the screen of 525 or 625 lines vertically is obtained

and with a circular spot and adequate bandwidth in the video signal the horizontal resolution is in the region of 1000 TV lines (500 cycles per picture width).

For signals of varying modulation level it is customary to use as a measure of horizontal resolution the modulation transfer function (MTF), which is generally a decreasing function of spatial frequency, see section I.1.3.2. Experimental measurements of CRT MTF frequently show that the useful spatial frequency for low contrast signals is significantly less than expected from simple resolution calculations. An example of the use of MTF methods is provided in [1.37], this paper being of particular interest in indicating the resolution obtainable from the small CRTs used in helmet-mounted displays. Under favourable conditions up to 600 cycles per horizontal line can be obtained at 50 % MTF on a 25 mm diameter tube. Figure 1.14 shows MTF curves for larger monochrome tubes, and is taken from Banbury's paper [1.38] which provides a useful guide to the measurement of MTF of CRTs. It should be noted that MTF is a characteristic which is valid only in linear systems, whereas phosphors exhibit significant non-linearity and hence some care needs to be taken in specifying the exact conditions in which the measurements are made. Thus the measured MTF is affected by the general level of illumination from the phosphor (the veiling glare) as well as by the local illumination caused by electron beam excitation at the point being tested.

For cursively-written CRTs the line width effectively defines resolution, and is principally a function of electron beam width and of spot-spreading in the phosphor. At the high current densities used in aircraft head-up displays or in monitors for outside use, spreading can be significant, but with well-focussed tubes and at lower current densities the width of a line-pair can be made below that which can be resolved by the eye.

Finally it may be noted that when lines are drawn at an angle to the scan lines on a raster display the line structure can cause an apparent series of steps in the line (jaggedness, aliasing, see section I.1.3.3). Although this is an inherent characteristic of raster displays the effect can be minimised in CRTs by shading each edge of the drawn line along the raster lines, so that the steps are less obtrusive and the display appears to have higher spatial resolution than it really has (center of luminance shifting). Nevertheless it is commonly accepted that cursively-written displays appear sharper than raster displays of lines or symbols.

1.7.2. Resolution-shadow-mask tubes

Measurements of the MTF and resolution of colour CRTs are given in [1.39] and [1.40]. For penetration colour tubes the performance is similar to monochrome, except that the size of phosphor granules tends to be a little larger than for

monochrome tubes and may be a little more apparent. For shadow-mask tubes the
phosphor pattern has a very real effect on resolution, especially in the case
of cursively-written symbology. If a narrow line is written across a
shadow-mask tube, coloured Moiré patterns are produced at low spatial
frequencies because of the discrete nature of the screen, and the electron beam
therefore has to be widened to about twice the triad pitch.

As mentioned in 1.2.2, very high definition shadow-mask tubes of large size
have recently been developed for graphical displays such as those used in
computer-aided design systems. Dot-spacing of 0.2 or 0.25 mm is typically used
[1.6,1.7], together with in-line guns to minimise electron beam defocussing so
that a horizontal resolution of over 1500 dots on a screen size of 360 mm x
270 mm is obtainable. Effective use of this capability depends, of course, on a
high resolution raster format, correspondingly high frequency deflection coils,
and a high bandwidth modulation capability. A more recent report [1.41]
describes a 20-inch square Trinitron CRT with a resolution of 2048x2048 display
elements operating at 60 Hz non-interlaced refresh rate. The horizontal scan
rate is 127 KHz and the video bandwidth is 300 MHz. It is clear from the
circuit descriptions provided that the high resolution obtained is firmly
dependant upon recent advances in amplifier technology.

Rugged shadow-mask tubes for use in high vibration environments (terrain
vehicles, aircraft) have similar del sizes, but because of their smaller
overall size their resolution in del numbers is much reduced. Armstrong [1.9]
reports that the latest avionics tubes have a dot-spacing of 0.22 mm and an
effective linewidth of 0.3 mm.

1.7.3. Luminance and contrast

The high luminous efficiency of most CRT phosphors, combined with the high
energy of the electron beams, results in a phosphor luminance which is adequate
for most applications. In the absense of any significant ambient lighting, the
veiling glare produced within the tube is very low so that the contrast is
correspondingly high.

However, if the CRT is to be used in a high illuminance ambient, a major
problem is created by the reflection characteristics of the phosphor screen.
The screen acts as a high efficiency, approximately Lambertian, reflector such
that from 25 to 75 % of the light incident upon it is scattered back from its
surface, and the light emitted from the phosphor is perceived against this
uniform background. In a light ambient of 10^5 lux, which is typical for bright
sunlight conditions, the display surface can have an apparent luminance of
3.10^4 cd m^{-2} so that to obtain a display contrast of 10 the luminance of the
excited phosphor would need to be $3x10^5$ cd m^{-2}, a very high figure even for

high efficiency phosphor. Only in circumstances in which the flexibility and size of CRT displays are specifically required is it worthwhile to pay the considerable cost of developing and producing the special types of display needed, for example in the case of military aircraft with totally transparent cockpit canopies. In such applications it is usually necessary with both raster and cursively-written displays to insert a filter between the screen and the viewer to attenuate the incident light before it reaches the screen; this reduces the required luminance by a factor of 10 or more. When a phosphor with a narrow band emission characteristic is used, the filter can similarly have a narrow band to achieve maximum attenuation of incident light with minimum attenuation of emitted light; see section I.1.4.2.

For colour displays the situation is further complicated by the lower luminance levels generated by both shadow-mask and penetration types of CRT: TV raster operation allows contrast ratios of less than two for all colours in a 10^5 lux ambient [1.40]. A shadow-mask display unit has been developed for military aircraft which is able to display stroke, raster or hybrid formats in full-colour, but with contrasts of only 1.15 for red, 1.5 for green and 1.06 for blue [1.9]. However by use of cursive writing it becomes possible to generate displays on a penetration tube having contrasts of 3.5 in the green and 1.7 in the red, which indicates that the luminance levels of colour CRT displays are near to being practicable for many bright ambient applications.

The contrast required for any particular display depends upon the characteristics of the displayed informaton and particularly upon whether the format is symbolic or pictorial. Colour displays generally require less luminance contrast because they impart information via colour contrast. Head-up displays create unique problems because the symbology is perceived against a real-world background; making allowances for the optical characteristics of the optical combiner, the required symbology luminance is approximately equal to that of the real-world scene.

1.7.4. Colour

The spectral distribution of the light emitted by the CRT is determined by the screen phosphors, and a considerable number of phosphors have been developed which cover most of the visible spectrum. Because the eye sensitivity is not uniform across the visible spectrum but peaks in the green region, the luminous efficiencies of phosphors also vary considerably and tend to be a maximum for the green colours. The characteristics of most of the available phosphors are registered and published in the WTDS register, and it predecessor the P register, operated by the US Electronic Industries Association [1.3]. A description of some recent developments in CRT phosphors has been given by Woodcock and Leyland [1.42].

Distinction   may   be   made   between   broad   band   and   narrow   band   spectral
distribution phosphors. Although they are not distinguisable by the   human   eye
when   viewed   directly,   the use of filters can have quite different effects in
the two cases. Thus narrow band phosphors may be used with narrow band   filters
to   improve   the   contrast under high ambient illumination conditions without a
significant change in perceived colour. However filters may be used with   broad
band phosphors to select a part of the emitted spectrum which may then appear a
different colour.

Figure   1.15   shows   a   number of the more
common   phosphors   in   their   appropriate
positions   on   the   colour   triangle,
including   an   example   of   a   penetration
phosphor   (P49)   whose   apparent   colour
depends upon the excitation energy of   the
electron   beam.   For   the   green phosphors
shown   on   the   triangle,   efficiencies   of
better   than   50 ℓm W^{-1} are obtainable at
low current density. However it should   be
noted   that   all   phosphors   show   non-
linearities in their   characteristics   and
the   high   efficiencies   obtained   at   low
current   densities   may   fall   off
unacceptably at higher current levels. For
example, the new rare-earth P53 can accept
much   higher currents than some earlier P1
and   P43   phosphors   although   it   is   less
efficient up to 3000 cd m^{-2}.

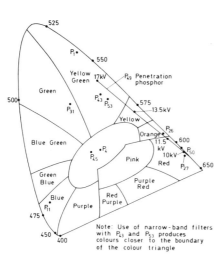

Fig. 1.15 Colour triangle
showing locations of CRT
phosphors.

For   multi-colour   displays it becomes necessary to use phosphors of efficiency
significantly lower than the 50 ℓm W^{-1} quoted above, which partly accounts   for
the   brightness   problems   of avionics tubes described in section 1.7.3. Colour
displays of both the shadow-mask and penetration types have the   characteristic
that,   because   the emissions from the separate phosphors are added together at
the eye, the maximum luminances achievable at   the   intermediate   colours   are
greater than at the primary phosphor colours.

1.7.5. Flicker

The range of refresh frequencies over which an intermittently energised display
appears   to   flicker   varies   with   a   number   of   parameters, including display
luminance, brightness of surroundings, size of   display,   position   of   display

relative to the visual (fixation) axis of the eye, and persistence of the phosphor. Thus it is not possible to quote exact figures, but it is generally accepted that for most operating conditions there is no perceived flicker provided the refresh frequency is kept above a value of 50–60 Hz [1.43], which largely accounts for the choice of field rate adopted for the normal TV standards. If the display is particularly bright, or is perceived peripherally, it may be necessary to increase the frame rate. On the other hand, if the persistence time of the phosphor is greater than a few milliseconds it can have a useful effect in reducing perceived flicker (section I.2.5), although for dynamic displays it is not possible to rely on a long persistence phosphor as this will result in 'smearing'.

Care must be taken if the display is used in a vibrating environment, since with short persistence phosphors it is possible to produce an aliasing effect due to the interaction of the refresh frequency and the vibration frequency. Particular care has to be taken to match the persistence characteristics of the phosphors used in colour tubes when these are viewed in a vibration environment, otherwise a 'colour break-up' may be perceived by the observer. Short persistence phosphors also create problems when used in displays incorporating cameras, and synchronisation of the camera with the display scan becomes necessary.

1.8. State of development

The cathode ray tube is the firmly established leader in electronic displays for all but the simpler types of character and symbolic displays. As quoted by Castellano [1.44], the annual market value of CRTs of all types in 1982 approached $ 5 billion, and although over 90 % of this was accounted for by domestic TV application, the industrial/commercial markets were also very large.

The domestic TV CRT is technically well established and mature, and is manufactured in enormous numbers. Although the Sony 'Trinitron' has taken a small share, the market is mainly satisfied by shadow-mask tubes very similar to those developed by RCA and sold as production items from 1954 onwards. Improvements have been in the detailed design rather than in operating principles, and have included the use of the black matrix phosphor screen and larger deflection angles, resulting in wider but shorter tubes.

As the newer types of display, such as the liquid crystal and electroluminescent matrix panels, challenge the CRT at the lower end of the performance scale, the CRT is being developed to satisfy new requirements which are appearing at the high performance end. The two main directions for these developments are the high definition display needs for graphics terminals used in computer aided design and for tabulation of many characters on a single screen, and the high luminance high definition displays required in aircraft; both those types require full-colour display, which currently implies a shadow-mask or similar technique, although for some aircraft applications such as head-up displays, monochrome tubes are still assessed as being most suitable.

For the large high definition CRT for graphics and characters, the principal task has been the generation of very narrow beams and the production of masks and screens of pitch size down to 0.2 mm. As described in section 1.2.2, and in [1.6,1.7] and [1.41], high definition shadow-mask CRTs are now available with this pitch size in screen sizes up to 360x270 mm, giving a horizontal resolution of over 1500 dels. Such definition requires standards of production and assembly much tighter than in domestic TV tubes, with a corresponding increase in manufacturing costs.

For aircraft use, monochrome tubes of adequate performance are now in widespread use. They are typically used in cursive mode for head-up displays for daytime use, and in either raster mode or cursive mode for head-up displays in night use and for head-down panel displays. The desire for head-up display of pictorial information from forward looking infra-red (FLIR) sensors in aircraft has led to the development of CRTs which can operate at high performance in both raster and cursive modes.

Other current developments in monochrome tubes are aimed principally at improving the contrast ratio and MTF under high illuminance ambient conditions; these improvements are being achieved through narrower beamwidths from better guns, improved electron optics and better phosphors. The dispenser cathode is able to provide a higher beam current density, and because it can be manufactured to a precise shape allows the beam to be very accurately positioned within the electron optics. Care has to be taken to prevent the cathode from being poisoned by residual gas molecules, and the complications which this produces in the CRT production process have meant that tubes employing dispenser cathodes are only recently emerging from the development phase. Particular care must also be taken with the thermal design because of the considerable heating at the electron gun and at the shadow-mask.

Great improvements have been made in phosphors in recent years, and narrow band rare-earth activated phosphors are becoming more common. Their principal advantages are a better match with the narrow band filters used to increase contrast, and their use in combination with diffractive optic elements such as used in head-up displays. Another interesting development is the recent P53 phosphor which can accept much higher current densities than previous phosphors without saturation or damage and is therefore well suited to be used with dispenser cathodes in the next generation of tubes.

Colour CRTs are now also being widely adopted for aircraft use, building on the technology base established for airborne monochrome CRTs, together with that of high definition computer shadow-mask display tubes. Following their adoption in aircraft such as the A320 757, 767 and F100, it can confidently be predicted that all future civil aircraft of any complexity will use them. A cursive format has frequently been used, but as tube performance is improved, hybrid and raster displays will become more common.

Because of the difficulties in designing colour tubes appropriate for military aircraft, applications in this market have been later than for civil aircraft. Initially the penetration phosphor technique was used, but the need for commonality in overall design of the display suite is likely to mean that raster shadow-mask displays are now becoming the preferred solution. The performance quoted in [1.9] suggests that significant progress has been made in establishing the required standards of brightness and ruggedness, although whether this is yet adequate for high resolution colour pictorial information is not yet clear. The beam index tube also looks likely to make a significant penetration of the military aircraft market.

1.9. Health implications for CRT users

In parallel with the increasing use of CRTs for a wide range of applications, has been an increasing concern over the health implications of this usage. The main concerns have centred over CRTs in video display terminals (VDTs), primarily because these are used by dedicated operators for many hours of each working day, and also because the operator typically sits very close to the screen. The primary concerns seem to centre around three possible hazards:

(a) Visual impairments and fatigue.
(b) X-ray radiation associated with emission from the screen.
(c) High frequency radiation associated with driving and scanning circuits.

Experimental evidence to support the view that CRT-based VDTs can impair health has proved to be very difficult to obtain, since the population of users is very diverse and difficult to compare with a parallel "control" group. An

overall summary of the situation regarding radiation hazards is that these are
generally well understood, and that no danger will result from using VDTs which
conform to generally accepted standards of screening. In the case of visual
hazards, it is certainly true that fatigue and possibly permanent eye damage
can result if levels of brightness, contrast and flicker are manifestly outside
the range for comfortable viewing, and even within that range it is considered
desirable for operators to take rests at periodic intervals in order to allow
time for eye relaxation. Standards for visual characteristics are being
established in a number of countries, but the possibilities for abuse of such
standards is bound to be considerable.

CRTs operated at higher levels of EHT voltage unavoidably produce some X-ray at
the screen, but it is rarely necessary to incorporate within the display a
specific filter to absorb this radiation, since the use of lead glass
faceplates ensures that the emitted radiation remains within the permissible
level.

For further information on radiation effects, the reader is particularly
referred to [1.45], and on visual effects to [1.46,1.47,1.48] and [1.49].

REFERENCES

[1.1]   Versnel, L.J.W., High resolution monitor tube, Electronic Components
        & Applications, 5(2), (1983), pp. 108-112.

[1.2]   Infante, C., Denham, D. and McKibben, B., A 230 MHz bandwidth high
        resolution monitor, SID Digest, (1983), pp. 124-125.

[1.3]   Keller, P., Recent phosphor screen registrations and the worldwide
        phosphor type designation system, Proc. SID, 24(4), (1983), pp. 323-328.

[1.4]   Doi, K., Cathode ray tubes: recent trends for character-graphic display,
        Displays, 4(4), (1983), pp. 197-200.

[1.5]   Chen, H.Y., An in-line gun for high resolution colour display, Proc.
        SID, 23(3), (1982), pp. 123-127.

[1.6]   Hayashi, K., High resolution CRT display is more than a picture tube,
        Journal Electronic Eng., 20(202), (1983), pp. 34-37.

[1.7]   Hayashi, S. and Yamaguchi, Y., Improving the resolution of CRT display
        equipment, Journal Electronic Eng., 20(202), (1983), pp. 27-29.

[1.8]   Dietch, L., Palac, K. and Chiodi, W., Performance of high resolution
        flat tension mask colour CRTs, SID Digest, (1986), pp. 322-323.

[1.9]   Armstrong, J. and Spencer, R., Performance of a five-inch by five-inch
        very high resolution full-colour avionic CRT display, Proc. NAECON,
        (1983), pp. 1373-1378.

[1.10] Galves, J.P., Multi-colour and multi-persistence penetration screens, Proc. SID, 20(2), (1979), pp. 95-104.

[1.11] Chevalier, J. and Galves, J.P., CRTs with phosphor and impregnated cathodes for avionics displays, SID Digest, (1982), pp. 60-61.

[1.12] Ohkoshi, A., Takeuchi, O., Kusama, H., Kambayashi, K. and Yukawa, T., Current-sensitive multi-colour CRT display, IEEE Transactions Electron Devices, vol. ED-30, no. 5, (1983), pp. 484-489.

[1.13] Woodcock, S. and Leyland, J.D., High resolution CRTs and their application to helmet-mounted displays, Proc. SID, 20(2), (1979), pp. 105-109.

[1.14] Aiken, W.R., A thin cathode ray tube, Proc. IRE, 45(12), (1957), pp. 1599-1604.

[1.15] Gabor, D., Tothill, H.A.W. and Smith-Whittington, J.E., A fully electrostatic, flat, thin television tube, Proc. IEE, 115(4), (1986), pp. 467-478.

[1.16] ..... Flat screen mini-TV from Britain, Electro-optical systems design, 13(5), (1981), pp. 28-29.

[1.17] Woodhead, A.W., Washington, D., Mansell, J.R., Overall, C.D., Knapp, A.G. and Schagen, P., Channel multiplier CRT, Proc. IEE, 131(1), (1984), pp. 2-5.

[1.18] Knapp, A.G., Washington, D., Guest, A.J., Gill, R.W.A., Pook, R. and Francis, L.H., Large area channel electron multiplier for CRT applications, Proc. IEE, 131(1), (1984), pp. 6-9.

[1.19] Lamport, D.L., Woodhead, A.W., Washington, D. and Overall, C.D., Flat deflection system for a channel multiplier CRT, Proc. IEE, 131(1), (1984), pp. 10-12.

[1.20] Mansell, J.R., Woodhead, A.W., Knapp, A.G. and Stone, H.D., Colour selection in the channel multiplier CRT, Proc. IEE, 131(1), (1984), pp. 13-16.

[1.21] Emberson, D.L., Caple, A., Field, R.L., Jervis, M.H., Smith, J. and Lamport, D., A thin flat high resolution CRT for datagraphics, SID Digest, (1986), pp. 228-229.

[1.22] Credelle, T.L., Large screen flat panel television; a guided beam display, Electro-Optical Systems Design, 14(1), (1982), pp. 31-42.

[1.23] Goede, W.F., A digitally-addressed flat panel CRT, IEEE Transactions Electron Devices, vol. ED-20, no. 11, (1973), pp. 1052-1061.

[1.24] ..... Flat terminal combines plasma and CRT techniques, Displays, 4(3), (1983), pp. 172-173.

[1.25] Haven, D.A., Electron beam addressed liquid-crystal light valve, IEEE Transactions Electron Devices, vol. ED-30, no. 5, (1983), pp. 489-492.

[1.26] Schwartz, J.W., Beam index tube technology, Proc. SID, 20(2), (1979), pp. 45-53.

[1.27] Ando, K., Inoue, F., Jitsukata, H., Eto, M. and Yamazaki, E., A beam index colour display system, SID Digest, (1983), pp. 74-75.

[1.28] Beck, V.D., Electron optical array corrections in the MBCRT, SID Digest, (1986), pp. 50-52.

[1.29] Odenthal, C.J. and Quick, R.G., A Gatling-gun multi beam CRT, SID Digest, (1986), pp. 53-55.

[1.30] Shanks, I.A., Multi-colour displays using a liquid crystal colour switch, AGARD Conference proceedings, CP-167, Paper 18, (1975).

[1.31] Baggen, E.A., Hunter, M.W. and Snyder, H.L., Visual performance evaluation of liquid crystal shutter and shadow-mask CRTs, SID Digest, (1986), pp. 161-164.

[1.32] Bos, P., Buzak, T. and Vatne, R., A full-colour field sequential colour display, Proc. Eurodisplay Conference, (1984).

[1.33] Hunt, G.H., Optical screens, U.S. Patent 4231068, (1980).

[1.34] Koehler/Beran, R. and Bos, P.J., A unique active contrast enhancement filter using liquid crystal Pi-cell technology, SID Digest, (1986), pp. 436-438.

[1.35] Aspen, W.M., COMED a combined display including a full electronic facility and a topographical moving map display, AGARD Conference proceedings, CP-167, (1975), Paper 29.

[1.36] Byrd, J.C., New video standards, Proc. NAECON, (1983), pp. 322-326.

[1.37] Bedell, R.J., Modulation transfer function of very high resolution miniature cathode ray tubes, IEEE Transactions Electron Devices, vol. ED-22, no. 9, (1975), pp. 793-796.

[1.38] Banbury, J.R., Evaluation of MTF and veiling glare characteristics for CRT displays, Displays, 3(1), (1982), pp. 23-29.

[1.39] Kojima, A., An analysis of horizontal MTF in colour CRTs, SID Digest, (1983), pp. 66-67.

[1.40] Brun, J. and Martin, A., Comparative evaluation of high resolution colour CRTs, Thomson-CSF Paper NTV 6210, based on paper given at SID symposium, (1980).

[1.41] Awata, Y., Sumiya, H., Shibata, Y. and Uemura, S., A new large-screen high resolution Trinitron colour display monitor for computer graphics application, SID Digest, (1986), pp. 459-462.

[1.42] Woodcock, S. and Leyland, J.D., The choice of phosphor for modern CRT display applications, Displays, 1(2), (1979), pp. 69-82.

[1.43] Turnage, R.E., The perception of flicker in cathode ray tube displays, Information Display, 3(4), (1966), pp. 38-52.

[1.44] Castellano, J.A., Current US world markets for electronic displays, SID Digest, (1982), pp. 24-25.

[1.45] Terrana, T., Merluzzi, F. and Giudici, E., Electromagnetic radiations emitted by visual display units. In: Ergonomic Aspects of Visual Display Terminals, by E. Grandjean and E. Vigliani (Taylor & Francis), (1960), pp. 13-21.

[1.46] Laubli, T., Hunting, W. and Grandjean, E., Visual impairments in VDU operators related to environmental conditions. In: Grandjean and Vigliani, loc cit, pp. 85-94.

[1.47] Daindoff, M.J., Visual fatigue in VDT operators. In: Grandjean and Vigliani, loc cit, pp. 95-99.

[1.48] Sauter, T., Gottlieb, M.S., Jones, K.C., Dodson, V.N. and Rohrer, K.M., Job and health implications of VDT use: Initial Results of the Wisconsin-NIOSH Study, Communications of the ACM, 26(4), (1983), pp. 284-294.

[1.49] Knave, B. e.a., Proceedings of the conference 'Working with visual display units', Stockholm (1986).

DISPLAY ENGINEERING: D. Bosman (Editor)
© Elsevier Science Publishers B.V. (North-Holland), 1989

II.2: VACUUM FLUORESCENT TUBES

G.H. HUNT

Royal Aircraft Establishment, United Kingdom

2.1. Historical survey

The vacuum fluorescent device (VFD) is a simplified descendant of the earlier magic eyes which displayed tuning accuracy in vacuum tube radios by means of a varying area of cathodo-luminescence. Since in the VFD the area remains constant, the device is only required to modulate the density of electrons hitting the phosphor. Therefore its construction is that of a triode, the anode of which is coated with the phosphor. The source of electrons (filament, cathode) can be common to all the triodes in the glass envelope and, if the addressing takes place by switching anode potential, also the grid. With such tremendous simplification vast numbers of dels can be manufactured in one VFD. The triode VFD was first marketed in 1967. Its subsequent history has been described in [2.1,2.2] and in [2.3] where particularly comprehensive reviews are given of both the history and the various types of display which have been developed (mainly in Japan) up to 1982.

Several distinct phases of development can be identified starting with the first generation of displays, which comprised single 7-bar numerics each encapsulated in a cylindrical glass envelope. By 1975 multi-digit tubes were available capable of displaying up to 13 digits, and these were followed by alpha-numerics and symbols. The third generation of VFDs was physically simplified by dispensing with a separate ceramic substrate for the anode and using one side of the glass vacuum envelope for this purpose, thus achieving lower manufacturing cost and improved appearance. The fourth generation comprises the development of area or matrix displays, which are now of sufficient resolution to approach the requirements of TV or computer screens, together with bi- and tri-colour options. Looking to the future, a fifth generation appears likely which will incorporate active substrates to fully exploit the possibilities of large-size matrix arrays.

2.2. Principles of operation

Vacuum fluorescent tubes form a type of display in which light is emitted from phosphor-coated anodes when these are subjected to electron bombardment, using the principle of cathodo-luminescence. As compared with cathode ray tubes the electron energies are much lower, typically no more than 50 eV, and the display pattern is formed of discrete anode elements rather than by a focussed beam

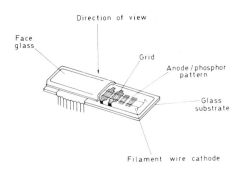

Fig. 2.1 Elements of a simple
vacuum fluorescent display.

striking one part of a continuous
anode surface.

Figure 2.1 shows the layout for a
typical simple tube. Although diode
operation is possible, tubes are
usually constructed as triodes to
allow easier control of luminance.
The cathode is a thin tungsten
filament which is run at dull-red
heat, and the grid is a thin metallic
mesh. The anode is commonly formed of
a pattern of thin-film electrodes
deposited onto the top surface of the
rear glass wall of the vacuum
envelope, and on each of the electrode elements a further deposit of phosphor
is laid down. This fixed pattern of anode/phosphor elements may be in the form
of a series of 7-bar numerals, 5x7 dot-matrix patterns, or sometimes a series
of bars forming an analogue one dimensional scale indicator.

An important difference between this type of tube and a CRT is that the anode
phosphor is viewed from the side which is impacted by the electron currents.
Thus the cathode and grid structures must be as transparent as possible to the
emitted light and, moreover, have very low reflectance for light from the
outside. This layout is used in order that the emission efficiency of the
phosphor layer can be maximised, particularly in view of the low energies of
the electrons; [2.3] quotes a luminous efficiency of about 6 ℓm W^{-1} for a
ZnO:Zn phosphor when used in a VFD at about 35 V, as compared with 25 ℓm W^{-1} in
CRTs operated at about 10 kV. In order to improve the appearance, however,

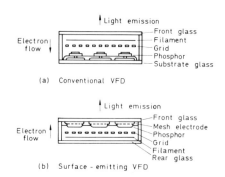

Fig. 2.2 Alternative structures for
VFDs.

some recent 'top view' VFDs have been
developed with a layout more similar
to the CRT, having the phosphor
excited from the rear. Experiments
[2.4,2.5,2.6] show that by careful
choice of phosphor thickness and of
the geometry of the anode mesh upon
which the phosphor is deposited, the
loss of luminous efficiency can be
kept reasonably low. Figure 2.2b
shows the layout of this type of
tube, as compared with the
conventional type in figure 2.2a.

Nakamura et al [2.3] provide a detailed description of the typical simple VFD tubes shown in figure 2.1. Their subsequent development into larger display tubes of essentially similar principles of operation led to the capability to display 240 characters each formed of a 5x7 dot-matrix [2.7]. Further expansions to a fully flexible 256x256 dot-matrix display having a dot size of 0.2x0.2 mm^2 and a dot pitch of 0.4 mm; and a 512x512 dot-matrix display are reported in [2.8,2.9].

The luminance of each dot or bar can be controlled by a combination of the grid and anode voltages; in practice it has usually been found desirable to operate the grid and anode at the same potential to avoid focussing problems, so the control of electron current and luminance in each area of phosphor is through simultaneous variation of grid and anode voltage. For the smaller matrix tubes the grid voltages of the individual dels are operated in a time-multiplex mode; duty factors of 1/20 to 1/50 are typical for these displays, with a consequent loss of luminance as compared with continuously operated displays. For this reason the anode voltage may be raised to 120 V to restore the luminance to acceptable levels.

For large matrix displays the conventional time-multiplexing techniques may also be used, but the loss of luminance and the difficulties of rapidly switching the electrode voltage levels have in the early 80s led to experiments with active substrates. References [2.11,2.12] describe the progressive development of vacuum fluorescent matrix displays of 241x246 and 216x246 dels respectively, with full addressing flexibility at TV frequencies. These incorporate active elements on a silicon substrate to isolate the phosphor display elements from the address circuits, and the display size is thus constrained to a relatively small substrate size appropriate to conventional silicon technology.

2.3. Physical characteristics

Although there is a range of different anode layouts which determine the size and arrangement of the displayed symbology, the essential physical arrangement of VFDs is that of a three-element (triode) vacuum tube as shown in figures 2.2a and 2.2b. In each case the tube is formed of a flat-walled glass envelope, the dimensions of the rectangular front and rear faces being determined by the required size of alpha-numeric array or the size of graphic matrix. Character sizes range from 5 mm to 15 mm in height, and overall display sizes from 20 mm x 50 mm to 140 mm x 270 mm. Typical tube thickness is in the range 10 mm to 14 mm.

The tube is fitted with a large number of wire lead-outs through the glass envelope, providing for signals to be applied to the cathode, grid and anode electrodes. In most cases the display tube is supplied in the form of a complete module consisting of the tube mounted on a circuit board together with a number of addressing and driving circuits, thereby minimising the number of separate connections required in assembly or replacement. Figure 2.3 shows this arrangement for a matrix display.

The display devices are very rugged and can withstand a broad range of environmental conditions; they are therefore suitable for a wide range of industrial and consumer applications, and using well-established vacuum tube technology there are few difficulties in their production. The production process was greatly simplified and reliability improved in 1974 when thick-film printing techniques on a glass substrate were utilised to replace ceramic substrates, and allowed a greater variety of display patterns to be produced.

Fig. 2.3   240-character VFD with integral driving circuits.

An example of this is provided by the 240-character display in [2.7] which describes that, to improve production yield and reliability, it was decided to minimise the number of electrical overlapping points in the anode matrix wiring array. It describes the production process which produced a 3-layer anode substrate having only two overlapping points per dot.

The grid and cathode structures are required to be robust and yet, especially in the arrangement of figure 2.2a, be unobtrusive when observed from the front.

The cathode is normally formed of a pure tungsten filament wire of 7–15 μm diameter, coated with alkaline earth carbonates. The grid is constructed by photo-etching with hexagonal holes a stainless steel plate of 50 μm thickness to give a resultant light transparency of about 85 %. To obtain uniformity of performance, the spacing between cathode and grid must be maintained within close tolerance, and this involves pre-tensioning the cathode wire and maintaining the tension by supporting it on a spring mount. An analysis of the design factors for the cathode filament is given in [2.10].

Construction of the glass envelope uses conventional techniques. Ordinary window glass is found to be quite adequate for the front and rear elements, which are then joined by a fritting process. Details of the complete manufacturing process of the envelope, anode, cathode and lead assembly are given in [2.3]. The surface-emitting tube has significant differences in design of anode assembly. As described in [2.5], the phosphor is deposited on an anode structure which has the form of a thin mesh; experiments have shown that if the ratio of open area to total area of the mesh is about 80 %, maximum brightness is obtained. Problems in design of the grid and cathode are reduced because they are not required to be transparent.

2.4. Addressing/driving and system interface

Except for very simple single-numeral displays, individual dots or bars are excited by time-multiplexing; [2.7] describes the method used for a 240 character display, each character being formed by a 7x5 dot-matrix plus a cursor. This is one of the most complex displays which is commercially available, but the addressing technique used on simpler displays is similar in principle.

The 240 characters are divided into six lines of 40 characters, with 40 grids each covering a column of six characters. The anodes for each column have 216 individually connected dots (six characters x 36 dots). Complete freedom for addressing individual dots is provided by time-multiplexing the voltages on the 40 grids; allowing for a blanking time of about 50 μs between grid pulses a duty cycle of 1/45 is obtainable when the complete cycle is repeated at 50 Hz.

A micro-processor is used for the addressing system. The CPU executes control of driving circuits and I/O data, and the driving program and all character patterns are stored in ROM. A shift-register is used to assemble a column of six characters before these are input to the drivers, allowing the pattern for one column to be built-up while the previous column is being excited. Anode and grid voltages are maximum 50 V, the luminance of the display being reasonably linear with voltage. Anode and grid currents per column are in the 0–30 mA range, so that the driver circuit requirements are compatible with IC characteristics. A single IC package incorporating shift registers, latches and drivers is described in [2.7].

The low duty cycle which is an inherent characteristic of large matrix time-multiplexed displays significantly reduces output luminance. To improve this, and to allow the display tube to better match to TV input signals, active addressing of the anodes has been developed. An example of this is provided by a TV matrix display using CMOS transistors embedded within the anode substrate,

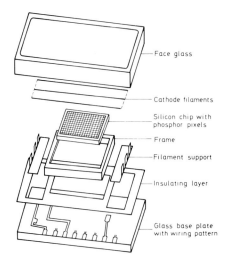

Face glass

Cathode filaments

Silicon chip with
phosphor pixels

Frame

Filament support

Insulating layer

Glass base plate
with wiring pattern

Fig. 2.4 Structure of the on-chip
VFD.

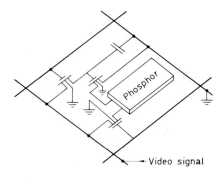

Phosphor

Video signal

Fig. 2.5 Driving circuit of active
matrix VFD.

see figure 2.4. As describe in [2.12, 2.15], and shown in figure 2.5, each del is decoupled from the input video and scanning signals by a transistor circuit which contains CMOS components formed within the silicon substrate under the phosphor. To obtain screen brightness up to 300 cd m^{-2}, high voltage (20–30 V) transistors must be used. The size limitation inherent in this type of construction results in a display area of only 6.91 mm x 9.10 mm, so that some optical magnification technique will be necessary in order to provide a display appropriate for most applications.

2.5. Visual characteristics

Many vacuum fluorescent tubes are fixed format displays in which the patterns of the figures or lines which can be displayed are defined by the shapes of the phosphor layer printed on the anode. These shapes often take the form of 7x5 dot-matrix or 7-bar numerals, although other more complex fixed shapes are used in analogue strip and pointer displays for car dashboard and other uses. However, with the development of larger area matrix displays, graphical formats are becoming available; the displayed graphic image produced by the 256x256 dot-matrix display described in [2.8] is shown in figure 2.6. With such a matrix display, and with suitable driving cicruits, it is possible to "scroll" the format either vertically or horizontally.

The colour of light emission from the phosphor is an important visual

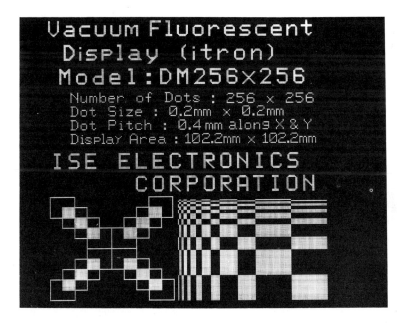

Fig. 2.6 Graphic image on 256x256 matrix VFD.

characteristic. The phosphor normally used is ZnO:Zn, which has a broad band emission peaking near 0.5 μm, and having a half-peak width of about 0.1 μm. A range of individual colours can be selected from this broad band by using suitable filters, so that colours from blue to orange can be seen, but this decreases the perceived luminance by a factor of 3 or more. Much work has therefore been done to develop other phosphors which themselves have narrower bands of emission. Although a range of such phosphors is available for CRTs, most of these are not suitable for VFDs because of their poor efficiency at low electron energies. However, as reported in [2.2,2.3], and shown in table 1 below, a range of phosphors has been developed for use in VFDs, though all have

efficiencies significantly lower than that of the familiar blue-green of
ZnO:Zn.

Table 1   Phosphor colour characteristics

Ingredients of phosphors	Emitting colour	Peak wavelength (nm)	Colour coordinate X	Y	Emitting efficiency (ℓm W^{-1})
ZnO:Zn	Bluish-green	505	0.25	0.44	5 ~ 7
(ZnCd)S:Ag+In$_2$O$_3$	Red	665	0.67	0.33	0.5 ~ 1
ZnS:AuAℓ+In$_2$O$_3$	Lemon	555	0.39	0.56	1 ~ 1.5
ZnS:Mn+In$_2$O$_3$	Yellow	585	0.52	0.47	0.3 ~ 0.6
ZnS:Ag+In$_2$O$_3$	Blue	450	0.16	0.12	0.1 ~ 0.3
ZnS:CuAℓ+In$_2$O$_3$	Green	520	0.25	0.59	1 ~ 1.5

A full-colour VFD has been described in [2.15] which incorporates a 160x120
pattern of red, green and blue phosphor dels. To obtain good colour purity the
individual dels must be printed with great accuracy, and when energised there
must be minimum crosstalk between adjacent dels. No figures are given in [2.15]
of the purities achieved, nor of the brightness of the display which is likely
to be low because of crosstalk problems and because of the fundamental
limitations of matrix addressing.

An alternative method of achieving variable colour in VFDs has been proposed
[2.5] using the same technique as the penetration-phosphor CRT, i.e. using a
phosphor whose colour of emission is a function of electron energy. Although
samples have been produced which operate in the green-orange region, the scope
for producing a really useful multi-colour display with this technique does not
appear good, principally because of the low voltages which must be used and the
correspondingly low phosphor efficiencies.

The use of filters can, as well as providing a choice of colours, also improve
contrast by reducing the level of light scattered back from interior surfaces
of the display; the worst effect of this scattered light is caused by the
printed phosphor patterns on the anodes, the grids and the individual
conducting leads to the anodes which all contribute to a fixed pattern which is
very apparent in bright ambient illumination. The use of narrow band filters as
in some CRT displays to minimise ambient light scattering is not feasible with

vacuum fluorescent tubes because of the broad band emission from the zinc oxide phosphor.

Phosphor luminance is particularly important in displays which are to be viewed in high illuminance ambient, for example in car dashboards, see Part III, section 3.3.5. Experiments described by Morimoto [2.4] show that response times for correct recognition of numeric displays in an ambient of 50.000 lux were acceptably short (i.e. less than 0.15 secs.) provided the display luminance was about 2000 cd m^{-2}, with surface-emitting tubes showing some advantages compared with conventional VFDs. It is also reported in [2.3,2.4,2.5] that satisfactory performance has been obtained with respect to variation of luminance with temperature and with length of time of usage. For most applications, in which displays are not being driven continuously at very high luminance, a lifetime of 10.000 to 50.000 hours can typically be obtained before there is any significant loss of luminous efficiency.

A visual characteristic of importance in some applications is the permissible angular range over which the display must be seen. This can be a particular problem in conventional VFDs having the cathode/grid structure between the emitting phosphor anode and the observer. Care must therefore be taken in designing this structure to minimise the visual screening, particularly with high density matrix displays; Uchiyama et al [2.8] describe the layout used in their 256x256 dot-matrix display. Such problems are avoided in the surface-emitting type of VFD, and it is claimed by Watanabe and West [2.13] that the front-luminous 640x400 VFD which they describe eliminates the problem of obscuration at different viewing angles.

2.6. State of development

The vacuum fluorescent tube is in large-scale production for a wide range of display applications including calculators, radio and communication equipment, clocks and other domestic and industrial applications. Such applications are principally of alpha-numeric displays based on simple 7-bar or 7x5 fonts, together with dedicated numbers, letters and symbols. Further market penetration into computer terminals, word-processor and other applications in the rapidly-expanding area of information processing will follow as the performance of high density matrix displays approaches that of existing CRT displays. In these areas the importance of production cost, including that of the associated addressing and driving circuits, will be paramount.

For the current range of VFDs, excellent reliability, life and environmental characteristics are claimed, and production cost is generally relatively low. For these reasons the display is potentially attractive to the automobile market, and indeed some market penetration has already been made [2.14]. In this area the display formats can generally be kept relatively simple, so that multiplexing duty cycles can be kept high, with correspondingly high luminance as required in the bright ambient light conditions. For aircraft or other applications in which a combination of large matrix arrays and high luminance are required, an active matrix substrate method of addressing will be required. Uchiyama [2.8] indicates, on the basis of laboratory experiments, that luminance levels of up to 15.000 cd m^{-2} should be obtainable for individual dels fabricated with transistors and capacitors and operated at TV refresh frequencies. However much further work will be necessary to achieve this.

The prospect of active matrix addressing also improves the possibilities for obtaining acceptable luminance levels from less efficient phosphors, such as those in the blue and red and this, together with higher density matrices, could lead to the development of full-coloured matrix VFD displays which could be used for a wide range of applications.

REFERENCES

[2.1]   Kiyozumi, K. and Nakamura, T., Vacuum fluorescent displays: from single digits to colour TV, Displays, vol. 4 (4), (1983), pp. 213-219.

[2.2]   Kazuhiko, K. and Nakamura, T., The present and future of vacuum fluorescent display device, Proc. First European Display Research Conference, (1981), pp. 156-159.

[2.3]   Nakamura, T., Kiyozumi, K. and Mito, S., In: "Advances in Image Pickup and Display", vol. 5, Ed. B. Kazan, Academic Press, (1982), pp. 199-280.

[2.4]   Morimoto, K. and Dorris, J.M., Front luminous vacuum fluorescent / display, SAE Paper 830044, (1983).

[2.5]   Yoshida, Y., Kadota, Y., Miyazaki, T. and Chin, B., Latest technology in fluorescent indicator panel, SAE Paper 830045, (1983).

[2.6]   Shimizu, H.D., The itron VFD with reverse view configuration, SAE Paper 830497, (1983).

[2.7]   Kasano, K., Masuda, M., Shimojo, T. and Kiyozumi, K., A 240-character vacuum fluorescent display and its drive circuitry, Proc. SID, vol. 21/2, (1980), pp. 107-111.

[2.8]   Uchiyama, M., Masuda, M., Kiyozumi, K. and Nakamura, T., High resolution vacuum fluorescent display with 256x256 dot-matrix, Proc. SID, vol. 23/3, (1982), pp. 163-167.

[2.9]   Kiernan, R.P., Watanabe, T., Daishaku, E. and Suzuki, S., High resolution X-Y dot-matrix VFD, SID Digest, (1982), pp. 214-215.

[2.10] Kishino, T., Mizohata, T. and Pykosz, T.L., Multi-function large scale glass vacuum fluorescent displays for automotive applications, SID Paper 830043, (1983).

[2.11] Uemura, S. and Kiyozumi, K., Flat VFD TV display incorporating MOSFET switching array, IEEE Transactions Electron Devices, vol. ED-28, (1981), pp. 749-755.

[2.12] Yoshimura, T., Fujii, K., Tanaka, S., Uemura, S. and Horie, M., High resolution VFD on a chip, SID Digest, (1986), pp. 403-406.

[2.13] Watanabe, H. and West, R.A., A 640x400 graphic front luminous VFD, SID Digest, (1986), pp. 407-409.

[2.14] Pagel, E.O. and Sterler, G., Electronic dashboard for the audi quattro, Proc. IEE/I Mech E International Congress on Automotive Electronics, (1983), pp. 189-191.

[2.15] Tamara, M., Vacuum fluorescent display gains large capacity and multi-colours, Journal Electronic Eng., vol. 23 (231), (1986), pp. 82-84.

II.3: LIQUID CRYSTAL DISPLAYS

A.J. HUGHES

Royal Signals and Radar Establishment, United Kingdom

3.1. Introduction

The term "Liquid Crystal" was first used by Lehmann in 1890, who applied it to a substance which flowed in the manner of normal liquids, but whose optical behaviour was similar to that of an anisotropic crystal. It designates a state of matter intermediate between solids and liquids, having some of the properties of each. Liquid crystal phases are called "mesophases" because of this intermediate nature.

The early commercial developments of Liquid Crystal Displays (LCDs) concentrated on small numeric and alpha-numeric displays which rapidly replaced LEDs and other technologies in applications such as digital watches and calculators. More recent developments have produced larger and more complex displays for use in many applications including public information displays, portable computers, miniature TVs, etc. The main attractive features of LCDs include:

a)   the display does not emit light, it merely modulates the ambient light and retains good appearance in even the brightest conditions. Low power subsidiary illumination is readily provided for night-time viewing;

b)   display power consumption is minimal, a few microwatts per square centimetre, so continuous battery-powered operation is feasible over long periods;

c)   displays operate at low voltages, from 2 to 20 V, compatible with low power integrated circuit drivers;

d)   life-times in normal environments can be very long, up to 50,000 hours or more;

e)   manufacturing costs in large quantity production are very competitive;

f)   displays may either be viewed directly in transmission or reflection, or may be projected onto large screens.

There are disadvantages as well, including:

a)   each liquid crystal phase exists over only a limited temperature range and the physical properties of materials vary considerably with temperature, so there have been difficulties in making materials and displays to operate over the wide temperature ranges encountered in severe environments;

b)   at low temperatures the display response becomes noticeably sluggish;

c)   the most frequently used display effect, the "Twisted Nematic" effect, requires polarisers which restrict the display brightness;

d)   there are considerable difficulties in making very complex displays (e.g.
     for TV and computer terminals) that have acceptable brightness, contrast
     and angle of view.

The aims of this chapter are to provide a background understanding of liquid
crystals, the effects used in displays and their limitations; to summarise the
capabilities of the currently available production displays; to describe a
selection of experimental prototypes potentially capable of extending the range
of application of LCDs. Section 3.2 deals with the background information on
liquid crystals relevant to display applications, followed in section 3.3 with
a discussion of a range of electro-optical effects that are being exploited in
displays. Finally, section 3.4 discusses some of the difficulties of addressing
LCDs and describes a number of solutions and their limitations.

3.2. Physical characteristics

Detailed references are not given in this section; for more comprehensive
reviews and bibliography see references [3.1] to [3.3].
Liquid crystals have been classified in two distinct classes known as
"Lyotropics" and "Thermotropics". Lyotropic phases occur as a function of the
concentration of solutions of suitable organic molecules, e.g. the slime on a
wet bar of soap. Although lyotropic phases are of considerable importance in a
wide range of situations they have not as yet been utilised in displays and
will not be discussed further here. Thermotropic phases occur as a function of
temperature between the true solid phase at low temperatures and the isotropic
(clear) liquid phase at high temperatures. Two principal types of molecules are
known to form thermotropic liquid crystal phases. The original liquid crystals,
and all those used in the display effects discussed here, consist of long,
thin, organic molecules. More recently, however, liquid crystalline properties
have been observed with flat, disc-like molecules, though these have not yet
been exploited in display applications.
In a perfect crystalline solid the relative positions and orientations of the
molecules are well defined, whereas in a liquid the relative positions and
orientations are virtually random. In a liquid crystal, however, the relative
orientations of the molecules remain well defined, but various aspects of the
positional ordering of the crystal are lost.
Liquid crystals are classified according to the symmetry of their positional
ordering. In "Nematic" liquid crystals all positional order is lost, only the
orientational order remaining. "Cholesteric" materials are closely related to
nematics and are sometimes called "chiral nematics". Here the asymmetry of the
molecules causes a small angular twist between molecules which results in a
macroscopic spiral structure of well-defined helical pitch.
In "Smectic" materials, of which there are several types, the molecules tend to

exist in layers, but there is no long-range positional order within the layers. The "Smectic A" (S_A) phase has the molecules oriented perpendicular to the plane of the layers, whereas the "Smectic C" (S_C) phase has the molecules tilted uniformly from the perpendicular. A particularly interesting version of the S_C phase occurs when the molecules are chiral. This produces a helical structure, with the helical axis perpendicular to the smectic layers, known as a "Chiral Smectic C" (S_C*) phase. This has the correct symmetry properties to show ferro electric and pyro electric effects. Still further smectic phases exist with various degrees of short range positional order, but they will not be discussed further.

Any particular material may pass through several of these mesophases as temperature is varied, but usually the more highly ordered, smectic, phases are found at lower temperatures than the less ordered nematic or cholesteric phases, and the transparent isotropic phase, as shown for a hypothetical material in figure 3.1. In this diagram no attempt has been made to represent thermal fluctuations which must affect both the positions and orientations of the molecules to an extent which depends on the temperature.

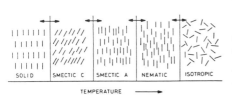

TEMPERATURE ⟶

Fig. 3.1 Schematic phase diagram of hypothetical liquid crystal with phase structures included.

Two important concepts related to molecular orientation are the "director" and the "order parameter". The director is a unit vector representing the local alignment direction of the long axes of the molecules, obtained from the average alignment of a group of molecules. The order parameter, usually called "S", measures the degree of alignment of the molecules relative to the director and is defined by

$$S = \tfrac{1}{2}\,(3\,\overline{\cos^2\theta} - 1)$$

Measured values of S are strongly temperature dependent and range from about 0.4 at the nematic to isotropic transition temperature to values approaching 1 at low temperatures.

In spite of all this ordering, however, liquid crystals are clearly liquid, the nematic phase in particular flowing with quite low viscosity. Consequently the orientational order is not preserved automatically over indefinite distances as in a solid crystal. Typically the correlation length of the director orientation is a few tenths of a millimetre in the absence of external constraints.

A most significant aspect of liquid crystals is the large anisotropy of most of
their physical properties when measured in directions parallel and
perpendicular to the director. For example, the anisotropy of dielectric
constant means that electric fields can be used to control the orientation of
the director. This effect is utilised in all the displays discussed later. This
dielectric anisotropy may be either "positive" (i.e. maximum dielectric
constant parallel to the director) or "negative", and both types of material
have been exploited in displays. The anisotropy of magnetic susceptibility is
also significant and permits orientation control by magnetic fields. This is
used extensively in research but has not yet been exploited for display
purposes. The anisotropy of refractive index, Δn, usually between 0.05 and 0.2,
is much greater than in most crystalline solids and is the basis of most
optical effects used in displays. Other physical properties, such as electrical
conductivity, elasticity, viscosity, etc., are also strongly anisotropic and
have significant effects on the static and dynamic behaviour of materials and
devices.

Finally, liquid crystal materials may also interact strongly with solid
surfaces. These effects are also important since, in the absence of electric or
magnetic fields, the structure and orientation of thin layers of liquid
crystal are largely determined by surface interactions. Methods have been
developed of treating glass surfaces with organic or inorganic films,
possibly followed by controlled mechanical abrasion, which align the
director either perpendicular to the surface ("homeotropic" alignment), or parallel to the surface ("homogeneous" or "planar" alignment), or at some intermediate angle as shown in figure 3.2. This control of alignment via surface forces permits very large areas of uniform orientation and texture to be produced which is essential for the uniform appearance of displays.

Although many electro-optic effects in liquid crystals have been used for display purposes, the construction of the display cell is normally based on the outline shown in figure 3.3. The figure, which is not to scale, shows

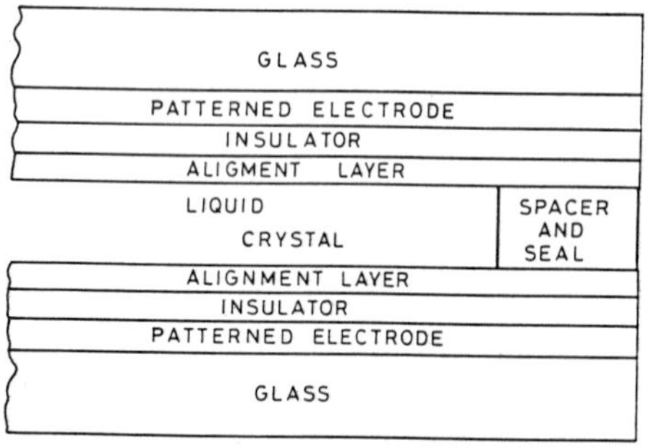

Fig. 3.2 Representation of various
surface alignments.

Fig. 3.3 Schematic representation
of a typical liquid crystal cell,
not drawn to scale.

the two flat glass substrates which are separated by a uniform space, typically between 5 μm and 20 μm thick, which is filled with the liquid crystal. The inner walls of the glass are covered by the electrode patterns which define the active areas of the display. In transmissive cells both electrode layers are made of a transparent conductor such as an indium tin oxide mixture, whereas in some reflective cells the rear set of electrodes may be metallic. Covering the electrodes are insulator layers of sufficient thickness to protect the liquid crystal from inadvertent exposure to DC. Finally the insulators are coated with the appropriate alignment layers if needed. The spacing of the cell is sometimes controlled by a spacer around the periphery, though rather better spacing control is obtained using inconspicuous spacers distributed over the whole area. These spacers may either be short lengths of glass fibre or plastic spheres. The cell is sealed around the edge by either a thermoplastic bond, a UV-curable epoxy or a higher temperature glass frit seal. The liquid crystal material is introduced through gaps in the peripheral seal, either by capillary injection or vacuum processes, after which the filling holes are plugged. Finally the polarisers and reflectors if required are attached to the outside of the glass, and electrical connections are made around the appropriate edges of the cell.

In order to be generally acceptable there are many requirements placed on the LC materials. Firstly, it is vital that they should be non-toxic, both for ease of handling during manufacture and to avoid risks caused by breakage during use. Secondly, they should be highly stable, not only to ensure long life in the operating environment, but also to avoid the need for difficult and costly manufacturing processes. Many of the early liquid crystal materials were suspect on account of either toxicity, susceptibility to atmospheric oxidation or to degradation caused by blue or ultraviolet light. The development of the cyano-biphenyl family [3.4] and [3.5] provided the first satisfactory solution to all of these problems and enabled the production of displays with claimed operating life-times in excess of 50,000 hours. Subsequently, continuous materials development has been required to keep pace with the demands for wider operating temperature ranges, closer control of electro-optic characteristics, faster response speeds, novel electro-optic effects, etc.

Temperature affects all the physical properties of liquid crystals, but the first consideration must be the range over which the chosen LC phase exists. This is bounded at high temperatures by a transition to a less ordered LC or isotropic liquid phase, and at low temperatures by a transition to a solid or more highly ordered LC phase. The nematic to isotropic transition is usually well-defined and reversible, but solidification is often accompanied by extensive super-cooling. It is most important that the quoted minimum temperature for a material represents melting from the solid and does not rely on supercooling: the super-cooled state is only metastable, and damage to the surface alignment layers is possible if repeated solidification occurs.

Single chemical compounds are rarely liquid crystalline over a usably wide temperature range, but multi-component mixtures have been developed [3.6] and [3.7] which have mesophases over very wide ranges. One basic principle of mixture development is illustrated by the two-component phase diagram shown in figure 3.4, where the eutectic composition is nematic from 20 C to 70 C. By adding precise quantities of other materials a multi-component eutectic can be made with even wider range. For example, the four-component mixture E7 (BDH nomenclature), used for many years in watch displays, operates from − 10 C to + 60 C. More recent mixtures have considerably extended this range and at the same time afforded improvements in other physical parameters. For example, manufacturers are now able to offer nematic materials operating from below − 50 C to over 110 C, with a considerable range of other properties [3.8].

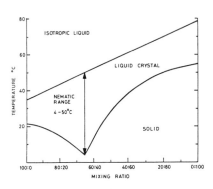

Fig. 3.4 Phase diagram of a two-component LC mixture, showing the wide temperature range of the nematic phase at the eutectic composition.

3.3. Principles of operation

3.3.1. Introduction

A wide range of electro-optical effects in liquid crystals have been devised and exploited for display purposes. Effects have been developed in many different LC classes, including nematics, cholesterics, smectic A, chiral smectic C, and others. Some of the optical effects obtained depend solely on the LC birefringence, such as optical scattering, rotation of the plane of polarisation, variable birefringence colours, and may or may not require external polarisers. Other optical effects rely on the optical absorption in anisotropic dyes dissolved in the LC material. The electrical control of these optical effects may be obtained solely by interactions between the applied electric field and the dielectric anisotropy of the LC, or may rely on a combination of this with local heating or electrical conduction caused by ionic dopants.

In this chapter a selection of electro-optic effects is discussed in more detail. The selection, which is by no means exhaustive, is intended to include the historically and commercially important effects plus those effects which are showing considerable promise in research and development laboratories. Almost inevitably the "twisted nematic" (section 3.3) dominates the discussions since this has been the mainstay of LC display production for many years and is still being developed and improved. For many applications, however, its limitations are clearly apparent, and the need to develop and assess alternatives is all too obvious.

3.3.2. Dynamic scattering in nematics

The dynamic scattering effect [3.12] was the first LC effect to be used in displays, some 80 years after Reinitzer first observed a liquid crystal phase in 1888, and is included here for historical reasons.

The effect usually uses a nematic material of negative dielectric anisotropy containing a dopant to increase its conductivity. Homogeneous alignment layers ensure that clear transmission occurs in the undriven (off) state. An applied electric field produces both current flow and a dielectric torque on the molecules. When a critical threshold field is exceeded a turbulent flow condition occurs. In this state the director alignment is lost and spatial variations of refractive index occur on a scale suitable for strong light scattering. The on-state then appears cloudy.

No polarisers are required so the display brightness can be quite high, but contrast depends critically on the illumination conditions and may be poor.

This display effect has never received widespread popularity, its premature commercial exploitation causing considerable public mistrust in LCDs for several years. It was unfortunate that the use of unstable LC material, non-ideal ionic dopants and rather poor optical effects in those early displays combined to produce a totally inadequate device. Subsequent research has produced stable LC materials and a better understanding and control of ionic dopants and of scattering displays in general, so that quite satisfactory dynamic scattering displays can now be produced. For most applications, however, there are alternative LC effects available with generally superior properties, so the use of dynamic scattering in nematics has been restricted to a very few special purposes. More recently a similar effect has been discovered in smectic A materials (see section 3.8).

3.3.3. The twisted nematic effect

This is the effect used in the vast majority of liquid crystal displays produced over the last few years, following the pioneering work of Schadt and Helfrich [3.13].

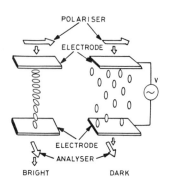

POLARISER
ELECTRODE
V
ELECTRODE
ANALYSER
BRIGHT   DARK

Fig. 3.5 Twisted nematic cell in on-
and off-states, used in transmission
between crossed polars.

A schematic twisted nematic (TN) cell is shown in figure 3.5, where a material with positive dielectric anisotropy is used. The diagram of the off-state shows homogeneous alignment on both surfaces of the cell, with these two alignments mutually at right angles. The director then spirals uniformly from one surface to the other giving the 90° twist that is used to name the effect (in practice various defects in performance are avoided if the surface alignment is not precisely homogeneous, tilt angles of up to 30° from the surface being used in some instances, but this does not materially affect the description given here). The front polariser produces linearly polarised light whose polarisation direction is either parallel or perpendicular to the director at that face. Because of the large refractive index anisotropy of the liquid crystal the plane of polarisation is guided through the cell, following the rotation of the director. It thus emerges polarised orthogonally to the incident polarisation. If the analyser is perpendicular to the polariser this emerging light is transmitted.

When a voltage above threshold is applied, dielectric reorientation causes the director to rotate parallel to the electric field in all places. No guiding of polarisation occurs, so the transmitted light is absorbed by the analyser. Of course, by rotating the analyser through 90° the opaque and transmitting states are reversed.

When the applied voltage is reduced below the threshold value the surface forces then re-establish the original twisted structure.

The display may be used either in the transmissive mode with an independent light source illuminating the rear of the display, or it may be used with a suitable reflector stuck to the rear polariser to reflect the ambient light. In the latter case a <u>diffuse</u> reflector which does not depolarise the light is required to maximise the display brightness and contrast.

An elegant, low power, solution to night-viewing of reflective TN displays is to use a "transflective" rear reflector, i.e. a reflector that transmits say 10 % of incident light. A very weak light source, possibly a beta-light or a low power filament bulb, placed behind this transflector then gives good transmissive mode viewing in the dark, with a smooth transition to reflective mode viewing at higher light levels.

For low power, portable applications,

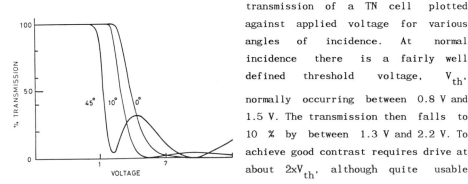

Fig. 3.6 Typical normalised optical transmission of a TN cell between crossed polars as a function of applied voltage at three angles of incidence. The 10° and 45° data apply to the "low voltage" quadrant.

drive voltage is a most significant consideration. Figure 3.6 shows the transmission of a TN cell plotted against applied voltage for various angles of incidence. At normal incidence there is a fairly well defined threshold voltage, V_{th}, normally occurring between 0.8 V and 1.5 V. The transmission then falls to 10 % by between 1.3 V and 2.2 V. To achieve good contrast requires drive at about $2\mathrm{x}V_{th}$, although quite usable contrast is obtained at slightly lower voltages. Since power dissipation is roughly proportional to the square of the drive voltage it is clearly beneficial in terms of battery life to operate with low V_{th} materials.

Batteries, however, are available only at certain voltages, so the major consideration may often be to match the required liquid crystal drive voltage to that produced by 2 or 3 cells of a specified battery.

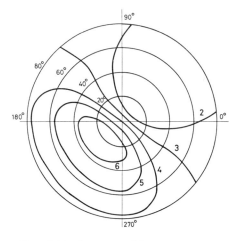

Fig. 3.7 Stereographic representation of the angular dependence of the transmission contrast ratio of a hypothetical TN cell.

The appearance of a TN cell driven slightly above threshold is a strong function of the direction from which it is viewed. This arises because the molecules are constrained to rotate in a particular direction in going from the off- to the on-state. The threshold voltage for the optical effect is lower than that at normal incidence in one particular viewing quadrant, known as the "low voltage quadrant". From all other directions the threshold is higher. Figure 3.7 shows a useful way of depicting the appearance of displays when viewed from arbitrary angles and at various voltages. In this diagram is plotted contrast ratio, defined as on-state transmission divided by zero voltage transmission at the same angle. This type of diagram shows clearly the low voltage quadrant and the variation of threshold with viewing direction, and emphasises the importance of obtaining the optimum relative positions of display, observer and illumination.

Unfortunately figure 3.7 does not give a complete description of display appearance for several reasons. Firstly it applies strictly to displays used in transmission. With reflective displays the ambient illumination may throw shadows of the displayed data onto the rear reflector. Usually these shadows increase the observed contrast, but if the reflector is a significant distance behind the liquid crystal layer then there may be disturbing parallax between the data and its shadow. In this case the intensity of the shadows may be reduced by illuminating from directions well outside the low voltage quadrant. Secondly, the perceived visual quality of a display does not depend solely on the contrast ratio as defined here – it depends equally strongly on the luminance of the transmitting state. It is indeed possible to use very dense polarisers that give a very large photometric contrast ratio, but whose transmission is so low that the bright state is unacceptably dull. In general display manufacturers choose polarisers and reflectors very carefully to achieve the optimum balance between brightness and contrast.

The threshold voltage, V_{th}, of most liquid crystals is temperature dependent, so the temperature coefficient, dV_{th}/dT, is a parameter of interest. Typical values range from -0.4 to -1.0 % C^{-1}. This effect is not too important for directly driven displays, providing that enough voltage is available to turn the display fully on at the lowest temperature, but it is a very significant parameter for multiplexed displays as discussed in section 3.4.

Perhaps the most striking effect of temperature is on the turn-on and turn-off times, $t(on)$ and $t(off)$, of the display, which may vary by more than 2 orders of magnitude over the operating temperature range; $t(on)$ is determined by 3 principal factors, the cell spacing d, a material viscosity coefficient η, and the amount by which the applied voltage V exceeds V_{th}:

$$t(on) \div \frac{\eta\ d^2}{V^2 - V_{th}^2} \qquad (3.1)$$

Similarly, $t(off)$ is determined by η and d, but the driving force is provided by the surface alignment forces and not an external field, so:

$$t(off) \div \eta\ d^2 \qquad (3.2)$$

Clearly $t(on)$ can be reduced by increasing V, but there is usually little advantage in making $t(on)$ much shorter than $t(off)$. Reducing d is obviously beneficial but two limitations are encountered. Firstly, the quality of the "guiding" of the polarised light in the twisted off-state is determined by the factor Δn.d. As Δn.d is reduced the emerging light becomes progressively more elliptically polarised giving significantly reduced contrast and possibly a coloured off-state. The minimum acceptable value is called the "Mauguin Limit" [3.14] defined by

$$\Delta n.d \geq 2 \lambda \tag{3.3}$$

Secondly, it is difficult to reliably manufacture cells with spacings much less than 5 μm over large areas. This leaves η which must be minimised to achieve small values of t(on) and t(off). In recent years a great deal of research has concentrated on finding materials which not only have low viscosities at room temperature, but also have small temperature coefficients. One result of this search is depicted in figure 3.8, where switching times of less than 0.5 s are obtained right down to − 20 C [3.15].

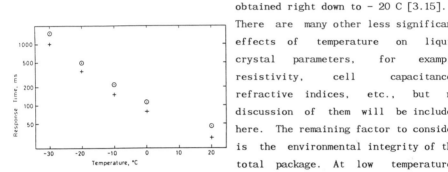

There are many other less significant effects of temperature on liquid crystal parameters, for example resistivity, cell capacitance, refractive indices, etc., but no discussion of them will be included here. The remaining factor to consider is the environmental integrity of the total package. At low temperatures there are no severe problems, but at high temperatures, particularly accompanied by high relative humidity, both the hermeticity of the cell seal and the adhesion of the polarisers has

Fig. 3.8 Variation of response times with temperature of the low viscosity LC mixture E200 (ref. 3.15), 6 V drive.

caused problems in the past. Considerable progress has been made recently, and now displays can be manufactured that perform reliably in extreme environments without further encapsulation [3.16].

Most of the discussion so far has applied to both directly driven and matrix TN displays. We must now consider the extra factors that ensure the steep threshold characteristic and low M-values (see section 4.1) required for multi-way multiplexing.

When no special precautions are taken the nematic mixture E7, used for so many years in directly driven displays, gives an M-value of about 1.8, which is only adequate for 3-way multiplexing with full contrast. It can be shown, however, that the steepness of the transmission voltage curve can be increased if certain elastic constants of the LC material are correctly chosen.

In this way materials have been developed [3.7] giving M < 1.55 and M' < 1.22.

From table 3.1, M = 1.5 implies excellent performance with 7-way multiplexing and M′ = 1.2 implies adequate performance at 30-way.

Table 3.1

Optimum RMS Voltage Ratios, V_{on}/V_{off}, for n-way Multiplexed LCDs

n	V_{on}/V_{off}	n	V_{on}/V_{off}	n	V_{on}/V_{off}
3	1.932	20	1.255	200	1.073
4	1.732	30	1.203	300	1.060
5	1.618	40	1.173	400	1.051
7	1.488	50	1.153	500	1.046
10	1.387	70	1.128	700	1.039
15	1.302	100	1.106	1000	1.032
		150	1.085		

Many other factors can have a significant effect on the measured M-values. These include the surface tilt angle, which should be very small, the accuracy of alignment of the polariser transmission axes and surface alignment directions, and the thickness of the insulator layers. The most important other factor though is the product of Δn.d. Gooch and Tarry [3.17] have analysed in detail the guiding properties of 90° twisted layers and have found that the graph of off-state transmission against Δn.d contains a damped oscillatory component. The first extremum of this function occurs below the Mauguin limit at Δn.d = $\frac{\sqrt{3}}{2}$ λ. This gives the best compromise between response time, which improves with reduced d, and optical performance. To capitalise on this effect requires excellent control of cell spacing in order to avoid variations in off-state colour caused by varying birefringence, and recently manufacturing methods have been developed which have achieved the required spacing control.

Several manufacturers are now offering a range of large area, highly multiplexed, TN displays. One example, chosen at random, is the Epson Model EG7001A-AR which has a 254x82 mm viewing area. This contains 640x200 dots, multiplexed 100-ways as a double matrix. It is capable of displaying up to 25 lines of over 100 characters, approximately equivalent to an A_5 page, and can also display high resolution graphics. With a total consumption of 120 mW this type of display is well suited for portable computer applications. Unfortunately the quest for maximum complexity has meant that visual appearance has been severely compromised. This particular display achieves a contrast ratio of 3 or better only between viewing angles 10° and 30° from normal incidence. Many potential users are severely critical of displays with such limited viewing angles and poor contrast!

3.3.4. Electrically controlled birefringence (ECB) effects

The large anisotropy in the refractive index of aligned nematic liquid crystals permits electrical control of birefringent effects [3.18]. An example is shown in figure 3.9, where a thin layer of planar aligned liquid crystal of positive dielectric anisotropy is placed between parallel polarisers. The incident polarisation is set at 45° to the director. By direct analogy with birefringence in solid crystals the transmission may be analysed in terms of ordinary "o" and extraordinary "e" rays polarised perpendicular and parallel to the director respectively. For certain wavelengths of incident light the optical path difference between o and e rays through the liquid crystal will be an integral number of wavelengths and the resultant will pass unhindered through the analyser. For all other wavelengths the light

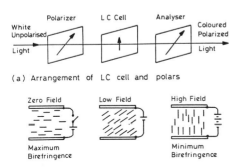

(a) Arrangement of LC cell and polars

(b) LC director configurations

Fig. 3.9 Display construction and liquid crystal director configurations for electrically controlled birefringence.

emerging from the liquid crystal will be elliptically polarised, to greater or lesser extent, and will therefore be partially or even completely absorbed by the analyser. Thus, in general, the transmitted light will be coloured. When an electric field is applied the director rotates changing the degree of birefringence and consequently altering the transmitted spectrum. In the limit, for very high fields, the director becomes normal to the plane of the cell, birefringence is reduced to zero, and white light is transmitted. If the polarisers are crossed then at each voltage the complementary spectrum is transmitted, giving a dark state at high voltage.

A similar effect is obtained with negative dielectric anisotropy material in a cell with nearly homeotropic surface alignment. With applied voltage zero there is no birefringence and therefore no transmission between crossed polarisers. As the applied voltage is increased the molecules tilt away from homeotropic and the birefringence rises. Provided this tilt is in a plane at 45° to the polarisation direction this birefringence results first in increasing transmission of white light, followed by a series of colours at higher fields. A further variation uses planar alignment on one surface and homeotropic alignment on the other - the "hybrid aligned nematic" (HAN) effect.

These effects have two principal potential uses. Firstly, the sharp threshold characteristic obtainable with the negative anisotropy material is in principle excellent for moderately complex matrix displays. Secondly, the electrically controllable colour effects are the most promising techniques for producing a range of colours from each pixel. Unfortunately, the electro-optic effect at each point depends on the applied field, $F = V/d$, at that point, and the resulting birefringence depends on the local value of $\Delta n.d$, which puts severe constraints on the uniformity of cell spacing. This has precluded exploitation of these effects in all but the smallest cells until very recently. However, now that large area cells can be manufactured reliably with spacing uniformity of better than 0.2 μm, it is possible that ECB will be further exploited.

3.3.5. The cholesteric nematic phase change effect

This effect [3.19] makes use of cholesteric material which is switched electrically between a scattering off-state and a transparent on-state, an optical effect which does not require polarisers.

The cell is filled with cholesteric material of positive dielectric anisotropy, the cholesteric pitch being a small multiple of the wavelength of light to optimise the scattering effect. When a field above threshold is applied the positive dielectric anisotropy causes all molecules to align parallel to the field, the cholesteric twist is lost and the molecular ordering and alignment are similar to the on-state of the twisted nematic cell. Since no polarisers are used the cell is completely transparent. When the field is removed the cholesteric twist is rapidly re-established throughout the bulk of the material. There is then no preferred direction for the orientation of the cholesteric spirals so a quasi-polycrystalline structure results which is strongly scattering.

This effect has a threshold field, rather than a threshold voltage, which is determined largely by the cholesteric pitch. It is generally found that cells which give strong scattering effects have threshold fields ~ 1 V/μm or higher, resulting in drive voltages above 10 V. The detailed behaviour of the off-state is very complex. At zero field the strongly scattering state is only metastable, but relaxation to the stable, non-scattering, twisted structure may take hours or may be very rapid. At slightly higher fields, but still below the threshold field, the scattering state is stable. Furthermore, if the field is slowly reduced from a strength above threshold there is considerable hysteresis in the characteristics; the rate of nucleation of the scattering state can be very slow at field strengths just below threshold. The extent to which these various effects occur can be altered both by choice of pitch and materials, and also by choice of surface alignments.

The visual appearance of this effect is somewhat similar to dynamic scattering, since the visual contrast is obtained between a clear, transparent state and a scattering state. Although high brightness displays are possible, the appearance depends strongly on illumination conditions and it is only in projection that good contrast is reliably achieved.

There has been no significant commercial exploitation of these effects, although Tani et al [3.20] produced a series of high complexity matrix displays using an electrical drive method that exploited long term hysteresis effects.

3.3.6. The dyed phase change (DPC) effect

This is a direct extension of the above effect which includes dyes, dissolved in the liquid crystal, to give optical absorption rather than scattering [3.19] and [3.21].

The dye molecules must be highly anisotropic, both physically and optically. They must align accurately with the director of the liquid crystal at all times; this is known as the "Guest-Host" effect. Also, they must be "pleochroic"; that is, their absorption spectrum should depend strongly on the relative orientations of the molecules and the polarisation of the light. Ideally, absorption should be zero when the optical polarisation is perpendicular to the long molecular axis, and strong when the polarisation is parallel to that axis, as shown in figure 3.10.

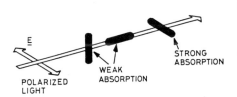

Fig. 3.10 The anisotropic optical absorption of pleochroic dyes as used in the dyed phase change effect.

Structurally the on- and off-states of the DPC effect are similar to those of the phase change effect above, but the presence of the anisotropic dye produces a totally different optical effect.

In the driven on-state of a DPC cell all the liquid crystal and dye molecules are forced to align perpendicular to the plane of the cell, so light passing through the cell is only weakly absorbed. In the undriven (off) state the twisted cholesteric structure ensures that all polarisations of incident light encounter sufficient dye molecules whose axes are suitably aligned to give strong absorption. As in the previous effect, the electric threshold is field rather than voltage dependent.

With a suitably designed diffuse reflector this effect is visually very attractive. It should be noted, however, that the displayed information appears bright on an absorbing background, the direct inverse of the conventional TN display.

Choice of the dye components involves many considerations. Their alignment in the liquid crystal host is described in terms of an optical order parameter [3.21] which must be as high as possible to minimise absorption in the on-state. Several dye components may be used simultaneously to achieve the required aggregate absorption spectrum, and care must be taken to ensure that the perceived hue does not depend strongly on the illumination spectrum. They must be sufficiently soluble in the LC host to give adequate absorption and contrast without risk of the segregation of dye particles at low temperatures. They should not degrade the host's physical properties, such as phase transition temperatures, viscosity, etc. Finally, they must be highly stable when exposed to solar radiation.

Exploitation of this effect was initially hampered by inadequacies in the dyes, either poor solubility, low order parameter, poor UV stability or unsuitable colour. Recently, however, a range of anthraquinone based materials has been announced [3.22] which combine excellent values of all properties.

The overall design and performance of a DPC display is subject to many compromises, since each variable affects more than one of the observable features. For example: a large cell spacing provides good contrast, but at the expense of either slow turn-on or high drive voltage; short cholesteric pitch improves contrast and gives rapid turn-off, again at the expense of high drive voltage; high concentration of dye gives good contrast but reduces on-state brightness and may involve low temperature solubility problems. Clearly, the achievable speed, contrast, drive voltage etc., will depend strongly on the external design constraints.

Temperature affects the threshold voltage in much the same way as for a TN display. Values of dV_{th}/dT are of the order of -1% per C. Similarly response times are affected by changes of viscosity with temperature. Although there is in principle the possibility of rapid response when short pitch cholesterics are used, this necessitates much higher drive voltages than TN cells. Furthermore, the presence of dye molecules can have an adverse effect on viscosity, degrading the response speed particularly at low temperatures.

The angle of view of a reflective DPC display is determined largely by the properties of the diffuse reflector. By control of the surface texture of this reflector the on-axis brightness of the display may be played off against the wide-angle appearance. Furthermore, since no rear polariser is required, the rear reflector can be located inside the LC cell, thus eliminating shadow parallax problems and further increasing the luminance and contrast of the display. In this case care must be taken to ensure that none of the diffusely reflected light is trapped within the display by total internal reflection at the front glass-air interface.

The same technologies for cell sealing can be used with DPC cells as with TN cells. The absence of polarisers, however, means that the DPC cell should be less susceptible to degradation in extreme environments.

In principle DPC displays have two major advantages over TN displays. Firstly, they do not require polarisers and therefore should appear much brighter and crisper than TN displays. Secondly, their optical properties are far less anisotropic – there is no "low voltage quadrant" – so they can be viewed clearly over a much greater angular range. Their one major disadvantage, however, is in the poor steepness of their transmission versus voltage characteristic, which is totally unsuitable for multiplexed drive schemes (section 1.5). To date the only successful demonstrations of high complexity matrix displays have used drive schemes which make use of the hysteresis of the electro-optic characteristics [3.23].

There is, at long last, a rapidly growing use of dyed liquid crystal materials in directly driven, large area, public informatior displays. Figure 3.11 shows

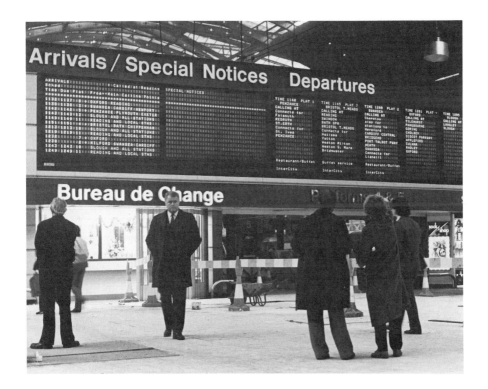

Fig. 3.11 Photograph of 60x7 ft. annunciator display at Paddington Station, London, made using 3500 single character modules. Single polariser, dyed nematic effect used in transmission (Courtesy Racal Microelectronic Systems Ltd., Reading, UK).

a photograph of a display measuring about 60x7 ft and containing 3500 single character modules. Each module contains a 7x5 array of display elements.

The modules are backlit, so act as electro-optic shutters. All the LC cells, which have 71x51 mm characters, contain the same black dyed LC mixture, but different colours are obtained using simple filters. Strictly speaking, the

modules employ a "single polariser, dyed nematic" effect which has some
advantage over the true dyed phase change effect when used in transmissive
mode.

3.3.7. "Supertwist" and related effects

The TN effect, with its $\pi/2$ twist angle, and the phase change effect, with its
"many π" twist angles, represent the two extremes of twisted LC displays. As
has been shown, their electro-optic behaviours are very different. Not
suprisingly, a considerable range of different electro-optic effects can be
produced with intermediate twist angles.

One interesting example is the "π-cell" [3.24], which can be switched
repetitively between two states of different birefringence with sub-millisecond
switching times. Unfortunately, the off-state is metastable so the rapid
switching can only be achieved with fairly frequent, repetitive switching. This
is ideal for some shutter applications, but is difficult to apply to dot-matrix
display devices.

Another example, discussed in [3.25], uses a type of bistable twist cell filled
with a cholesteric material. When the correct values of cell spacing,
cholesteric pitch and surface alignment are chosen, there are two stable
configurations of the LC director having markedly different optical properties.
Visual contrast is obtained either using crossed polarisers as in the TN
effect, or using dissolved pleochroic dyes. The bistability ensures that in
principle an indefinitely large matrix may be addressed without crosstalk
problems, although laboratory demonstrations have been restricted to quite
small matrices so far. Cell spacing uniformity is important and dust particles
have had a serious effect on stability in some examples, so it is difficult to
assess the potential of this approach.

The final example is known as the "supertwist" effect, [3.26] and [3.27], in
which twist angles between π and $3\pi/2$ are used. With careful choice of the
ratio of cholesteric pitch to cell spacing, LC elastic constants, surface tilt
angle, etc., it is possible to obtain a very steep and reversible optical
transition. This is ideal for multiplexed drive, giving M-values significantly
better than with the TN effect. The original descriptions of this effect [3.26]
and [3.27] identified two distinct devices, both using pleochroic dyes. One
case used liquid crystal with large Δn and required a single polariser. The
other case used LC with small Δn and did not need any polarisers. The great
attraction of this effect is that it employs cell construction methods and
electrical drive schemes very similar to conventional TN displays, but achieves
significantly better appearance. The angle of view is considerably wider, a
feature which is particularly apparent when high level multiplex drive is used.

The major disadvantage, however, is the increased response times which become significant below about 0 C.

More recently [3.28] an alternative optical effect has been demonstrated using birefringence rather than pleochroic dyes, which has been named the "Supertwisted Birefringence Effect" (SBE). This shares the same advantages of low threshold voltage, excellent M-value and wide angle of view. Figure 3.12

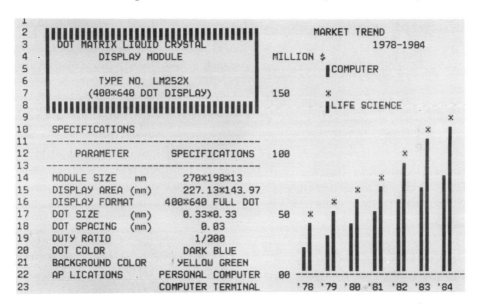

Fig. 3.12 Example of a commercial supertwisted nematic display using 200-way multiplex drive (Courtesy Hitachi Electronic Components, Ltd., UK).

shows an example of a commercial SBE display of nearly A$_4$ size with 400x640 dels. The drive scheme uses 200-way multiplexing.

3.3.8. Smectic A effects

As described in section 2, smectic A materials are more ordered than nematic or cholesteric materials. Not only do the molecules tend to align parallel to each other, but they also tend to lie in layers, the director being perpendicular to the layers. One common feature of most smectic materials is that their viscosity is very high compared to nematics. Consequently, any molecular alignment tends to be stable, since the realignment forces due to surface treatments do not propagate far into the material – the basis of memory effects. Furthermore, very large electric fields are needed to realign smectics, except at temperatures very close to the nematic or isotropic phase boundary. Mixtures of materials have been produced with very wide smectic temperature ranges, and recently these materials have been exploited in a variety of ways for display purposes.

In this section two effects are described, firstly a thermal-electric effect, secondly a dynamic scattering effect.

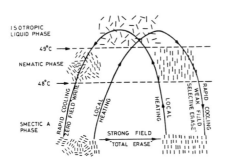

ISOTROPIC
LIQUID PHASE

49°C

NEMATIC PHASE

48°C

SMECTIC A
PHASE

RAPID COOLING
"ZERO FIELD WRITE"

LOCAL
HEATING

STRONG FIELD
"TOTAL ERASE"

LOCAL
HEATING

WEAK FIELD
"SELECTIVE ERASE"

RAPID COOLING
"SELECTIVE ERASE"

Fig. 3.13 Schematic diagram illustrating operating principles of "Thermal-Electric" Smectic A devices.

The thermal-electrical effect is illustrated in figure 3.13. At temperatures well below T_{SN} (the smectic to nematic phase boundary) the material can exist in either of two states; a "homeotropic" aligned state which is clear and transparent, or a "focal conic" state which is strongly scattering. Reversible transitions between these states are obtained by a combination of heat and electric fields. When the temperature is raised the material passes through the nematic into the isotropic phase, where all molecular ordering is lost. If it cools rapidly into the smectic phase this disorder tends to be frozen in and the strongly scattering, focal conic state is established. If, however, an adequate electric field is applied during cooling, the molecules are realigned in the nematic state, and this realigned state is frozen in. All regions of the cell which are not heated are virtually unaffected by this realignment field, so stored data may be selectively changed without the need to re-write all the information.

In one type of application a focussed laser beam is scanned across the cell to provide the local heating. Both cell plates are coated with transparent electrodes for application of the "erase" field. Total erase of all information is obtained by applying a large electric field. These devices are generally used in projection displays, having the advantage that write and erase can be performed while the cell is in the projector. The laser-written lines can be very narrow, 10 microns or less, so very high resolution is possible. Systems using various lasers and projection systems have been described in references [3.29] to [3.31], giving up to 10,000 lines resolution; [3.32] describes a 3 colour system with 2000 lines resolution using 2 smectic cells. Generally these systems are bulky and consume a lot of power, but moderately compact, transportable versions have been developed.

Figure 3.14 shows the screen of a 3000 line resolution, transportable system with a 0.75 m square screen [3.31]. This uses a cursively-scanned (vector) laser beam to write and erase the information stored in the LC which is

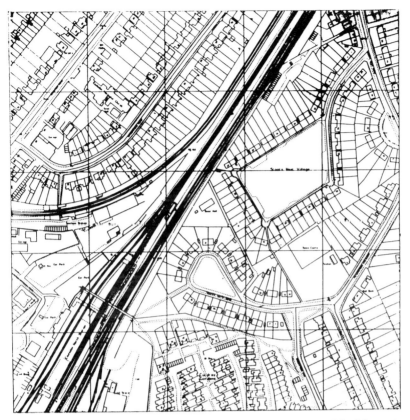

Fig. 3.14 Photograph of part of the screen of a transportable LC projection
display with 3000 lines resolution. For clarity the overlayed background
picture is omitted (Courtesy Laserscan Laboratories Ltd., Cambridge, UK).

simultaneously projected onto the screen to overlay the projected map picture.
Other implementations use raster scanning, particularly for high density data
displays. In all cases the rate of information change is limited by two
factors, the laser power available and the dynamics of the mechanical beam
deflection system, resulting in rather long page re-write times.
An alternative thermal-electrical approach uses matrix addressing [3.33]. The
line electrodes act as resistive heaters, so when a current pulse is passed
through one line that entire line of picture points is heated into the
isotropic phase. Voltage pulses are applied during cooling to selected column
electrodes to realign appropriate elements. The column pulses of course affect
only those elements that have been heated, so crosstalk is absent. In [3.34] a
laboratory demonstration is described using a 1 cm x 1 cm silicon slice as
substrate. The display was used in projection and the 256 lines were refreshed
at TV frame rate. A larger version, measuring 94x90 mm with 250x240 dels, has
been developed on a glass substrate for computer terminal applications. The

line address time of 20 ms implies a total frame time of 5 s. The operating
temperature range was 5 C – 40 C, with peak heating power of 15 W. A somewhat
similar device [3.35] used a dichroic dye dissolved in the smectic material to
improve the visual appearance. The operating principles were very similar, but
a bright diffuse reflector was placed behind the cell so that visual contrast
was obtained between the clear state (bright) and be scattering/absorbing
state (dark), as shown in figure 3.15. More recently [3.70] a 720x500 pixel

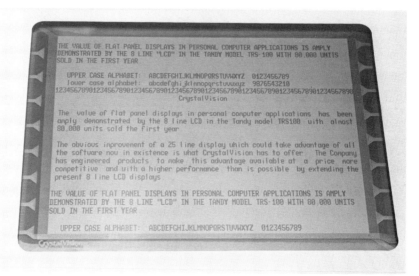

Fig. 3.15 Photograph of 200x480 pixel matrix addressed storage display using
dyed smectic A materials and thermal–electric drive (Courtesy CrystalVision,
USA).

display has been demonstrated which contained the additional facility of laser
"light pen" write–in and capacitative read–out of data.
The dynamic scattering effect has also been used in a smectic A display [3.36].
When dopants were added to reduce the material's resistivity it was found that
low frequency AC fields of adequate amplitude induced scattering, whereas high
frequency fields induced homeotropic (clear) alignment. Similar transitions
were also obtained with carefully controlled DC pulses, suitable for matrix
displays. It was shown that pulse conditions could permit addressing of an
infinite matrix in the line–at–a–time manner, and the addressing of a 5000 line
display was successfully simulated on a small matrix.
Figure 3.16 shows a prototype display [3.37] with 420x780 dels in an active
area of approximately A$_4$ size. The picture shows 25 lines of 80 characters,
based on a 9x16 dot character font. The total page re–write time is 0.8
seconds, during which the power consumption is about 5 W, but the permanent
memory ensures a flicker–free picture and zero power consumption subsequently.

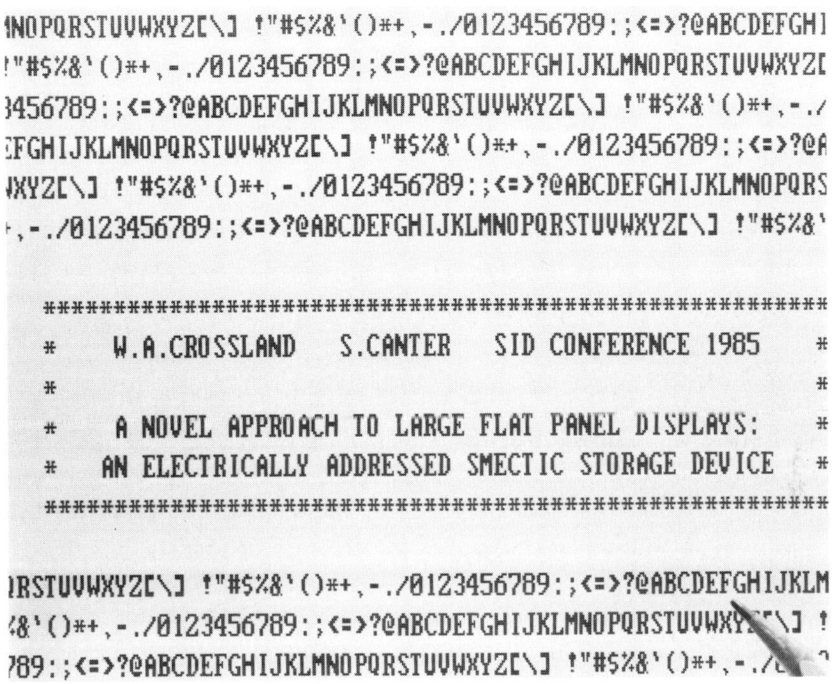

```
!NOPQRSTUVWXYZC\] !"#$%&'()*+,-./0123456789:;<=>?@ABCDEFGHI
!"#$%&'()*+,-./0123456789:;<=>?@ABCDEFGHIJKLMNOPQRSTUVWXYZC
3456789:;<=>?@ABCDEFGHIJKLMNOPQRSTUVWXYZC\] !"#$%&'()*+,-./
:FGHIJKLMNOPQRSTUVWXYZC\] !"#$%&'()*+,-./0123456789:;<=>?@A
!XYZC\] !"#$%&'()*+,-./0123456789:;<=>?@ABCDEFGHIJKLMNOPQRS
+,-./0123456789:;<=>?@ABCDEFGHIJKLMNOPQRSTUVWXYZC\] !"#$%&'


*****************************************************************
*     W.A.CROSSLAND    S.CANTER    SID CONFERENCE 1985     *
*                                                                *
*     A NOVEL APPROACH TO LARGE FLAT PANEL DISPLAYS:      *
*     AN ELECTRICALLY ADDRESSED SMECTIC STORAGE DEVICE    *
*****************************************************************


!RSTUVWXYZC\] !"#$%&'()*+,-./0123456789:;<=>?@ABCDEFGHIJKLM
!&'()*+,-./0123456789:;<=>?@ABCDEFGHIJKLMNOPQRSTUVWXY \] !
789:;<=>?@ABCDEFGHIJKLMNOPQRSTUVWXYZC\] !"#$%&'()*+,-./
```

Fig. 3.16 Photograph of a section of a 420x780 pixel storage display with electrical matrix addressing and smectic A material (Courtesy STL Ltd., Harlow, UK).

3.3.9. Chiral smectic C effects

The smectic C phase (S_c) has a layer-like structure with the director tilted away from the layer normal. When the constituent molecules have chiral properties a helical rotation of this tilt angle occurs between layers, somewhat analogous to the cholesteric structure. The resulting chiral smectic $C(S_c*)$ structure has the necessary symmetry to show ferro electric properties.

The significance of this has recently become apparent [3.38] when a bistable device was demonstrated which could be switched reversibly in sub-microsecond times. This discovery has sparked off a world-wide R & D aimed at capitalising on the apparent great potential of this effect. The difficulties, however, are not trivial.

The speed of the effect is attributed to the strong coupling between the spontaneous ferro electric polarisation and the applied electric field. But this bulk polarisation can only occur if the natural helical twist of the S_c* material is "unwound", possibly by the cell surface forces or by external fields. Unfortunately this unwinding is not easily achieved except in very thin cells, possibly < 2 μm thick, which are not easy to manufacture over large areas. Further difficulties include achieving good alignment, optimisation of materials to achieve good optical and electrical properties over a wide temperature range, and design of suitable drive schemes for complex matrix displays.

In spite of all these difficulties rapid progress is being made on all fronts. One of the most complex displays described to date [3.39] measured 260x162 mm and contained 640x400 dels. The line-write time, however, was 400 μs, so 320 ms were required to change the whole page of data.

3.3.10. Colour in LCDs

The discussions in earlier sections have shown how switchable colour effects can be obtained by using dissolved pleochroic dyes or by using various birefringence effects. Unfortunately none of these approaches gives the full-colour and grey scale capability required to generate realistic pictures in full-colour. In order to compete with shadow-mask CRTs for full-scale TV applications it is essential that such a capability should be developed. Ideally, each display element of a complex matrix display should be capable of showing all colours and luminance levels. The nearest that has been achieved emulates the shadow-mask CRT by arranging dels in triads, each triad containing individual pixels that are electrically switched black-to-red, black-to-green and black-to-blue respectively. Several displays with this type of colour capability are referenced in table 3.2. The colours are obtained by generating micro-dot colour filters, desposited on one internal surface of the LC cell, and registered with the conducting electrodes that define the pixel areas. The liquid crystal acts as a switchable shutter, usually using either the TN or DPC (with a black dye) effects. If the LC shutter effect has a grey scale capability then full-colour is possible. Considerable development has occurred in several technologies for producing the colour filters, resulting in well balanced hues with reasonably good saturation, but two major problems inevitably occur. Firstly, as in the CRT, with three coloured pixels required for each display element the spatial resolution of the display is drastically reduced. Secondly, since each colour can be transmitted by only one third of the display area at most, the resulting luminance is greatly reduced. Furthermore, if the TN effect is used the transmission is reduced by a further 50 % owing to absorption in the polarisers. Consequently full colour displays of this type are unacceptably dark when used in reflected ambient light, so are almost invariably used in transmission with a backlight. With large area displays this backlight usually consumes considerably more electrical power than all the display circuitry.

Table 3.2

A list of all the Active Matrix LCDs discussed at the International Display Research Conference, October 1985. Displays are listed in order of decreasing diagonal size (Note that "graphics" and "data" imply lack of grey scale or full-colour).

Maker	Diagonal in inches	Number of dels	Device material	LC effect	Comments	Reference
Toshiba	9.5	480x640	a-Si	TN	Colour, graphics	[3.51]
Suwa Seikosha	8.9	400x640	MIM diode	TN	Monochrome, data	[3.52]
Hosiden	8.9	400x640	a-Si	TN	Monochrome, data	[3.53]
Hosiden	7.24	520x520	a-Si	TN or dyed nematic	Colour, graphics	[3.54]
Hitachi	6.3	600x640	a-Si	TN	Colour, graphics	[3.55]
Suwa Seikosha	5.1	440x480	p-Si	TN	Full colour, TV	[3.52]
Imperial College RSRE	4.9	80x96	CdSe	Dyed nematic	Monochrome, data	[3.56]
CNET	4.5	320x320	a-Si	TN	Monochrome, TV	[3.57]
NEC	4.3	378x400	a-Si	TN	Full colour, TV	[3.58]
Seiko	2.24	220x240	a-Si	TN	Full colour, TV	[3.59]
Toshiba	2.2	220x240	a-Si	Hybrid aligned nematic	Birefringence colours, graphics, projection	[3.60]
ECD	1.8	32x32	a-Si diodes	TN	Monochrome, data	[3.61]

3.4. <u>Addressing/driving</u>

3.4.1. **Direct addressing of matrix displays**

Good liquid crystal materials have high resistivity, $> 10^{10}$ ohm cm, so that cells normally present a resistance of $> 10^9$ ohms mm^{-2} in parallel with 10 to 30 pF mm^{-2}. The resistive and capacitive impedances are therefore comparable in the low frequency, 10 - 50 Hz, range. If the electro-optic effect requires say 3V to drive it fully on, then the mean current density required is ~ 10 $nA.mm^{-2}$, and the power dissipation is a few tens of $nW.mm^{-2}$. All these factors are ideally suited to long term operation with small, low voltage batteries and low power CMOS drive circuits.

DC voltages should not be applied to LC cells since, although most electro-optic effects can be induced by DC drive, this entails the risk of electro-chemical degradation of the LC material. The risk is minimal at frequencies above about 10 Hz, and is further reduced if the electrodes are covered by insulator layers as shown in figure 3.3.

Many LC electro-optic effects respond to the RMS value of the applied waveform over a very wide range of waveforms and frequencies (section I.1.5), and this response will be assumed in the rest of this section. It must be noted, however, that other LC effects exist which show hysteresis or memory, and these effects require addressing schemes which are closely tailored to the precise details of the LC behaviour.

For very simple displays, such as 7-bar numerics with a small number of characters, it is possible to provide an individual electrical connection for each element of each character on one plate of the display, with a common electrode on the opposite plate. This of course requires an individual drive circuit for each display element. The simplest drive scheme applies zero volts to the back-plane and to the off-elements, and applies an AC sqare wave of adequate amplitude, $\pm$ V_D, to the on-elements. This simple scheme, however, requires a power supply and drive circuits giving 3 levels, + V_D, 0 and - V_D. A more economic scheme applies an AC square wave of amplitude ½ V_D to the backplane and off-elements, and applies a similar waveform 180° phase-shifted to the on-elements as shown in figure 3.17. This achieves the same effect at each display element, but

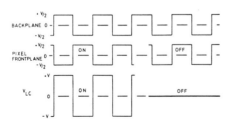

Fig. 3.17 Electrical waveforms suitable for directly driven TN displays.

requires a much simpler power supply.

This "direct" drive scheme is only economic for very simple displays. A numeric display of 10 digits would require at least 70 drivers and connections, and a single alpha-numeric character, based on a 7x5 font, has 35 elements. For these and for more complex formats a matrix approach is used. The transparent conductors on both cell plates are patterned so that each electrode is shared between several display elements. The behaviour of this arrangement is electrically equivalent to that of a rectangular matrix of n-lines and m-columns, the cross-points representing the display elements. The minimum number of connections is obtained when n = m, but even with n = 3 the number of connections to the above 10-digit example is reduced from over 70 to less than 30.

A simple way of addressing a matrix display is discussed in I.1.5, figures 1.26,1.28. The lines of the display are scanned ("strobed") repetitively in sequence by a "line select" pulse of amplitude V_S. While each line is selected the appropriate "select" and "non-select" data pulses, of amplitude $\pm V_D$, are applied to the columns. Net AC drive is achieved either by reversing drive polarity after each scan or by replacing each pulse by an alternating waveform. The problem now arises that supposedly "off" elements experience a significant drive voltage, largely composed of the data pulses to all other elements in the column. For a display consisting of N lines, using the optimum ratio of V_S to V_D, it can be shown that the ratio of "on" to "off" element RMS voltage is given by

$$\frac{V_{(on)}}{V_{(off)}} = \left\{ \frac{N^{\frac{1}{2}}+1}{N^{\frac{1}{2}}-1} \right\}^{\frac{1}{2}} \tag{3.4}$$

and values of this function are given in table 3.1. The importance of this optimisation, and the suitability of any RMS responding LC effect for matrix addressing, can be appreciated using figure 3.18 which shows the variation in transmission of a hypothetical LC with applied voltage at one temperature and two different viewing directions. The point on the 45° curve at 90 % transmission is conventionally taken to

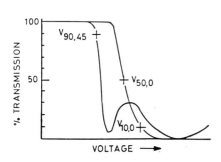

Fig. 3.18 Use of transmission versus voltage curves for calculating M and M', the multiplexing figures of merit.

define an acceptable off-state voltage, $V_{90,45,T}$. Similarly the point on the 0° curve at 10 % transmission defines the on-state voltage, $V_{10,0,T}$. The ratio of these voltages can be used as a figure of merit or M-value which, when compared with the drive waveform voltage ratios of table 3.1, determines the maximum usable multiplex ratio. In practice, this M-value is found to be an excessively harsh criterion. A more practical figure of merit, known as M', is defined as the ratio of $V_{50,0,T}$ to $V_{90,45,T}$. M'-values are inevitably much smaller than M-values, so, when they are compared with the data in table 3.1, much larger multiplexing ratios are suggested. In practice it is found that displays using multiplex ratios based on M'-values produce contrast ratios and angles of view that are just acceptable for many applications. It should be emphasised that the smaller the value of M or M', the larger is the matrix that can be addressed.

A more complete analysis must include effects due not only to viewing angle but also to temperature. Viewing angle depends strongly on the LC effect being used and can only be assessed when transmission versus voltage curves are available for a range of angles of incidence. It is generally true, however, that for a particular drive scheme and voltage there will be an optimum viewing direction and that display appearance will be degraded from all other directions. Changes in temperature cause two principal effects. Firstly, the electro-optic characteristic usually shifts bodily with temperature by amounts that preclude multiplexed operation with fixed drive voltages over wide temperature range. Accurate methods for temperature compensation of drive voltage have been developed which can maintain the visual appearance of displays over practically the entire temperature range that the material is liquid crystalline. These methods either involve a thermistor or other temperature sensor mounted close to the display cell, or may use the liquid crystal material as its own temperature sensor. This latter method [3.10] makes use of the dependence of liquid crystal capacitance on temperature and voltage, and has the advantage of sensing temperature at exactly the right place and with no time delays. In spite of the achievable accuracy of temperature compensation it is still advantageous to use materials with small temperature coefficients in order to minimise the effects of temperature gradients across the cell. Secondly, the curve shape may change slowly giving M and M' values that vary with temperature. This implies that the matrix complexity must be determined by the worst M-value encountered over the entire temperature range of operation.

From this analysis it is clear that the optimum electro-optic effects and materials for matrix addressing must have steep electro-optic characteristics that are as insensitive as possible to changes in temperature and viewing direction.

This discussion has dealt only with the maximum acceptable number of lines in the matrix. The number of columns, however, is not affected by these considerations, merely by the manufacturer's ability to provide connections and drivers. A simple way of effectively doubling the size of the matrix is to split the column electrodes at their mid-point and drive them independently from the top and bottom. In this way a matrix which appears to consist of N lines by M columns is electrically behaving as a matrix of ½ N lines and 2 M columns. Even more complex schemes, such as the Hitachi "quad-matrix" [3.11] have been proposed, but require excessively complex electrode structures.

3.4.2. Addressing of displays with restricted information

The analysis of Alt and Pleshko [3.9] was based on two fundamental premises, firstly that all combinations of "on" and "off" dels must be possible, and secondly that the lines and columns should be driven by essentially two-level waveforms. More complex waveforms, possibly depending in a much more sophisticated way on the data to be displayed, were not considered. Recently the general theory of matrix addressing of LCDs has been discussed [3.40] and [3.41]. The conclusions of these theories are that, in general, improvements are possible, but these improvements are only significant either when the number of scanned lines is very small or when the number of elements per column which are different from the background is very small. Clearly, neither of these conditions pertains to complex alpha-numeric displays, but there are many situations where the restricted information content of the display permits novel approaches, two of which are described here.

Firstly, a digital bargraph display consisting of a single column of dots would normally display data by having all dots below the indicated level on, and all dots above that level off. A liquid crystal display of this type has been described by Kmetz [3.42] in which the connections between elements were configured to be electrically equivalent to a conventional matrix. This matrix required few external connections and very few distinct drive waveforms. Nevertheless, the 3:1 voltage ratio between on- and off-elements ensured high contrast viewing over a wide angular and temperature range. Further extension to the multiplexed drive of double and triple bargraphs was demonstrated with only slightly degraded visual appearance.

Secondly, there are several types of display which may be configured to be equivalent to a matrix with only one element per column different from the rest. The first LC display of this type was the oscilloscope of Shanks et al [3.43], shown in figure 3.19, which introduced an addressing scheme now known as "Waveform Identity Addressing" (WIA). In this technique it is convenient to turn all the background dels "on" and to hold the data elements, one per column, "off". The method uses a set of distinct waveforms, one per line,

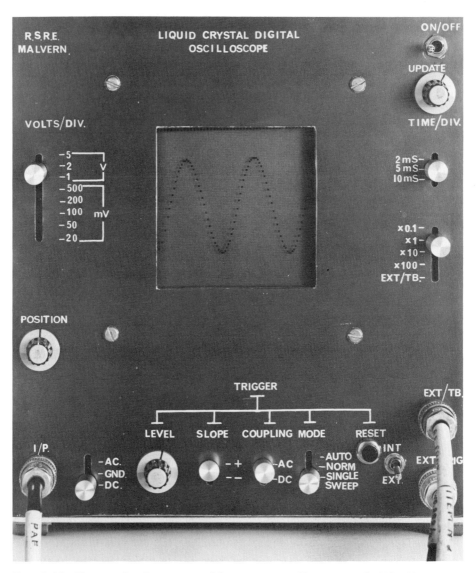

Fig. 3.19 **Photograph of prototype LC storage oscilloscope with 100x100**
dels using waveform identity addressing.

driving   the   matrix   continuously.   To obtain an "off" element at any required
line and column intersection it is merely necessary to apply to the column   the
same   waveform as is applied to the appropriate line. To achieve a good display
makes several demands on the set of waveforms. Firstly,   the   RMS   differences
between   any   two   members   of the set should be equal, to ensure an even back-

ground contrast level. These difference voltages should be a large proportion of the applied voltage to ensure the background areas are driven fully "on" without needing large supply voltages. Also the waveforms should be easy to generate and select and should have repetition times less than the LC response time to avoid flicker. The oscilloscope display used pseudorandom binary sequences [3.44], which provide an optimum solution in many respects. The 100x100 pixel prototype oscilloscope and display required only 60 integrated circuits and consumed less than 500 mW. Versions were demonstrated in reflection with both TN and DPC LC cells, while one model used a TN cell in transmission for large scale projection. A commercial version, the Scopex "Voyager", with 128x256 element DPC display was briefly available, and an independently developed device, the Brown Boveri "Metrawatt M2050" was marketed with a 64x128 TN display.

A number of other laboratory prototypes have also been demonstrated for other applications. For example, a low resolution radar plan position indicator (PPI) display was produced [3.45] with 120 bearings and 60 range gates. This used a combination of radial electrodes on one plate and spiral electrodes on the other (figure 3.20). WIA has also been applied [3.46] to analogue meters – see figure 3.21.

In this case the 10 reference waveforms are applied to sector electrodes on one plate, while the 10 high resolution electrodes meander around the other plate. The effective resolution of the meter is increased to 0.1 % by including a ten sector "vernier" in the display centre. Reference [3.46] also describes the application of WIA to numeric displays, where the economies of matrix addressing for multiple digits are realised with the same visual appearance as directly driven displays.

3.4.3. **Active matrix addressing**

The advantage of this approach is that it makes only minimal demands on the steepness of the transmission–voltage characteristic of the liquid crystal. The complexity now resides in the cell substrate, where an electrical (control) device, either active or non–linear passive, is placed in series with each liquid crystal display element (section I.1.5).

These devices are used exclusively in matrix addressed cells, where line select pulses are applied to the lines in sequence while data pulses are applied to the columns. The details of these select and data pulses of course depend on the type of control device and LC effect used. In general, though, LC elements are regarded as capacitors and the control devices must achieve three main objects. Firstly, they must prevent all partial select pulses intended for other dels from affecting the charge on non-selected dels. Secondly, when a full select pulse occurs it must permit the capacitance of the selected display

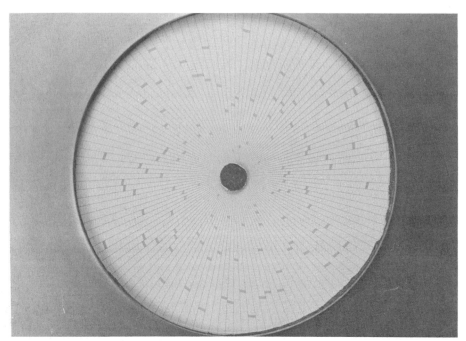

Fig. 3.20   Photograph of prototype display using waveform identity addressing.

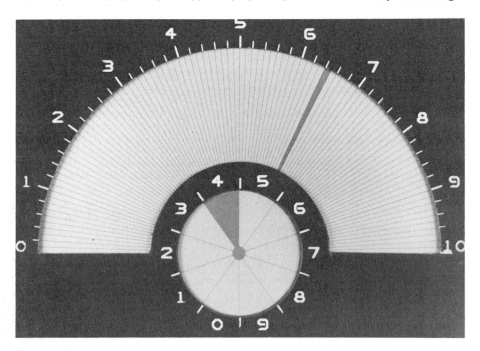

Fig. 3.21   Photograph of prototype "analogue" meter using waveform identity addressing to achieve 0.1 % resolution.

element to charge up rapidly to the applied data voltage. Thirdly, when the full select pulse is terminated, it must prevent the charge accumulated from leaking away until the next full select pulse occurs.

Several different types of device have been investigated for this application, and many are still being actively developed. Some are two terminal devices, including zinc oxide varistors, metal-insulator-metal (MIM) devices, and amorphous Si diodes of various types. Perhaps more versatile, however, are the three terminal transistor approaches. These include MOSFETs on large area Si substrates and several thin-film transistor (TFT) devices using poly-crystalline silicon (p-Si), amorphous silicon (a-Si), poly-crystalline cadmium selenide (CdSe), etc., as the semi-conductor material.

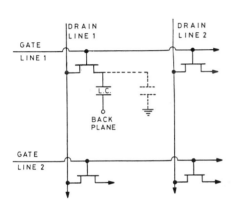

Fig. 3.22 Equivalent circuit of a small portion of a liquid crystal matrix with integral TFT addressing. The additional capacitor is not always needed.

The detailed operation of an active matrix display will be described using the Si MOSFET approach as an example, all other approaches following the same general principles. In this Si substrate approach the rear plate of the LC cell consists of a large Si slice operating as a single integrated circuit. The front plate on the cell is glass, coated on the inside with a layer of transparent conductor which acts as one side of each LC capacitor and does not need to be patterned. A simplified equivalent circuit of a few display elements is shown in figure 3.22, where the additional storage capacitor, C_s, shown dotted, is not always required. The FET gates are connected to line bus-bars and the sources to column bus-bars. Transistors are normally held "off" by applying appropriate voltages to all gate lines, and are then turned on one line-at-a-time by a strobe pulse on the desired gate line. Simultaneously the corresponding data pulses (or video levels for grey scale displays) are applied to the columns. The capacitance of the LC and the storage capacitor then charges to video voltage, and this charge is stored while all other lines are addressed. This addressing cycle is continuously repeated at some appropriate frame time T_f, often around 50 Hz. AC drive to the LC is obtained by periodically reversing the video polarity, usually on alternate frames.

Two simple criteria can be established for successful FET and LC operation.

Firstly, the time constant for charge leakage from the LC capacitance must be much longer than the frame time T_f. This leakage occurs both through R(off), the off resistance of the TFT, and through R(LC) the LC resistance itself.

Secondly, the time constant for charging and discharging the LC capacitance through the on-state resistance of the TFT, R(on), must be much less than the line-time, T_ℓ. For a simple data display, where dels are either fully "on" or fully "off", a significant range of RMS voltages can be tolerated without degrading the visual appearance of the display. So, providing sufficient supply voltage is available, a wide range of C and R values may be acceptable. For displays with grey scale, however, very precise control of RMS voltage across the LC is required. All dels must be charged up to tightly controlled voltages, and little charge leakage can be tolerated. The charge leakage problem is aggravated by high temperatures, where both R_{LC} and R_{off} are reduced. This can be alleviated either by reducing T_f or by increasing C_s, but both solutions require increased "on" currents from the transistors. Furthermore, the inclusion of C_s may increase the manufacturing complexity and reduce the yield of good devices. The ideal solution, therefore, requires transistors with a very large ratio of R_{off} to R_{on}, liquid crystal with very high leakage resistance, and the ability to select the transistor currents to suit the display element capacitance and data frame and line-times.

The early development of these devices on Si substrates included work by Lipton et al [3.47], who demonstrated a 45 mm square prototype having 175x175 dels. Because Si substrates are opaque the TN effect could not be used, so dynamic scattering was used instead.

Subsequent developments in various places have incorporated some of the decode and drive circuitry around the periphery of the display, thus reducing interconnection problems and the complexity of the external circuits. For example, Kasahara et al [3.48] presented a 220x240 element TV display with 2" diagonal that included integrated gate bus drivers. Hosokawa et al [3.49] made use of pleochroic dyes to improve the appearance of a similar display. Ymasaki et al [3.50] developed a 240x240 element TV display requiring only 12 external connections. This type of display will undoubtedly find many applications and has clearly not reached its development limit, but it will always be limited by the maximum size of Si slice available. Attempts have been made to produce larger displays by butting together 4 slices in a single display, but this has not been very successful.

The thin-film transistor (TFT) approach is electrically very similar in that each del is addressed by an individual transistor, but it offers several potential advantages over the Si slice approach. Firstly, the active devices can be deposited on a glass sheet, so the size limitation of the Si slice is avoided. Secondly, the transparent substrate means that almost any LC effect can be used, but particularly the well characterised TN effect. Thirdly, backlighting is easily achieved, and the LC cell can be constructed using well established techniques.

Several different semi-conductors have been exploited for this type of display, including amorphous silicon (a-Si), poly-crystalline silicon (p-Si), poly-crystalline cadmium selenide (CdSe), etc. It has been well established that all these materials can produce transistor characteristics suitable for displays with many hundreds of lines. Instead of an exhaustive review of all work on this topic, table 3.2 contains a summary of all the active LCD papers presented at the International Display Research Conference (IDRC), Oct. 1985. This illustrates clearly the range of technologies available, the complexity already achieved, the large area capability and the variety of useful LC effects. Figure 3.23 shows the first commercial active matrix display. This used CdSe TFTs and had 128x192 dels, each 0.5 mm square. The TN effect was used in the "transflective" mode with an electroluminescent backlight, giving a total package about 12 mm thick. Figure 3.24 shows a monochrome picture of a full-colour display with poly-Si TFTs. The display [3.52] has 440x480 dels and measures 88x96 mm. Again the TN mode is used, but the presence of colour filter triads necessitates the use of a permanent backlight.

At this stage it is not possible to predict which, if any, of these approaches will achieve long term commercial success, since all have their strong and weak points. For example, to achieve good semi-conductor mobility poly-Si requires

Fig. 3.23 Photograph of an early commercial LCD with "active matrix addressing" using CdSe TFTs (Courtesy PanelVision Corp., USA).

processing temperatures of around 600 C, which is rather high for cheap glass substrates. It also requires a rather complex transistor structure to achieve adequately low "off" currents. However, it does produce fast enough devices for the line and column shift registers and drivers, and complete drive circuits have been demonstrated, as in figure 3.25 [3.52].

Amorphous silicon is completely compatible with glass substrates, requiring process temperatures of between 300 C and 400 C. Unfortunately the resulting semi-conductor mobility is low, usually less than 1 cm^2 V^{-1} s^{-1}, so very wide, short transistors are needed to obtain adequate "on" currents. Low "off" currents are easily obtained, and a considerable technological investment in the material for solar cell applications has been highly beneficial. The majority of TFT displays currently being developed are based on a-Si.

Poly-crystalline cadmium selenide can also be desposited onto glass substrates at low temperatures. The resulting semi conductor mobility, however, can be as high as 100 cm^2 $V^{-1}s^{-1}$, permitting fast, high current transistors of modest dimensions. Furthermore, with correct processing, very low "off" currents can be achieved without the complex structures required in poly-Si. Apparently, the major factor against CdSe devices is that the material has few other applications and there is some commercial reluctance to develop it further. In spite of this, one of the earliest and largest TFT addressed displays to be marketed uses CdSe, as shown in figure 3.23. Further development of this approach has resulted in a 640x400 display measuring 250x180 mm [3.68].

Fig. 3.24 Photograph of 440x480 active matrix display using poly-crystalline silicon TFTs (Courtesy Epson-Seiko, Japan).

The two terminal device approaches offer the possibility of simpler manufacturing methods and no electrical cross-overs on the device substrates. The most successful of these to date is the metal-insulator-metal (MIM) diode approach based on tantalum oxide and pioneered by Streater et al [3.62]. Furher work by Morozumi et al [3.63] and [3.52] has resulted in a very reliable device, known as the "lateral MIM", used in the large area, high complexity display shown in figure 3.26. Unfortunately the published literature on this type of display says little about its grey scale capability or about its susceptibility to temperature change.

The final approach discussed here returns to the single crystal Si substrate, this time using a charge coupled device (CCD) shift register to accept video inputs directly [3.64]. In this device the CCD shift registers are on one face of the high resistivity Si slice, the stored charge being coupled through the slice to excite the LC cell on the opposite face. A prototype device measuring 5x5 mm with 256x256 elements has been developed, but at present the observed resolution is only 125x125 elements. Larger devices, up to 1000x1000 elements, are planned for use as light valves in various projection displays and optical processing applications. This display is in fact the third in a series of

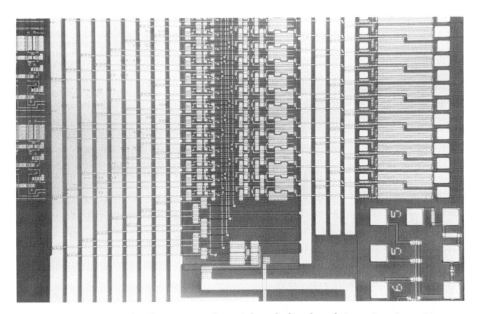

Fig. 3.25 Photograph of section of peripheral display drive circuit using poly-crystalline silicon TFTs (Courtesy Epson-Seiko, Japan).

developments for the same applications. The first two types used LC cells that included photoconductive layers, the state of the liquid crystal being locally controlled by the illumination of the photoconductor. The original device [3.65] and [3.66] used CdS as the photoconductor which was illuminated by an image of data displayed on a high luminance, high resolution CRT. Various liquid crystal optical effects were used to modulate the illumination from the projection lamp. The second device [3.67] used a Si slice both as the photoconductor and as one plate of the liquid crystal cell. In both cases high resolution was obtained and high luminance image projection was possible.

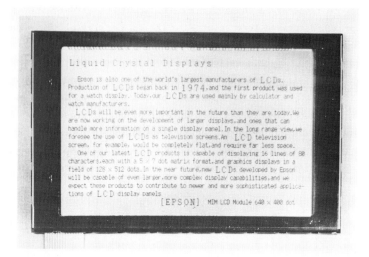

Fig. 3.26   Photograph of a 640x400 dot-matrix display incorporating "lateral MIM diodes" (Courtesy Epson-Seiko, Japan).

3.5. Conclusion

In the few years that LCDs have been commercially available they have had an enormous impact on the display market. At the low complexity end the LCD has secured total dominance of the market for small numeric and alpha-numeric displays. This has been achieved through low cost, mass-produced products that have adequate visual performance and make minimal demands on electronics and batteries. As display complexity increases, however, so does the competition with other flat panel technologies. Multi-way multiplex operation of TN displays results in poorer optical performance, so that the more costly light emitting technologies (e.g. plasma panels, electroluminescence, vacuum fluorescence, etc.) are able to offer greater visual appeal. The use of backlighting with transmissive mode LCDs can give improved visual appeal, but detracts from the intrinsic low power advantage of the LCD. At the highest complexity, LCDs must compete with CRTs which exploit so effectively the simplicity of "beam-addressing", can be made in such a wide range of sizes and have a long established capability for high resolution and colour.

Looking to the future a number of questions may be asked. For instance, how much will "supertwist" effects be capable of exceeding the complexity of TN displays? What size and complexity of TFT driven display can be manufactured at sufficiently high yield to be competitive? What can be achieved with storage

effects, particularly in smectic materials? What is the true potential of ferro
electric devices? Which LC effects will be most appropriate for colour and grey
scale displays? Can an acceptable colour display be obtained operating in
reflection? What is the size limit for LCD's? Will future developments retain
the low voltage, low power and low cost characteristics of reflective TN
displays? All these and many other questions present challenges to research,
development and production workers. The answers are not obvious, but we may be
confident that the enormous diversity of electro-optic effects obtainable with
liquid crystals gives good grounds for optimism. Given adequate investment in
research and production technology the momentum of LCD advance will surely be
maintained.

REFERENCES

[3.1] Blinov, L.M., Electro-optical and magneto-optical properties of liquid
      crystals, Wiley, UK (1983).

[3.2] Sprokel, G.J., (Ed.), The physics and chemistry of LC devices, Plenum,
      NY (1980).

[3.3] Chandrasekhar, S., Liquid crystals, Cambridge University Press, England
      (1977).

[3.4] Ashford, A., Constant, J., Kirton, J. and Raynes, E.P., Electro-optic
      performance of a new room temperature nematic liquid crystal,
      Electronics Letters, vol. 9, no. 5, (1973), pp. 118-120.

[3.5] Gray, G.W., Harrison, K.J. and Nash, J.A., New family of nematic liquid
      crystals for displays, Electronics Letters, vol. 9, no. 6, (1973), pp.
      130-131.

[3.6] Hulme, D.S., Raynes, E.P. and Harrison, K.J., Eutectic mixtures of
      nematic 4' - substituted 4-cyanobiphenyls, J. Chem. Soc., Chem Commun.
      no. 3, (1974), pp. 98-99.

[3.7] ...... Liquid crystals, Data Sheets, BDH Ltd. (1986).

[3.8] ...... See for example material E 209, BDH Ltd.

[3.9] Alt, P.M. and Pleshko, P., Scanning limitations of liquid crystal
      displays, IEEE Transactions Electron Devices, vol. ED-21, no. 2, (1974),
      pp. 146-155.

[3.10] Hilsum, C., Holden, J.R. and Raynes, E.P., A novel method of temperature
       compensation for multiplexed liquid crystal displays, Electronics
       Letters, vol. 4, no. 14, (1978), pp. 430-432.

[3.11] Kawakami, H. et al, Brightness uniformity in LCDs, SID 80 Digest,
       (1980), pp. 28-29.

[3.12] Heilmeier, G.H., Dynamic scattering, Proc. IEEE, vol. 56, no. 7, (1968),
       pp. 1162-1171.

[3.13] Schadt, M. and Helfrich, W., Voltage dependent optical activity of a twisted nematic liquid crystal, Appl. Phys. Letters, vol. 18, no. 4, (1971), pp. 127-128.

[3.14] Mauguin, C., Bull. Soc. Franc. Mineral, vol. 34, (1911), pp. 71-117.

[3.15] See for example material E 200, BDH Ltd.

[3.16] Fahrenschon, K. and Wiemer, W., LCD modules for harsh environmental conditions, Eurodisplay '84 Proceedings, (1984), pp. 37-40.

[3.17] Gooch, C.H. and Tarry, H.A., J. Phys. D (Appl. Phys.), vol. 8, (1975), p. 1575.

[3.18] Scheffer, T.J., In: Nonemissive electro-optic displays, Ed: Kmetz and von Willisen, Plenum Press, (1975), pp. 49-78.

[3.19] Raynes, E.P., In: Nonemissive electro-optic displays, Ed: Kmetz and von Willisen, Plenum Press, (1975), pp. 25-43.

[3.20] Tani, C., Ogawa, F., Naemura, S., Ueno, T. and Saito, F., Storage-type liquid crystal matrix display, SID Tech. Digest, vol. 10, (1979), pp. 114-115.

[3.21] White, D.L. and Taylor, G.N., New absorptive mode reflective liquid crystal display device, J. Appl. Phys., vol. 45, no. 11, (1974), pp. 4718-4723.

[3.22] Saunders, F.C., Harrison, K.J., Raynes, E.P. and Thompson, D.J., New photostable anthraquinone dyes with high order parameters, Conference Record of International Display Research Conference, (1982), pp. 121-125.

[3.23] Perrin, A. et al, Characteristics of multiplexed dyed dot-matrix LCDs, Eurodisplay 84 Proceedings, (1984), pp. 179-182.

[3.24] Bos, P.J., Johnson, P.A. and Beran, K.R.K., A liquid crystal optical switching device (π-Cell), SID 83 Digest, (1983), pp. 30-31.

[3.25] Berreman, D.W. and Heffner, W.R., Fast LCDs with low power addressing and permanent memory, SID 82 Digest, (1982), pp. 242-243.

[3.26] Waters, C.M. et al, Highly multiplexed dyed LCDs, Proc. Japan Display Conference, Paper 10.1, (1984).

[3.27] Waters, C.M. et al, Design of highly multiplexed LC dye displays, Mol. Cryst. Liq. Cryst., vol. 123, (1985), pp. 303-319.

[3.28] Scheffer, T.J. et al, 24x80 Character LCD panel using SBE, SID 85 Digest, (1985), pp. 120-123.

[3.29] Dewey, A.G. et al, A 4 Mpel LC projection display, SID 82 Digest, (1982), pp. 240-241.

[3.30] Smith, M.R., Burns, R.H. and Tsai, R.C., Ultrahigh resolution graphic data terminal, SPIE, vol. 200, (1979), pp. 171-178.

[3.31] Harrold, J. and Steele, C., High resolution vector addressed LC light valve, Eurodisplay 84 Proceedings, (1984), pp. 29-32.

[3.32] Tsai, R.C. et al. High density 4-colour LCD system, J. SID, May (1981), pp. 3-6.

[3.33] Le Berre, S., Hareng, M., Hehlen, R. and Perbet, J.N., A flat smectic LCD, SID 82 Digest, (1982), pp. 252-3.

[3.34] Hareng, M. and Le Berre S., Liquid crystal flat display, Proceedings. IEDM, Dec. (1978).

[3.35] Lu, S. et al, Thermally addressed pleochroic dye switching LCD, SID 82 Digest, (1982), pp. 238-239.

[3.36] Crossland, W.A. and Ayliffe, P.J., An evaluation of smectic dynamic scattering, Proc. SID, vol. 23, no. 1, (1982), pp. 9-13.

[3.37] Crossland, W.A. and Canter, J., An electrically addressed smectic storage device, SID 85 Digest, (1985), pp. 124-127.

[3.38] Clark, N.A. and Lagerwall, S.T., Physics of ferro electric fluids: a high-speed electro-optic switching process in liquid crystals, Recent Developments in Condensed Matter Physics, vol. 4, Plenum, (1981), pp. 309-319.

[3.39] Harada, T., Taguchi, M., Iwasa, K. and Kai, M., An application of chiral smectic C LC to a multiplexed large area display, SID 85 Digest, (1985), pp. 131-134.

[3.40] Clark, M.G., Shanks, I.A. and Patterson, N.J., General theory of matrix addressing liquid crystal displays, SID Tech. Digest, vol. 10, (1979), pp. 110-111.

[3.41] Nehring, J. and Kmetz, A.R., Ultimate limits for matrix addressing of RMS-responding liquid crystal displays, IEEE Transactions Electron Devices, vol. ED-26, no. 5, (1979), pp. 795-802.

[3.42] Kmetz, A.R., A twisted nematic dual bargraph system, SID Tech. Digest, vol. 8, (1977), pp. 58-59.

[3.43] Shanks, I.A. et al, LC oscilloscope displays, SID 78 Digest, (1978), pp. 98-99.

[3.44] Shanks, I.A. and Holland, P.A., Addressing method for non-multiplexed liquid crystal oscilloscope displays, SID Tech. Digest, vol. 10, (1979), pp. 112-113.

[3.45] Glasper, J.L. and Smith, C.J.T., A liquid crystal radar display, RSRE Research Review, (1985), pp. 150-154.

[3.46] Clark, M.G. et al, New applications for waveform identity addressing of LCDs, Eurodisplay 84 Proceedings, (1984), pp. 21-24.

[3.47] Lipton, L.T., Stephens, C.P., Lloyd, R.B., Shields, S.E., Toth, A.G. and Tsai, R.C., A 2.5 inch diagonal, high contrast, dynamic scattering liquid crystal matrix display with video drivers, SID Tech. Digest, vol. 9, (1978), pp. 96-97.

[3.48] Kasahara, K., Yanagisawa, T., Sakai, K., Adachi, T., Inoue, K., Tsutsumi, T. and Hori, H., A liquid crystal display panel using an MOS array with gate-bus drivers, IEEE Transactions Electron Devices, vol. ED-28, no. 6, (1981), pp. 744-748.

[3.49] Hosokawa, M., Oguchi, K., Ikeda, M., Yazawa, S. and Endo, K., Dichroic guest-host active matrix video display, SID 81 Digest, (1981), pp. 114-115.

[3.50] Ymasaki, T., Kawahara, Y., Motte, S., Kamamori, H. and Nakamura, J., A liquid crystal TV display panel with drivers, SID 82 Digest, (1982), pp. 48-49.

[3.51] Oana, Y. et al, A 9.5 inch diagonal multi-colour amorphous Si TFT-LCD panel, IDRC Record, (1985), pp. 32-33.

[3.52] Morozumi, S., Active matrix LCDs, IDRC record, (1985), pp. 9-13.

[3.53] Sunata, T. et al, A 640x400 pixels active matrix LCD using a-Si TFTs, IDRC record, (1985), pp. 66-67.

[3.54] Sunata, T. et al, A large area, high resolving power, active matrix colour LCD addressed by a-Si TFTs, IDRC record, (1985), pp. 18-23.

[3.55] Suzuki, K. et al, Amorphous Si active matrix addressed colour LCD, IDRC record, (1985), pp. 14-17.

[3.56] Lee, M.J. et al, A 10x8 cm LCD addressed by CdSe thin-film transistors, IDRC record, (1985), pp. 59-61.

[3.57] Lebosq, Y. et al, An improved design of active matrix LCD requiring only two photolithographic steps, IDRC record, (1985), pp. 34-36.

[3.58] Saito, T. et al, A high picture quality LC-TV using triangle trio-colour dots addressed by a-Si TFTs, IDRC record, (1985), pp. 27-29.

[3.59] Sakai, T. et al, A colour LC using a very thin-film transistor, IDRC record, (1985), pp. 30-31.

[3.60] Hitoh, H. et al, Large area projection HAN-mode multi-colour TFT-addressed LCD, IDRC record, (1985), pp. 62-65.

[3.61] Yaniv, Z. et al, A novel a-Si switching device for driving active matrix LCDs, IDRC record, (1985), pp. 76-79.

[3.62] Streater, R.W., Este, G.O., Maniv, S., Maclaurin, B. and Miner, C.J., MIM addressed LCDs: status and prospects, SID 82 Digest, (1982), pp. 248-249.

[3.63] Morozumi, S. et al, A 240x250 element LCD addressed by lateral MIM, Japan Display 83 Proceedings, (1983), pp. 404-407.

[3.64] Little, M.J., Braatz, P.O., Efron, U., Grinberg, J. and Goodwin, N.W., CCD-addressed LC light valve, SID Digest, (1982), pp. 250-251.

[3.65] Grinberg, J. et al, Photoactivated birefringent liquid crystal light valve for colour symbology display, IEEE Transactions Electron Devices, vol. ED-22, no. 9, (1975), pp. 775-783.

[3.66] Hong, B.S., Lipton, L.T., Bleha, W.P., Colles, J.H. and Robusto, P.F., Application of the liquid crystal light valve to a large screen graphics display, SID Tech. Digest, vol. 10, (1979), pp. 22-23.

[3.67] Efron, U., Grinberg, J., Braatz, P.O. and Little, M.J., A silicon photoconductor-based LC light valve, SID 81 Digest, (1981), pp. 142-143.

[3.68] Luo, F.C., Patterson, J., Braunstein, T. and Leksell, D., A 640x400 CdSe TFT-LC display panel, SID 85 Digest, (1985), pp. 286-288.

[3.69] Nagae, Y., Kawakami, H. and Kaneko, E., Thermally addressed LC flat display with laser light pen, SID 85 Digest, (1985), pp. 289-292.

DISPLAY ENGINEERING: D. Bosman (Editor)
© Elsevier Science Publishers B.V. (North-Holland), 1989

II.4: LARGE AREA GAS DISCHARGE DISPLAYS OR PLASMA DISPLAYS

J.P. MICHEL

Thomson CSF, Division Tubes Electroniques, France

4.1. Introduction

Undoubtedly the oldest electro-optical phenomenon able to produce light is an electrical discharge in a gas. Although this phenomenon is at the present time supposed to be at the origin of life on earth, millions of years elapsed until this effect was identified, analysed and mastered by man.

The progress in the knowledge of the behaviour of discharges in gases and of the parameters which govern them has been essentially parallel to that of electronics and vacuum. It is through the study of these discharges (Crookes 1875) that the "cathode rays" – as Goldstein named them (1876) because they appeared as faint light beams in low pressure gas under an electric field – were discovered and that later on Perrin (1895) identified these beams as made of negative charges and finally as beams of electrons (J.J. Thomson 1897). Most properties were discovered and explained during the last decades of the 19th century and the first quarter of this century by Faraday, Townsend, Penning, Jeans, Compton, Langmuir, Seeliger to name but few among literally tens of them. The "Paschen law" describing the discharge relationship (for a given applied voltage between two electrodes, the product of their distance multiplied by the gas pressure is a constant) dates only from the end of the 19th century (1889).

The properties of gas discharge are numerous and have been, and still are, applied in many different devices like:
- rectifying and switching tubes with either cold or hot cathodes: mercury pool rectifiers used in the early days of broadcasting, phanotrons also for high DC voltage supplies, ignitrons for welding or plasma fusion experiments, thyratrons for radar or scientific research, surge arresters for line protection in telephone or electricity
- switching diodes and trigger tubes used in timing circuits and relaxation oscillators (e.g. oscilloscope time base), switching of high energies, logic gates
- counting (or stepping) tubes like the "Dekatron" used in ring counters and registers
- stabilisers and voltage reference tubes
- gaseous triodes used for signal amplification and which were called "soft tubes" due to their control characteristic which was "weaker" than that of their "hard" conterpart (the name "hard tube" is still used for high vacuum, high voltage switching tubes but few people still know why)

- gaseous photocells for sound track reading of early talking motion pictures,
  or Geiger tubes for counting ionising particles.

None of these devices do make use of the light emitted by the discharge, but
rather of the low voltage drop, its constancy, the short de-ionisation time,
the arc transfer capability and the multiplication effect in a discharge.
Except where very high powers, energies or voltages are involved most of the
above functions are nowadays better and cheaper performed by solid state
devices. There is one area however where gas discharges still have decisive
properties. It is that of luminous devices.
Very efficient light sources rely on them like fluorescent lamps, high pressure
mercury lamps, high and low pressure sodium vapor lamps used for street
lighting, neon and other gases signs and flash tubes. On-off indicators and
character or numeral display tubes still find wide spread use too.

It is in the early fifties that a numeric indicator tube which became a
standard for readout devices under the name of NIXIE was developed. More
recently, in the late 60s, other techniques began to replace it. In the NIXIE
tube a single anode and a stack of separate cathodes, shaped in the form of
numerals, are immersed in a gas mixture [4.1].
First attempts to produce a matrix display panel were made in 1954 (Skellet),
the ignition of a discharge taking place at any intersection of two sets of
electrodes. This technique was, however, limited to one cell at a time [4.2].
Since then, research has continued, the constraining and contributing factors
identified, and a host of different approaches have evolved.
All these are designated by the names of "Gas Discharge Displays" or "Plasma
Display Panels" (PDP), the latter expressing the physical fact that when
emitting light, matter in its fourth state, made-up of free ions and electrons.

4.2. Principle of operation

Before describing the operation of Plasma Display Panels it is necessary to
detail the characteristics of the gas discharge itself and how the different
parameters are related to each other.

4.2.1. Electrical characteristics (figure 4.1)

Consider two metal electrodes at a given distance in a glass tube at a reduced
gas pressure. If no ions or electrons are available, the residual gas behaves
like an insulator. However, ions and electrons can be produced by the
ionisation of the gas, due to cosmic rays or residual radio activity or by some

external means. If a small potential is applied, these charges will be collected and a current of the order of 10^{-15} Ampère will be measured. As the potential is increased the current grows in the range $10^{-10} - 10^{-9}$ A or more, where all these available charges are collected but no other effect has yet taken place (Region A) [4.3].

At this point it is important to note that this priming current strongly depends on the activity of the source of ionisation and that it has to be initiated by some external means.

If the potential is then further increased (AB' Region), avalanche phenomena take place and the current grows very rapidly. This is due to two effects:

- the electrons gain sufficient energy to ionise additional atoms of the gas by collision
- electrons are extracted from the electrodes by ions, photons produced in the gas or metastable atoms (a metastable atom is an atom which has been excited −not ionised− to an

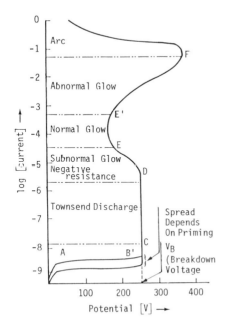

Fig. 4.1 I-V characteristics of a low pressure DC discharge (Arthur D. Little).

energy level from which it cannot return to the ground state without some external shock).

When the so called "break-down potential" or "firing potential" is reached (V_B; CD Region) the discharge becomes self sustaining and any attempt to raise it, results in a very strong increase of the current (over a range of almost 3 orders of magnitude for a potential variation of a few volts) while radiance varies accordingly. The name of Townsend discharge has been given to that state, which he studied extensively (1900). This self sustained discharge is due to a space charge and a high potential gradient at the proximity of the cathode. It is called self sustained because the priming current is no more necessary to maintain it.

Townsend showed that in the AB' (priming) region the current was governed by the equation

$$i = i_0 \exp \alpha d.$$

Where α is the first Townsend coefficient, which represents the number of electrons produced by a single electron, travelling unit distance under a given field E and at a given gas pressure p.

In the CD (break-down) region he also showed that the current relationship is

$$\frac{i}{i_0} = \frac{\exp \alpha d}{1 - \gamma (\exp \alpha d - 1)} \qquad (4.1)$$

Where γ is the second Townsend coefficient, which represents the total number of electrons produced per positive ion in the discharge (including those also produced by photons and metastables at the cathode).

The critical potential for break-down V_B occurs when $i/i_0 = \infty$, that is when:

$$1 - \gamma (\exp \alpha d - 1) = 0 \qquad (4.2)$$

As α/p and γ are both functions of the field and the pressure (E/p) one can finally write:

$$(V_B/pd)^{\frac{1}{2}} = \frac{D}{K + \log pd} \qquad (4.3a)$$

C and D are constants
depending on the gas

$$\text{where } K = \log \frac{C}{\log (1 + 1/\gamma)} \qquad (4.3b)$$

For large values of pd V_B rises almost linearly with pd.

For small values of pd V_B rises as pd decreases.

A minimum occurs when $\dfrac{dV_B}{d(pd)} = 0$

or $K + \log(pd) = 2$; where $V_{B_{min}} = D^2/4pd_{min}$ $\qquad (4.5)$

The resulting curve represents the already mentioned Paschen law, as in figure 4.2. It is U shaped, the left part being steeper than the right one. Thus it is possible to trade-off between the three governing parameters: p,d,V, for instance to optimise the break-down voltage or to decrease the distance so as to increase the resolution of matrix displays; or else to make the firing voltage independent of the applied external pressure. However, for a given gas, it will be impossible to reduce the break-down voltage below the minimum of the curve.

Beyond that region CD, the voltage drop across the discharge decreases due to a space charge increase while the current grows for 2 or 3 more orders of magnitude: the discharge presents a negative impedance, and is called the

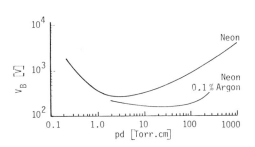

Fig. 4.2 Paschen curve for neon and neon + 01.% argon· Pennning mixture (Arthur D. Little).

"subnormal glow", region "DE" in figure 4.1. When the voltage reaches a minimum in the region (EE'), the glow is said to be "normal" and the potential across it is the "maintaining potential" or "sustaining potential" V_S. It is characterised by a glow around the cathode.

In the event that the applied voltage is reduced from that state, the current will fall into the priming range $(10^{-9} - 10^{-8}$ A) and the discharge will stop as well as the light emission and a closed on/off cycle has been described.

Still beyond, (E'F) when the current has been increased enough for the glow to completely cover the cathode, the current density increases and the glow becomes "abnormal". The potential in the glow rises considerably, the cathode heats-up, an "arc" discharge takes place and the current may be so high as to result in the destruction of the device, except if the material of the cathode has been chosen with a low yield in secondary electrons –Townsend 2nd coefficient– so as to stabilise the working conditions.

Therefore some means of current limitation must be employed, generally a resistance if only a steady state is considered, but a reactive impedance can be used as well if only transient states are to occur, the other means consisting in current pulse width control or in an arrangement such that the cell operates in the positive impedance abnormal glow region.

4.2.2. Luminous distribution

Figure 4.3 [4.3] shows the luminous distribution of a discharge operating in the subnormal and normal glow regions as it is the case with most gas display devices. The reason for this is quite obvious: as the luminance is directly related with the current flowing between the electrodes, it is

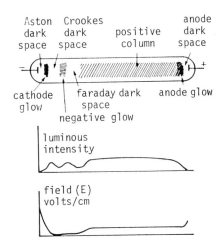

Fig. 4.3 Characteristics of a DC glow discharge (Arthur D. Little).

advantageous to operate in that region; with the additional benefit of the
possibility to go through the on/off cycle mentioned above.

Close to the cathode and closely following its shape is the negative glow.
Between that glow and the cathode is the cathode dark space where most of the
voltage drop takes place. There is a strong field resulting in numerous
excitations of neutral atoms and corresponding intense glow. The cathode glow
is followed by the Faraday dark space where the potential is almost constant
and the electrical field is low with almost no excitations and no light. Under
particular conditions the field can even be in the reversed direction with
respect to that at the cathode, with almost no excitations and no light.
In some instances a Crookes dark space and a negative glow may be observed too.

Beyond the Faraday zone is the positive column which extends to the anode,
whatever its distance. The field here is low too but the ion and electron
densities however are significant and almost equal. These conditions correspond
to what is called a plasma. The radiance output may be fairly high (it is the
part of the discharge used in fluorescent lamps).

Close to the anode, an anode glow may be observed separated from the positive
column by the anode dark space. In fact, only the cathode dark space and the
negative glow are important to the discharge and when the inter-electrode
distance is varied, only the positive column length changes. Although some
attempts have been made to use the anode glow, all existing plasma displays use
the negative glow.

To be noted also is the fact that the colour of both discharges may be
different. For instance with air the negative glow can appear milky blue and
the positive column pink.

4.2.3. Colour

It has been shown that several processes take place in the discharge. When the
energy of electrons exceeds a threshold which is characteristic for each gas,
the type of collision with neutral atoms may change. Instead of an elastic
collision which leaves the energy of the electron almost unaltered it may
transfer to the atom a significant part of its kinetic energy which, in turn,
raises the energy of an external electron of the atom.
These quantities of energy which may be transferred, form a series of discrete
values which, again, are characteristic of each gas.

Between the ionisation potential (21.6 V in the case of Neon) and the first
excitation potential (16.6 V for Ne) many excitation levels will be obtained.
From these excitation levels, where the atoms can remain only about 10^{-8}
seconds, they will relax to the ground state, or from 2p levels to 1s levels,

or still other levels, while emitting a photon of corresponding energy.

Therefore the spectrum will be very rich of lines at different wavelengths and in the case of Neon, will result in a characteristic orange negative glow with a high yield of 0.5 ℓm W^{-1}. Ultra violet, non-visible photons will also be emitted, some of which will result in electron emission from the cathode and corresponding avalanche contribution (Townsend second coefficient).

4.2.4. Penning mixture

However, two of the 1s levels of Ne are metastable states and, as already said, are not allowed to emit photons without some additional shock: typically they last 100 to 1000 times longer than excited species. Such atoms are not charged and like other "regular" atoms are displaced only by the slow process of diffusion.

It has been shown by Penning that for an accurately chosen proportion of additional gas, having a lower ionisation potential than the energy of that metastable atom, this additional gas will be ionised itself; thus contributing very strongly to lower the firing voltage.

In case of Neon such ideal Penning mixture contains 0.1 % of Argon. The metastable atoms of the first having an energy of about 16.6 eV and the ionisation potential of Ar being 15.8 eV there is a very high probability that a collision results in the ionisation of the Ar atom.

The Paschen curve is shifted to the right, the minimum voltage is lower and the voltage dependence of pd in smoother (figure 4.2). Therefore this mixture is used in most plasma displays.

4.2.5. Time dependence

The cumulative -avalanche- nature of the break-down process makes it obvious that some delay is involved for the discharge to reach its steady state. As has been stressed, there is a strong dependence on the priming current existing during the "off" state; the more charges available to initiate the avalanche, the lower the "statistical lag". It is also reduced by increasing the applied voltage well above the firing voltage.

The second parameter of the delay is "formative lag" or the time constant of the exponential current rise which, again, depends on the applied voltage.

Therefore, here also it will be necessary to trade-off between the desired rise time and the acceptable voltage - which is always too high, as seen by the driving circuits engineer.

Similarly if the applied voltage is removed, the recombination of free ions and electrons will take some time until the glow disappears and the gas returns to the completely non-ionised state, leaving an "after-glow" during the "de-ionisation time" (or "recovery time").

In plasma displays the build-up time can range from 0.1 to 100 microseconds and the recovery time from 5 to 50 microseconds.

However, after the discharge quenching, residual charges may remain free, thus modifying the priming conditions for sometimes up to some milliseconds.

These time dependences directly govern the maximum frequency at which a gas discharge device may be driven. For instance when thyratrons are to be used for radar applications, a very short de-ionisation time is requested and the proper gas will be Hydrogen whereas special low energy spark gaps will use Helium. In plasma displays where light yield and low break-down voltage are the prime requirements, a Neon Argon Penning mixture is used in spite of the relatively long rise and recovery times.

In summary, the key parameters which are unique to gas discharge are:
- the very sharp threshold which is so steep that matrix addressing of plasma panels (section I.1.5) poses no problem whatever the number of dels as opposed to most other phenomena,
- the initial conditions (priming current) are directly dependent on the availability of energetic particles: it will be possible to change these initial conditions by external or internal means,
- the firing and sustaining voltages are well defined and show little dependence on temperature (1 Volt change of V_S for 70 C temperature change [4.1]),
- the negative impedance characteristic, which leads to two steady states for the same applied voltage; the "on" state in the glow region or the "off" state in the priming current region: the discharge may be operated as a bistable element,
- the high luminous yield (0.5 ℓm W^{-1} for Ne/Ar Penning mixture) and the quasi proportionality of light output to current flowing,
- the dependence of spectral emission upon the gas (Xe will emit mainly Ultra-Violet wavelengths which may be used to excite phosphors and make them fluoresce).

All these parameters though, are dependent upon the nature of the gas and of the electrodes, their shape and distance etc.

4.2.6. Driving methods

Apart from the on-off status indicator, the most simple numerical displays are the tubes where each cathode has the shape of a numeral and which, when energised, is surrounded by the cathode glow and thus displays that numeral. In the same category are the segmented displays which, instead of a single complete figure, can display one or several rows of figures, characters or symbols in a stylised form. However, they all display a limited number of fixed patterns with no flexibility to change size, bias or location.

When more complex display capability is required the matrix addressing technique becomes a necessity and we will restrict our selves to such devices. This technique calls for a threshold for selectively addressing a given cell; quite fortunately the plasma discharge offers that property to an unparallelled extent [4.4,4.5,4.6].
Once the discharge has started, the impedance of the cell becomes very low or negative and some means must be used to limit the flow of current and to avoid coupling through "on" cells.

In 1963 Thomson [4.7] made a 10x10 array arrangement with one resistance associated to each cell, any of which could be selectively and concurrently turned "on" or "off". This was the precursor to DC gas discharge displays.
In 1964 Bitzer and Slottow [4.9] realised that the current limiting impedance could also consist of a capacitor.
This precluded the use of DC since the discharge could not be maintained in the cell; capacitive coupling is, however, compatible with AC. The AC plasma display panel was presented in 1966 [4.9] and since then has become the dominant solution.

Many different devices belong to the family of Plasma Displays. As shown in figure 4.4, gas discharge displays are indeed classed as AC displays or DC

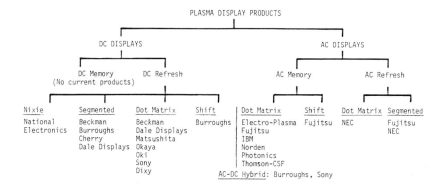

Fig. 4.4 Plasma displays products and suppliers [4.11].

displays [4.10,4.11]. Both can be operated in storage or non-storage modes, storage meaning that the memory is inherent to the display device, whereas in the non-storage or refreshed or cycling mode, an external memory is added to the display and the information is sequentially transferred and refreshed frequently enough to avoid flicker, like in any other "non memory" display.

4.3. AC plasma displays

4.3.1. General description

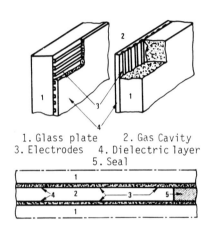

1. Glass plate   2. Gas Cavity
3. Electrodes   4. Dielectric layer
5. Seal

Fig. 4.5 AC plasma panel structure.

Most manufacturers produce AC PDPs of various sizes operating on the principle described hereafter [4.12].

Figure 4.5 shows a cutaway view and a magnified cross-section of a large AC plasma panel intended to display a great number of characters or elaborate graphics.

It consists of two identical glass plates (1) imprisoning a gas (2) in which the discharge takes place.

The inside surface of each glass plate carries rows of parallel conducting electrodes (3) insulated from the gas by a dielectric layer (4), made of materials which must also satisfy specific requirements, such as resistance to sputtering by ionic bombardment, lowered firing voltage, etc. In general it is not possible to have these electrodes made of transparent conductive material like Indium Tin Oxide due to the large surge current taking place when the discharge starts.

The plates are assembled with their electrode networks orthogonal to each other and with a small uniform gap between them. In some realisations a matrix of holes in a thin plate was used to localise the cells but in most products this has been avoided and the cells correspond to the intersections of the two sets of crossed rows of electrodes. The gap is first evacuated and then generally filled with a Neon-Argon mixture which emits a characteristic red-orange luminescence with each electrical discharge.

4.3.2. Operating principle

In operation, a square-shaped AC voltage called the sustaining voltage is permanently applied to all X and Y electrodes: figure 4.6a [4.10]. Its value is such that the electric field is not sufficient to initiate the discharge of the gas. In the absence of any other signal, the panel is in the "off" condition.

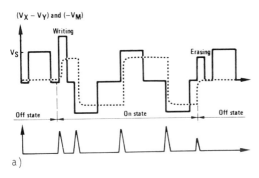

Ignition of a cell, defined by the intersection of a line and a column is achieved by applying between them a writing signal, in the form of an auxiliary instantaneous voltage of proper phase exceeding the firing voltage, which initiates the discharge: figure 4.6b.

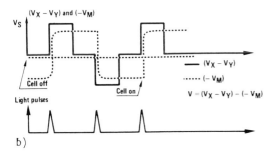

Fig. 4.6 Sustaining, writing, erasing voltages waveforms.

The ions and electrons generated by this discharge will build-up on the dielectric covering the electrodes, creating an opposite potential $(-V_M)$, and the actual potential across the gas will drop causing the discharge to be rapidly quenched. During the next half-cycle, when the sustaining voltage presents a reversed polarity, the potential of the walls has no longer an adverse effect, but on the contrary, will add to the sustaining voltage, so that the resulting voltage is then sufficient to exceed the firing potential. A new discharge will occur with corresponding electron and ion deposit and extinguish again; the addressed points, and only these, will then continue to fire twice per cycle (once per half-cycle) of the sustaining voltage. The luminance will therefore be proportional to the frequency of the sustaining voltage, up to an upper limit, where there is no longer enough time for the charges to build-up on the walls and the memory effect no longer exists.

Figure 4.7 gives the equivalent electrical model of this cell. Between the   two
external   electrodes   X   and   Y   are   two   capacitors   Cd, representing the two

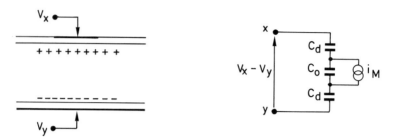

Fig. 4.7   AC PDP equivalent circuit (Thomson–CSF).

dielectric layers, and a capacitor, C_0, representing the cell itself. A current
generator I_M, at the terminals of capacitor C_0 represents the current flowing
at the time of gas discharge. The potential difference V at the cell   terminals
is   the   sum   of   the   externally   applied   voltage (V_x-V_y) and the voltage V_M
resulting from the charges stored in the   capacitor C_0.   This   voltage V_M is
called the internal or memory voltage. It follows that:

$$V = (V_x-V_y) \; \frac{Cd}{2C_0 + Cd} \; + V_M \qquad\qquad (4.6a)$$

$$V = \alpha \; (V_x-V_y) + V_M \qquad\qquad (4.6b)$$

The coefficient α is related the cell geometry, and is close to unity (≈ 0.95).
Thus:

$$V \approx (V_x-V_y) + V_M \qquad\qquad (4.6c)$$

Extinction of a cell is achieved similarly by applying a short pulse of the
proper phase to the appropriate pair of X and Y electrodes. A   short   discharge
(figure   4.6b)   takes   place   which   cancels   the   stored charges and hence the
corresponding memory potential $(-V_M)$. The next   half–cycle of   the   sustaining
voltage   will   find   the cell as if it had not been previously "on", and cannot
therefore ignite it. In other words, the panel has an inherent memory   and   any
point   can   be selectively written or erased. The written information is stored
without any need of "refreshing". The data input rate may be very low.

4.3.3. **Physical characteristics**

The shape of the display using a PDP is mainly determined by the dimensions of the panel itself. The thickness of the panel does not exceed 20 mm and its overall and useful areas are essentially related to the pitch of the cells and to their number. However operating constraints (high surge current, waveshape changes along long connections which affect the operation, etc.) impose that most of the drive electronics be placed immediately adjacent to the panel. The total thickness of panels with their drive electronics will be no more than 50 to 70 mm.

- Display size and capacity: the front dimensions of panels (in those where there is no requirement for rear optical access) can be arranged so as not to exceed the useful area by more than a few cm. Panels with 128x128, 128x256, 512x512, and up to 1212x1596; 1000x2000 or 1280x1728 cells are available from different manufacturers with cell pitches ranging from 0.57 mm to 0.27 mm and with associated electronics allowing alpha-numeric, semi-graphic, or full-graphic addressing.
  The dimensions of these panels range from 100x100 mm^2 up to 600x800 mm^2; the number of dels from 16,000 to almost 2.10^6 and in the alpha-numeric mode a display capability ranging from some several hundreds to 40,000 characters.
- Operating life and reliability: the operating life of AC panels is very long. Working hours in excess of 50,000 or 6 years of continuous operation and failure rates of less than 0,04 % having been demonstrated. This is more than enough for most applications.
- Ruggedness: the panels themselves are very sturdy and the associated electronics may be ruggedised without special difficulties. They -or rather their drive electronics- may be nuclear radiation hardened and high altitude operation is made easier than for CRTs for instance, by virtue of the much lower voltages involved.
  However, should the panel be exposed to low pressures, it would "inflate" due to the internal pressure of gas and some characteristics may change, resulting in erratic firing, or misfiring of cells. This effect remains limited provided a proper design has been made permitting the use of a lower pressure.
- Electromagnetic interference - security and shielding: like in other devices where high surge currents are switched, electro magnetic interference is produced. This EMI may be disturbing in some instances but it may also enable to decipher the flow of data being transmitted. In the case of AC plasma panels with their intrinsic memory where only new data are delivered in an acyclic form (as opposed to the periodic scanning of CRTs or scanned plasma

or EL panels) the information is written or updated only once and EMI is unintelligible.

If stringent requirements are to be met, special design of the panel and of the driver electronics may become necessary. This has been successfully achieved and the remaining radiation is so low that it is below most standards.

Another advantage of plasma panels is that they do not need shielding because the trajectories of the charged particles are too short to be deflected significantly enough to impair the display.

4.3.4. Addressing/driving

This description of operation just mentions the main principles. In fact, the firing voltage and the "memory margin" vary somewhat from cell to cell due to manufacturing tolerances. On the other hand, although the frequency of the sustaining voltage is only about 50 kHz, the waveshape, the impedance of the source and the multiplexing technique used have a direct impact on the performance of the panel.

Figure 4.8 shows the organisation used to drive an AC PDP. The circuit for generating the sustain waveform is separated from the address drivers. The address pulsers are a major system cost, since there must be one for each display electrode. In practice, they are packaged in an integrated circuit in groups of 32 (for example SN 75 500 and SN 75 501 from TEXAS or LIEB 4732 from THOMSON-CSF).

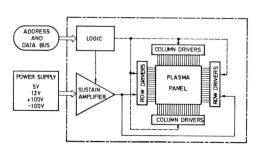

Fig. 4.8 Drive organisation of an AC PDP (Thomson-CSF).

These integrated drivers apply selection and sustain signals to logic signals and two high voltage the electrodes using low voltage input (100 V) amplifiers.

The general organisation of the electronics includes 3 main elements:
- The address IC drivers (for example 32 circuits on each face for a 1024x1024 PDP).
- A set of high voltage switches to generate the sustaining signals.
- A logic controller to receive the data and user commands and ensure their decoding and distribution to the above elements.

Connections between the edge of the panel and printed circuit boards can be made by a number of techniques. For AC PDPs the electrodes are usually connected by a pressure contact to a flexible circuit strip held in place by a spring clip.

In AC PDPs, the basic writing or erasing cycle is 30 μs long. During this 30 μs period, the addressing mode allows addressing of either an isolated del or all the dels of a complete line (line-at-a-time addressing). This last capability allows a 512x512 panel to be scanned in 15 ms, and entirely erased in 30 μs only.

The form of the <u>sustain</u> waveform has not significantly changed over the years. It swings between 0 and + 100 V and - 100 V with switching times much shorter than 500 nanoseconds and the current may reach 10 A peak for a 512x512 panel having 50 % of the cells lit. After the surge current has established the discharge, the sustaining voltage has to be maintained to 100 V (+ or - depending upon the phase) for several microseconds with no current.

On the other hand, the increase in size and resolution of PDPs has led to new waveforms for the <u>drive</u> signals. Instead of accurately calibrated width and amplitude square wave, the shape of the selective addressing signal is rather that of a ramp with an active edge slope if 10 V per microsecond with a resulting current of around 10 mA. The analysis of these characteristics emphasises the opportunity of isolating the sustaining circuits from the addressing circuits which operate at very much lower energy.

In the first generation of AC PDPs, multiplexing of these two signals was achieved with two diodes and a resistor associated with each electrode, plus a high voltage driving amplifier.

Integrated amplifiers with BIDFET output stages were used later, this output stage being common for the two signals. The impedance of that output stage limits however the current handling capability of the device and hence the size and resolution of the display, and to solve this problem on large size panels two circuits must be used in parallel, adding considerably to the cost while lowering the reliability.

In the integrated circuit described at the SID 1984 meeting [4.13] (figures 4.9 and 4.10), diodes D3 and D4, associated to switches AE + and AE - provide a separate path for the sustaining signal whereas the drive signals are supplied by DMOs

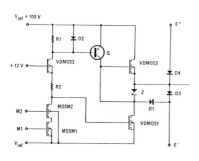

Fig. 4.9 Output stage of the driver for AC PDP.

transistors 1 and 2. The additional advantage of that solution is the use of
reduced silicon area.

A new possibility of simplifying the driver circuit would be the ISA
(Independent Sustained and Address) technique of Weber [4.14] which allows a
reduction by two –possibly by ten– of the number of drivers while retaining an
almost unchanged technology for the panel.

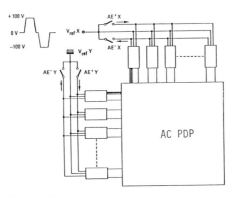

Fig. 4.10 Organisation of the
sustaining signal switches Open
VDMOS.

Another solution developed in several
places employs a coplanar structure in
which the electrode arrays are
deposited on the same (rear) glass
panel and are separated by a dielectric
layer. Therefore the discharge takes
place in their vicinity. In this
approach [4.15] each del is defined by
three electrodes; the address function
is ensured by two orthogonal
electrodes, independent of the sustain
electrodes, which are parallel to each
other. By combining this principle with
block multiplexing, substantial savings
can be made on the number of drivers so
that n del lines can be controlled by only $2\,n^{\frac{1}{2}}$ drivers. A panel with 240x80
dels controlled by 31 sustain drivers and 4 low power 20 bits ICs has been
presented.

The front glass sheet receives the phosphor dots which are excited by the U.V.
photons of a Xenon discharge but are not subject to ion bombardment which would
shorten their life. Additional advantages of that structure include the absence
of electrodes on the front glass, resulting in a better luminance and an easier
deposition of the phosphor.

4.3.5. System interface

Due to the close inter-relationship between panel and drive electronics, and
the highly significant part of the cost that the latter represents, most
manufacturers provide the user with a "display function" incorporating the
panel and factory-set drive electronics, which can be made very compact and
sturdy. The system electrical interface is then reduced to a power supply and
some TTL signals. In a typical alpha-numeric panel, for instance, these will
consist of:

• "line" and "column" address of the pixel (graphic panels) or of the character
  (alpha-numeric panels).

- ASCII code of the character.
- Control signals: "Writing", "Selective erasure" or "Total erasure".
- Luminance level control signal.

Graphic panels, of course, require more inputs.

Manufacturers offer a choice of panels with different sizes and organisation:
- alpha-numeric only where characters my be written in a predetermined format:
  (5x7; 7x9; 8x10; 10x10 display elements);
- graphic, where any element may be driven independently;
- or semi-graphic, where elements may be addressed independently but in a predetermined "block" of dels.

Typical specifications:

Input: TTL logic levels

Writing time or Selective erasure:

One dot	30 μsec.
One 5x7 character	150 μsec.
Bulk erasure	30 μsec.

Power consumption:

Discharge consumption	75	μW/cell
Capacitive consumption	10	μW/cell
Total power consumption		
lit cell	85	μW/cell
unlit cell	10	μW/cell

4.3.6. Visual characteristics

- Display quality: the most important visual feature of AC PDPs is the quality of the display: it is absolutely <u>flicker-free</u> thanks to the high supply frequency (2,000 times the eye's image fusion time, with two ignitions per cycle). No <u>image break-up</u> or stroboscopic effect (temporal aliasing) occurs in the case of relative motion between the observer and the display as opposed to "refreshed" displays (CRTs, scanned plasma panels or electro-luminescent panels). Each point is perfectly located so that no geometric distorsion occurs and the pattern remains steady and free from any jitter. Therefore, there is no noticeable fatigue of the eye even after a long observation time.

- Resolution: the commonly offered resolution (0.4 mm pitch) is already good
  enough to permit the display of alpha-numeric or graphic data observed from a
  typical distance of 400 mm without noticeable effect (spatial aliasing)
  resulting from the spatial sampling; this in accordance with the eye's
  characteristics. Large panels with increased resolution (0.27 mm pitch) have
  been designed and are in production.

- Viewing angle: the viewing angle is very wide, close to 180° with a small dip
  in luminance on the perpendicular to the panel. This is due, as for other
  matrix displays, to the presence of the column conductors on the front glass
  sheet which, even if transparent, partly hide the discharge.

- Rear access: by proper arrangement of the drive electronics it is possible to
  give access to the rear face for the addition of data, either by projection
  on a matte finish, or through direct superposition, by taking advantage of
  the transparent nature of both constitutive plates.

- Contrast ratio: as the glass sheets are transparent and as no diffusing
  material is in contact with them, like phosphor in CRTs or electroluminescence, the contrast is good even on bare panels. It may be further
  improved by the addition of an absorbing material at the rear, and an
  anti-reflection coating or a filter in front.

  Typical values are: under low level   illumination $C_r = 25$

  under 1000 lux illumination $C_r = 10$

  under 15000 lux illumination $C_r = 2.5$

- Luminance: the display element peak luminance is about 200 or 300 cd m^{-2}.
  This does not allow usage for the most demanding applications in e.g. the
  fighter aircraft cockpit, although it is quite sufficient for some
  applications in transport vehicles and aircraft like fuel management, area
  navigation, and of course all crew compartment use and in ground based
  systems.
  The device having an intrinsic memory, the luminance is totaly independent of
  the rate at which the information is delivered.

- Dimmability: dimming can be achieved by two ways: either by changing the
  sustaining voltage frequency, thus varying the luminance of the whole
  display, or erasing and re-writing selectively only a part of the display
  which gives "shades of grey". Dimmability can go up to a ratio of 10 to 1.

– Half-tones: the first half-tone images were produced with DC panels but the need for half-tones is specially apparent with large screens and therefore AC panels. Attemps have been made [4.16,4.17] on AC panels in spite of their intrinsic memory which apparently precludes the possibility of obtaining shades of grey with them.

Thanks to a new (patented) principle which relies on a temporal modulation of the on time of a cell so that the average brightness perceived by the eye varies as required, it is now possible to obtain 16 and even more half-tones. If the averaging time T is divided into N parts of proper durations, 2^N shades of grey may be obtained (figure 4.11).

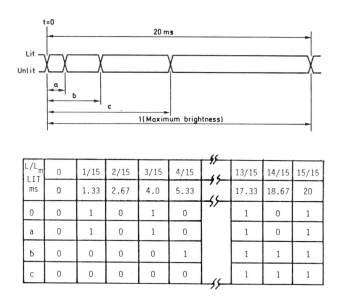

L/L$_m$ LIT ms	0 0	1/15 1.33	2/15 2.67	3/15 4.0	4/15 5.33		13/15 17.33	14/15 18.67	15/15 20
0	0	1	0	1	0		1	0	1
a	0	1	0	1	0		1	0	1
b	0	0	0	0	1		1	1	1
c	0	0	0	0	0		1	1	1

Fig. 4.11 Principle of digital grey shades on AC PDP.

During one image time T, each dot of the screen must be addressed N times. The access time to one line is T1, the number of lines of the panel N1. Then

T = T1.N1.N

T is defined by the human eye and must be shorter than 20 msec.

T1 is defined by the drive electronics and at present is 25 µsec.

Under these conditions, the addressable number of lines is 800 and for a 200 lines panel 2^4 = 16 levels are obtained.

This only requires the use of a RAM with a capacity equal to the product of the number of pixels of the panel multiplied by the number of bits for levels

coding (e.g. 512x512 dots; 16 levels = 2^4; requires 1 M bit capacity).
With an additional video-logic interface which writes in the memory the logic
signals, the display may be considered as a complete substitute to a
conventional video monitor (figure 4.12).

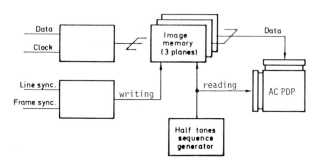

Fig. 4.12 Half-tones AC PDPs driving diagram (Thomson-CSF).

- Colour: the colour orange (λ_{avg} = 600 nm) is typical of the neon-argon gas
mixture, it is agreable and not tiring to the eye.

No full colour AC PDPs are available industrially although the feasibility of
multichrome panels has been demonstrated [4.18,4.19,4.20].
In the multichrome approach, sets of three properly locates pixels (triads or
stripes) are coated with different phosphors, and the sensation of hue
results from the addition, within the eye, of the three stimuli so obtained;
in the very same way as in a shadow-mask CRT. The price paid for colour
capability is a drop in resolution, since at least three luminescent sites
are needed to present one perceived
element like in other colour displays.

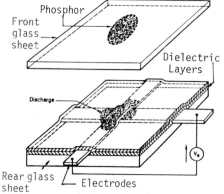

Fig. 4.13 "Single substrate"
AC colour plasma display panel
(Thomson-CSF).

One colour panel under development in
France employs a single substrate
structure for the X and Y electrodes
which are separated by a dielectric
layer and are addressed like
conventional AC PDPs, (see figure
4.13). Each "white" element is in fact
made of 4 dots: one green, one blue,
where the eye has a good sensitivity
and two red ones where the eye
sensitivity is lower as well as the

luminous efficiency of phosphors: see figure 4.14. This improves the colour balance and the overall luminance. With the 3 basic colours, 8 hues may be

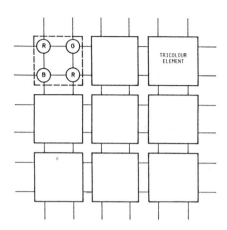

obtained, green, yellow, red, magenta, blue, cyan, white and black. Adding two half-tones for each basic colour leads to 4 hues. Taking advantage of the "split time" half-tone scheme described above results in 512 colours on a 256 lines panel at video rate.

4.3.7. State of development

Invented more than 20 years ago, AC panels have now reached industrial maturity and are extensively employed. First uses were found in non-military applications, like the PLATO automated teaching system, for which they were indeed first designed, and banking and air traffic control terminals. They are

Fig. 4.14 Principle of full-
colour matrix displays
(Thomson–CSF).

finding widening acceptance in the military field, mainly in Command and Control functions, such as in submarines, Fire and Air Support Systems, Computer Systems; and in several Electronic Counter Measures (ECM) systems.

One interesting application is the SAPHIR network of the french "Gendarmerie" (country police). In this system all the cars down to the small patrol cars will be linked to their platoon and up to the headquarters by radio and wire so that they will be able to consult in real-time the local and central computers where reported crimes and offenses are recorded. This programme will need about 10,000 PDPs.

Probably the largest PDP production facility in the world has a capacity of more than 50,000 panels per year. A typical office display has a resolution of 960x768 dels with a useful area of 340x274 mm^2 for a thickness of 6.4 cm including drive electronics; it can display letters, drawings, charts or photographs. Another product can display 9920 characters (62 rows by 160 columns) or 5250 characters (50 rows by 105 columns).

Plasma display panels generally are used in the alpha-numeric mode, although in some applications provision for semi-graphic or even full-graphic addressing is made, allowing cyrillic, arabic, katakana or kanji characters to be displayed, together with symbols and drawings. Some of these may be superimposed on a map placed at the rear face. No pictorial TV applications are known to have been developed. Once again, the limitations do not come from the panels themselves, but rather from the associated electronics.

To summarise, one can say that at present, AC panels with dimensions of up to 1 m wide are available [4.21], panels with resolutions of 0.3 mm (83 l/inch) are available and up to 0.2 mm have been demonstrated; alpha-numeric and graphic displays without half-tones are common-place, and video rate half-tone displays have entered production. Military environment extremes are, or can be fulfilled, and colour and full-image characteristics have been demonstrated and begin to be produced at the expense of somewhat complex electronics which remains the most difficult and expensive point of AC PDPs even now that high voltage integrated circuits are available in technologies convenient for both plasmas and Thin-Film Electroluminescence.

Although the memory AC PDP is largely dominant in the AC type, the "shift" versions have to be mentioned [4.22,4.23,4.24]. Like all AC panels they have memory and as a result there is no flicker and the luminance is independent of the rate of data arrival. They have been designed to reduce the number of external connections and use either the priming effect or the wall charge coupling effect.

In the first type, the two sets of X and Y electrodes are bussed for the whole panel. They are shaped in a meander form and shifted half a pitch with respect to each other. By properly changing the time at which the appropriate waveform is applied, the discharge is shifted along the lines thanks to the priming coupling effect similar to that described in the DC panels section. The initial discharge is produced by a third set of electrodes, the writing electrodes: one per line. Once written and shifted to the final position the display is maintained by application of another waveform.

The shifting process is somewhat slow and limits the data flow as well as the number of dels. Again the advantage in the number of drivers and connections has been swapped with increased panel complexity and performance loss.

4.4. DC Plasma displays

4.4.1. General description

Most of what has been said for AC plasma displays applies to DC plasma displays. The major differences with the description of the former, given in 4.3.1, are that the electrodes or resistive extensions thereof are directly immersed in the gas mixture, and that in most cases it is necessary to add a stencil plate. The holes of the stencil plate register with the crossings of the electrodes and thereby localise the cells. This is generally necessary to limit the consequences of electrode sputtering by ion bombardment, or for separating the display discharge from the transfer discharge (see below). This obviously adds to the difficulties in manufacturing the panel.

4.4.2. Operating principle

Various principles can be used to operate DC gas discharge panels, but basically they are supplied with a unidirectional voltage, either pulsed or sustained, the latter giving the memory mode [4.25,4.26,4.27,4.28].
Individual cells are struck "on" by application of a voltage pulse in the range 150-200 V. As soon as a stable discharge has been established, the cell impedance drops significantly, and it is therefore necessary to provide a series resistor to limit the current flow in the cell.
Most commercial matrix panels are multiplexed in the pulsed mode and because only one line of the display is addressed at any one time, it is necessary to provide a resistor in series with each column driver. This drive technique however, yields a rather low luminance level (<150 cd m^{-2}). Where higher display luminance is required, it is necessary to operate the display in the memory mode, in which an addressed cell is active for almost the whole of the frame time (section I.1.5), thereby yielding a much higher mean brightness, and considerable efforts have been made in that direction [4.29,4.30].
The memory effect is achieved by striking a cell "on" as described previously, and then reducing the voltage to a level high enough to sustain the discharge, but too low to ignite adjacent cells. The main disadvantage with this drive scheme is the necessity of a limiting resistor in series with each individual cell in the display.
The difference between the strike voltage and the maintaining or sustaining voltage is known as the working margin. The spread which occurs in these two voltages due to the manufacturing tolerances presents a further difficulty for memory operation. Degradation of the working margin results in non-uniform operation across the display.
In fact, no DC panels with memory are known to be in production.
A third problem is the statistical time lag between application of the trigger pulse and the establishment of a stable discharge in the cell – an effect known as jitter. This poses no difficulties for the AC plasmas thanks to the high frequency of the sustaining voltage but does pose problems with operation in the memory and the pulse modes of DC plasmas. In order to minimise jitter, it is necessary to prime the cells by introducing ions or electrons into the cell, prior to the application of the strike voltage. This is normally achieved by incorporating priming cells into the panel in close proximity to the data cells. The priming cells are blanked from view, so that their light emission cannot be observed.
The priming effect, which is the ability to use one discharge to affect a

neighbouring site, and thus manipulate the position of the discharge, is unique to gas discharge displays. Figure 4.15a illustrates how the striking voltage V_B

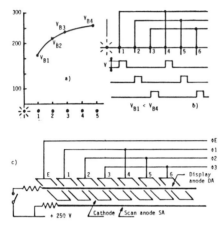

varies with the distance from an active site: the closer they are, the lower V_B is, which is understandable when one recalls that in the physics of initiation of the discharge, the larger the number of remaining charged particles, the earlier the break-down occurs. Therefore, by just using a three phase voltage variation applied on scan electrodes it is possible to shift the discharge from cell to cell: figures 4.15b and 4.15c [4.31].

The priming effect is also used in a device more recently described, able to display several rows of 80 characters and intended for small basic word-processors [4.32].

Fig. 4.15 Priming and shifting mechanism (Thomson-CSF).

The scan mode, however, suffers several limitations regarding display size, brightness and number of displayed elements, which restricts its application mainly to cash registers and banking terminals. In order to simplify the electronics, these panels are activated by a multiplexed scanning signal along the lines and colums of the matrix. The frame scan frequency must exceed about 50 Hz and a line cannot contain more than 300 dots, otherwise flicker becomes quite noticeable and the display dims. Therefore, these panels are generally limited to purely alpha-numeric display of about forty 5x7 dels characters per row.

To overcome some of these limitations. a modified type of operation has been described [4.33], combining AC 16.7 kHz operation for memory display and DC drive.

Figure 4.16 is a cutaway view of the design, and figure 4.17 shows the waveforms for the different voltages. As can be seen they are quite similar to those shown in figure 4.6.

A thin metal plate with holes in it, called the priming plate, separates the discharge space into a front memory section operated in the AC mode, and a rear, DC operated, scan section.

The front plate is coated with a transparent continuous conductive layer, itself coated with an insulating layer. The stencil priming plate is also covered with an insulating layer. The result is that all cells in the memory layer are connected in parallel. Each cell is isolated from its neighbours by the glow isolator mesh. These two insulator covered plates, when fed with an AC voltage provide the memory capability. Scan anodes are located at the bottom of

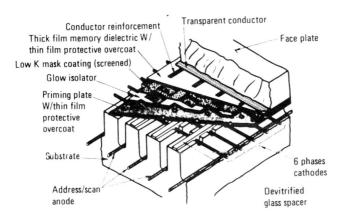

Fig. 4.16 Self-scan with memory display panel structure (SID).

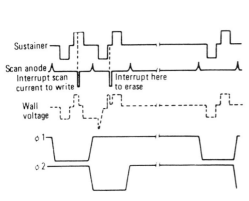

Fig. 4.17 Self-scan with memory addressing waveforms (SID).

grooves in the rear plate. Six phase cathode wires can be seen in the figure, orthogonal to the anodes and isolated from the priming plate by glass stripe spacers. Operation of the scan section is similar to that of earlier refresh type "periodic shift" displays, except that scan is initiated only when changes are required in the display content.

This configuration did claim to allow the display of over 96 characters per row without flicker, with improved luminance and mutiplexing possibility resulting in a reduction of the number of address lines to about one tenth of those required to address each del. Unfortunately, it has not been demonstrated that this economy was large enough to balance the added cost in manufacturing the panel, nor whether this design might be compatible with displays other than alpha-numeric, and the projected venture which aimed at producing it collapsed.

4.4.3. **Physical characteristics**

– The front dimensions of DC PDPs together with the associated electronics, do
  not significantly differ from the useful area, like for AC panels, and their
  thickness is also in the order of several centimetres.

– Display capacity: the commercially available panels have a display capability
  from 1 row of 16 characters up to 12 rows of 80, while an alpha-numeric and
  limited graphic unit contains 17 lines and 192 columns. A panel of 128x128
  elements has been reported. The number of characters then ranges from 96 to
  960 and the overall length is around 30 cm.

– Resolution: generally resolution is poor (see 4.4.6) typically 0.75 mm
  horizontal; 1 mm vertical. This may be considered adequate for alpha-numeric
  character display but is unlikely to give sufficient resolution for vector
  graphic display. Further development of the technology to provide finer
  pitches is likely to be limited by the mechanical tolerances of the panel
  components, crosstalk caused by the close proximity of adjacent cells, and by
  tolerances in the resistor arrays.

– Operating life: the operating life expectancy is 30,000 hours under normal
  conditions. However this can be drastically impaired to, say 100 hours for
  panels operating below 10 C, and which contain mercury, as it is usually the
  case for DC panels; and for these at least, no operation at all is possible
  below 0 C. The Hg vapor in the gas mixture is necessary to avoid sputtering
  of electrode material. At low temperatures, the Hg protective effect is
  greatly reduced, therefore bringing about increased electrode deterioration
  rate.

4.4.4. **Addressing/driving**

As has been seen in "Operating Principle" there are several different ways for
operating DC panels, and the requirements are still more closely related to the
panel design than for AC operation. Therefore the vast majority of DC panels
are supplied with the drive electronics (except perhaps for some single
character-row types) and the interface is reduced to supply and logic.

4.4.5. **Visual characteristics**

– Display quality: as in all other matrix displays, the picture is free of
  geometric distortion and each element is precisely located.
  The major pitfalls of these panels when operated in the pulse mode are the

flicker, which imposes a refresh frequency of typically 70 Hz thus limiting the number of cells per line as mentioned, and a display "break-up" under strong vibrations. This is particularly true for panels with no phosphor coating, but in spite of the decay time added by phosphors, it remains objectionable to DC panels where such a coating is used.

– Viewing angle: in most designs, holes in a stencil sheet define the cells. If the cathode, and hence the discharge, is not located on the front surface (or on top of the hole) the viewing angle may be lower than for AC panels, not exceeding a cone of 100° to 120°.

– Colour: the colour is neon–orange or green, the latter being obtained by the use of suitable U.V. activated phosphors. The phosphors are coated onto the cell walls and are activated by the U.V. emission in the gas discharge, in which case Xe is incorporated in the gas to enhance U.V. emission. Several manufacturers market Ne based displays incorporating phosphors which emit in the green. This however, further restricts the viewing angle, and produces a colour shift from green to reddish yellow and red when changing from the orthogonal to the lateral viewing position. A matched green filter should obviate this effect, and since the contribution from the glow to the total luminance is small, the reduction in luminance should be small too.

Work on multi–colour PDP began in the seventies almost only on DC panels [4.34,4.35]. These devices used the anode discharge "positive column" (like in fluorescent lamps) which yielded luminous efficiencies around 1 ℓm W^{-1}. The most impressive results obtained are those of NHK which demonstrated a 16 inch diagonal 240x320 dels trichrome display [4.36]. The complexity and poor resolution of DC PDP precluded further development.

– Luminance: the luminous efficiency (0.5 ℓm W^{-1}) yields a mean pulse mode driven panel luminance in the range of 100 – 150 cd m^{-2} which, as stated previously, is inadequate for good legibility in conditions of high ambient illumination. Some panels, which are pulse driven, have a quoted mean luminance of 60 to 100 cd m^{-2}.

For very high luminance displays, memory addressing techniques are used, as described in [4.25] for an experimental display. The panel can be operated at a luminance as high as that found in CRTs, but the current density required to do this involves excessive sputtering of the cathode material, and the life of the panel is significantly reduced.

Panels displaying 7 rows of 15 characters each having a 7x5 format, which can be operated in the memory mode at up to 1,000 cd m^{-2} have been produced. The basic inefficiency of the discharge mechanism, combined with energy

dissipation in the series resistors, means however that considerable heat   is
dissipated at these high luminance levels.

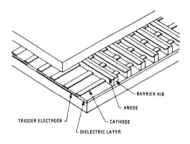

Fig. 4.18a   DC PDP with trigger electrode structure.

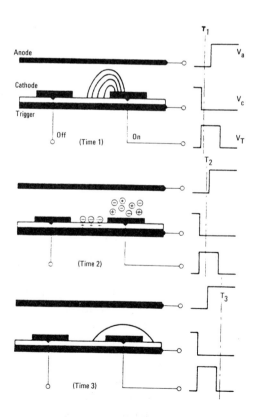

Fig. 4.18b   DC PDP with trigger electrode; principle of operation (SID).

- Dimming: a dimming ratio of 5:1 for the priming shift (self-scan) displays has been quoted, but because of the visibility of the scan glow, legibility in dark environments is impossible. On the other hand, to satisfy the goal of e.g. aircraft cockpit use, it is necessary that panels be sufficiently bright to be legible under 100,000 lux and should have a dimming ratio of 1,000:1. Drive circuits for high luminance panels have been developed in the U.K. which enable dimming ranges greater than 500:1 to be achieved.

4.4.6. State of development

Existing DC plasma panels have been specifically developed for the display of a few hundred alpha-numeric characters, and have found wide applications mainly in bank teller's terminals, industrial process management, etc. There are at present no commercial panels suitable for the display of vector graphic information.

A DC panel capable of a very high resolution (0.2 mm pitch) and operating with low driving (180 V) and switching (3 - 50 V) voltages has been described [4.36] and is now in production. Moreover the number of drivers can be significantly reduced.

It is based on the addition of a set of trigger electrodes on the back of the panel, isolated -and therefore capacitively coupled- to the cathodes (see figure 4.18a). When an appropriate voltage is applied between the trigger electrode and a cathode (figure 4.18b) charges build-up on the wall in a short time, up to a voltage where the trigger discharge stops in a way very similar to that of AC panels. But if an anode voltage is raised from its biased level to a value V_p high enough, the trigger discharge is transferred between cathode and the anode and a glow discharge takes place within a few microseconds.

V_p can be set at a value lower than the striking voltage and higher than the extinction voltage without trigger assistance. A panel with a single trigger electrode can therefore operate on low supply and switching voltages. In case of multiple trigger electrodes associated with several cathodes, multiplexed operation is possible with the additional advantage of fewer cathode drivers. As the panel structure is fairly simple high resolutions are also posible.

In an experimental 512x256 dels panel a resolution of 0,3 mm has been achieved, a luminance in excess of 80 cd m^{-2} and a contrast of 50:1 have been obtained at 60 frames per second; 512 anode (column) drivers, 32 cathode (lines) and 16 trigger drivers were used. Production types contain 640x400 dels on 9 or 12 inches (0.3 or 0.4 mm resolution). Viewing angle is 115° laterally and 140° vertically; luminance $\approx$ 100 cd m^{-2}; power consumption less than 20 W.

The luminance of displays presently available from the major manufacturers of DC panels is inadequate for full sunlight viewing, but it is hoped that panels under development will have sufficient luminance while maintaining the characteristics of life expectancy and gaining temperature independence.

Market surveys by Arthur D. Little or SRI estimated the total PDP market for 1985 at US $ 50 Millions with 1/3 to the DC family and 2/3 to the AC type.

This is understandable due to the importance of the military market in the US which can be satisfied by AC PDPs exclusively whereas DC PDPs are more "industrial appliance" oriented.

To summarise, DC PDPs have up to now been produced for non-military, alpha-numeric applications with limited size, number of characters, and moderate resolution. Although they have the inherent capability of displaying half-tones, this seems to have been little employed. Their high brightness capability in the memory mode may be hindered by heat dissipation and still more by life expectancy problems, particularly at low temperature.

4.4.7. Plasma electron excited CRT

To this point we have described display devices that make use of the light output of the discharge to present data. Different workers, coming back to the early days of the Crookes tube or the Braun oscillograph tube have tried to take advantage only of the availability of free electrons in a plasma to use the discharge not as a light source but as a large area cathode.

By combining plasma discharge electron generation, high luminous efficiency of cathodo-luminescent phosphors and digital addressing techniques, A. Schauer

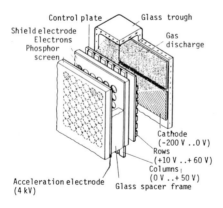

Fig. 4.19 Flat CRT plasma electron excited (IEDM).

[4.37] succeeded in producing a hybrid tube between the CRT and the plasma display panel.

It offers high luminance, contrast, resolution, and efficiency, colour compatibility, large size and slim profile (figure 4.19).

Basically, the laboratory model consists of a rear chamber where a DC gas discharge is maintained and a front acceleration chamber separated by a control plate bearing mullion and transom electrodes for line and column selection. The discharge produced between the back electrode (cathode) and the line electrodes on the control plate provides the electrons which will be accelerated towards the screen after selection through holes in the control plate at line-column electrodes intersections. They will impinge on phosphor dots which may deliver different colours. The distances between cathode and control plate (25 mm) on the one hand and control plate and screen (1 mm) on the other, are such that according to the Paschen law a discharge cannot take place in the front chamber whereas it does in the rear one at the selected voltages (200 V and 4 kV) and pressure (1 millibar) (figure 4.20).

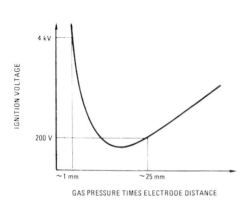

Fig. 4.20 Paschen law for plasma CRT (IEDM).

Many attempts have been made to excite phosphors by U.V. light emitted in gas discharges but luminous efficiency – 0.1 ℓm W^{-1} is rather low, whereas electron excitation of phosphors at 4 kV leads to an efficiency of 6 ℓm W^{-1} and a total power requirement of only 20 W. One of the models presented has a 350 mm (14") diagonal for a total thickness of 60 mm. It displays a matrix of 448x720 (322 560) dels with a horizontal pitch of 0.32 mm and a vertical one of 0.4 mm (3.1 and 2.5 dots per mm). Driving requires voltage swings of 50 V on both lines and columns and to avoid any flicker a refresh frequency of 80 Hz had to be choosen so that drive electronics has to run at 27.2 MHz.

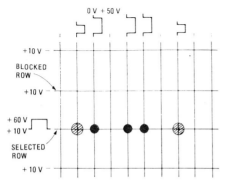

Fig. 4.21 Driving voltages for half-tones.

However this figure is divided by 2 by splitting each set of electrodes in two groups and putting the drivers at both ends. Special Dimos (double implanted MOS) ICs have also been developed that properly operate the panel. Luminance reaches 200 cd m^{-2} and contrast ratio is 20:1 due to the visibility of the faint rear glow through the holes. Two half-tones are obtainable by column pulse width modulation rather than amplitude modulation which would result in non uniform currents and hence luminance, see figure 4.21.

REFERENCES

[4.1]   Weston, G.F., Cold cathode glow discharge tubes, London IL-IEEE Books Ltd, chap. 1 + 9, (1968).

[4.2]   Maynard, F.B., Carluccio, J. and Poelstra, W.G., Grid switched gas tube for display presentation, Electronics, vol. 29, (1956).

[4.3]   Arthur D. Little, Large area flat panel display, Market & Technology 1982-1992 report.

[4.4]   Hirsh, M.M. and Oskam, H.J., Gaseous electronics, Electrical Discharges, Academic Press, vol. 1, (1978).

[4.5]   Deschamps, J., What does the future hold for plasma panel? Proceedings Eurodisplay, Munich VDE Verlag GmbH, Berlin (1981).

[4.6]   Jackson, R.N. and Johnson, K.E., Gas discharge displays, a critical review. Advances in Electronics and Electron Physics, NY and London, Academic Press, vol. 35, (1974).

[4.7]   Lear-Siegler Inc., Development of experimental gas discharge display, Progress reports on Contract NOBSR 89201 Bu. Ships. August 1963 - June 1965.

[4.8]   ..... Coordinated science laboratory progress report, Sept. - Oct. - Nov. 1964, University of Illinois (1965).

[4.9]   Bitzer, D.L. and Slottow, H.G., The plasma display panel-A digitally addressable display with inherent memory, 1966 Joint Computer Conference (San Francisco), AFIPS Conf. Proceedings, vol. 29, (1966), p. 541.

[4.10] Sobel, A., Gas discharge display: the state of the art, IEEE Transactions Electron Devices, vol. ED-24, no. 7, (1977), pp. 835-847.

[4.11] Weber. L., SID Seminar notes, vol. 1, (1983).

[4.12] Slottow, H.G., Plasma displays, IEEE Transactions Electron Devices, vol. ED-23, no. 7, (1976), pp. 760-772.

[4.13] Delgrange, L., A high voltage IC driver for large area AC plasma display panels, SID Symposium Digest, (1984), pp. 103-106.

[4.14] Weber, L.F., Younce, R.C., Independent sustain and address technique for the AC plasma display panel, SID Symposium Digest, (1986), pp. 220-223.

[4.15] Shinoda, T. et al, Green surface discharge plasma decode display, IDRC, (1985), p. 51.

[4.16] Judice, C.N., Jarvis, J.F. and Ninke, W.H., Using ordered dither to display continuous tone pictures on AC plasma panel, Proceedings SID, vol. 15, (1974), pp. 161-174.

[4.17] White. A.B., Johnson, R.L. and Judice, C.N., Animated dither images on AC plasma panel, Record SID Biennial Display Research Conference, (1976), pp. 35-37.

[4.18] Stredde, E., The development of a multi-colour plasma display panel, Coordinated science laboratory report R 730, University of Illinois (1967).

[4.19] Brown, F.H. and Zayac, M.T., A multi-colour gas discharge display panel, Proceedings SID, vol. 13, (1972), pp. 52-55.

[4.20] Hoehn, H.J., Recent developments on three colour plasma display panels IEEE Transactions Electron Devices, vol. ED-20, Nov. (1973), pp. 1078-1081.

[4.21] Willis, D.R., Johnson, R.L., Ernsthausen, R.E. and Wedding, D.K., Large area displays, Proceedings Eurodisplays Munich, VDE Verlag GmbH Berlin (1981), pp. 191-194.

[4.22] Umeda, S. et al, Self shift plasma display, SID Symposium Digest, (1972), p. 38.

[4.23] Andoh, S. et al, Self shift plasma display with meander electrodes, SID Symposium Digest, (1977), p. 78.

[4.24] Ngo, P., Temporal distribution of space charge on plasma cell discharge, IRDC, (1985), p. 86.

[4.25] Jackson, R.N. and Johnson, K.E., Address methods for gas discharge panels, IEEE Transactions Electron Devices, vol. ED-18, (1971).

[4.26] Smith, J., Experimental storage display panel using DC gas discharge without resistors, IEEE Transactions Electron Devices, vol. ED-22, (1975), pp. 642-649.

[4.27] Holz, G.E., Pulsed gas discharge display with memory, SID Symposium Digest, (1972), pp. 36-37.

[4.28] Lustig, C.D., Pulsed memory mode for gas discharge displays, Proceedings IEEE, vol. 61, (1973).

[4.29] Mikoshiba, et al, IEEE Transactions Electron Devices, vol. ED-26, no. 8, (1979), pp. 1177-1181.

[4.30] Murakami, H., et al, An 8-in diagonal pulse discharge panel with internal memory for a colour TV display, SID Symposium Digest, (1984) pp. 87-90.

[4.31] Cola, R. et al, Gas discharge panel with internal line sequencing ("Self-Scan" Displays), Image Pick-up and Display, vol. 3, (1978), p.83.

[4.32] Smith, J., A gas discharge display for compact desk-top word-processor, Record SID Biennial Display Research Conference, (1980), pp. 79-82.

[4.33] Holz, G., Ogle, J. et al, A "Self-Scan" memory plasma display panel, SID Symposium Digest, 28-30, New York (1981).

[4.34] Okamoto, Y, Mizushima, M., IEEE Transactions Electron Devices, vol. ED-27, no. 9, (1980), pp. 1778-1783.

[4.35] Kamegaya et al, IEEE Transactions Electron Devices, vol. ED-25, no. 9, (1978), pp. 1177-1181.

[4.36] Kojima T. and al, Proceedings SID, vol. 20/3, (1979), pp. 153-158.

[4.37] Schauer, A., A plasma electron excited phosphor flat panel display, IEDM Technical Digest, (1982), pp. 304-307.

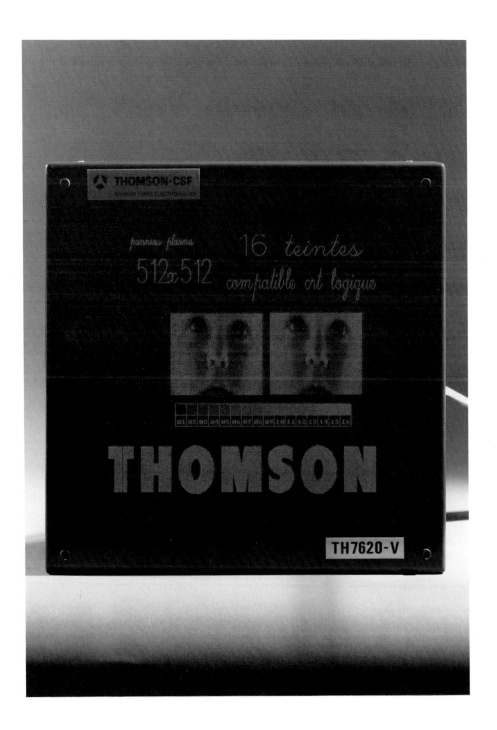

AC Plasma Panel, 512 x 512, 16 half tones. Photograph: Thomson-CSF.

DISPLAY ENGINEERING: D. Bosman (Editor)
Elsevier Science Publishers B.V. (North-Holland), 1989

II.5: ELECTROLUMINESCENT DISPLAYS

B. GURMAN
U.S. Army Avionics R&D Activity, U.S.A.

5.1. Historical survey

Electroluminescence (EL) is defined as the light generated by a phosphor under the influence of an electric field. It is the electric field, as the source of energy, that distinguishes electroluminescence from other light emission phenomena such as chemiluminescence (chemical energy), incandescence (heat energy), phosphoresence (light energy), etc. Electroluminescence occurs in two forms known as injection electroluminescence and high field electroluminescence. In injection electroluminescence, light energy is released upon recombination of minority and majority carriers across the band gap in a single crystal solid. This process is the basis of light emitting diodes and diode lasers (see chapter II.6). The remainder of this chapter will deal with high field electroluminescence in which the emission is due to impact excitation of luminescent centers by charge carriers which have been accelerated by high electric fields within the solid. Depending upon the phosphor material and electrode geometry, light may be generated by an alternating (AC) or continously applied (DC) field. The phosphor layer may be fabricated from powder EL materials by screening, spraying or sedimentation or it may be deposited as a thin-film by evaporation, sputtering or other vapor deposition technique. Therefore four types of high field EL devices are possible, namely, AC powder, DC powder, AC thin-film and DC thin-film. This variety of EL devices arises from different means of achieving the high field and protecting the device from destructive avalanching currents produced by the high field.

Significant advancements in the state of the art of EL displays have been achieved during the past decade. Although powders have much improved and are used in certain vehicle and household applications, and also for instrument and liquid crystal display backlighting, the main breakthrough has been in thin-film phosphors. The most stable thin-film devices utilise high quality dielectric layers on both sides of the phosphor to limit avalanche currents and therefore these devices must be operated with AC drive voltages. The thin-film structures are highly transparent allowing the use of an absorbing black rear layer for contrast enhancement. A highly non-linear luminance versus applied voltage relationship in the thin-film EL allows for direct matrix addressing. Because the AC thin-film EL (TFEL) devices currently represent the greatest investment in time and money, moreover to many as having the most application potential, sections 5.2 through 5.6 will focus principally on AC TFEL. The powders and DC thin-films will be discussed in section 5.7.

The electroluminescence (EL) phenomenon was first discovered by Destriau in 1936, in zinc sulfide (ZnS) with copper (Cu) used as an activator [5.1]. This discovery led to several efforts to apply the EL phenomenon to lighting. During the 1950s and 1960s, extensive research was carried out to fabricate EL displays using phosphors made by heating (fusing) ZnS:Cu powders. The powdered EL, however, could not be made daylight readable, nor could the brightness levels achieved be maintained for even moderate lifetimes. The limited life resulted from the migration of the activator (Cu) under the influence of the applied electric field. Further, the mechanism of destruction was fast, even at moderately high temperatures. On the other hand transparent thin-film phosphors were made as early as 1934 by DeBoer and Dippel [5.2], and later good quality films were made by Williams in 1947 [5.3], and Feldman and O'Hara in 1954 [5.4], though not for EL display applications. The earliest thin-film EL phosphors were devised by Thornton [5.5] in the late 50s, while the matrix addressing concept for EL panels was patented in 1955 by Piper [5.6].

Of the various thin-film devices being developed for display application, the two most advanced in terms of practical characteristics are the three layer, double insulated, AC coupled, ZnS:Mn thin-film device first developed by Soxman [5.7] (see figure 5.1). In this configuration, which is now the standard

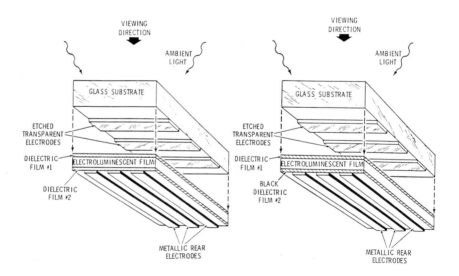

Fig. 5.1 Three layer, AC coupled ZnS:Mn thin-film device with 2 transparent dielectric layers.

Fig. 5.2 As figure 5.1, but with a black, light absorbing, rear dielectric film.

construction, the active light-emitting layer is sandwiched between two transparent high-dielectric-strength layers, and interposed between thin-film electrodes, at least one of which is a transparent electrode on a transparent glass substrate. When viewed from the front, the substrate is transparent through to the rear electrode. The front and rear electrodes are perpendicular to each other to form a crossed matrix pattern.

In the late 60s, Soxman reported on reproducible AC thin-film matrix EL devices having over 50 cd m^{-2} and several thousand hours of life [5.7]. However, because the rear electrodes were reflective, the contrast ratio was poor. It was suggested that if the rear electrodes could be made transparent and backed with a diffuse black material, or if the second dielectric layer were to be made highly absorbing, the unexcited device could appear quite black, thus providing good contrast in high ambient illumination (figure 5.2).

The advantages of the black layer were explored by several other researchers during the mid 60s using powders and thin-films. The first thin-film EL display with a black layer was produced in the 1960s by Soxman and Steel, [5.8] who reported on some samples having in excess of 160 cd m^{-2} luminance with flat maintenance (see 5.3.2). Major advances were reported at the 1980 SID International Symposium in San Diego. Inoguchi et al [5.9] presented data on a three layer thin-film ZnS:Mn EL display exceeding 10,000 hours of life at 340 cd m^{-2} luminance with flat maintenance. Mito et al [5.10] presented a second paper describing the reproduction of TV imagery on a 108x81 line TFEL panel. However, these displays did not include a black layer. Because they used reflective aluminum rear electrodes with a front-mounted circular polariser for contrast enhancement, the visual performance was marginal. The performance of thin-film EL devices using ZnS:Mn, as demonstrated by Soxman and Inoguchi et al, has since been confirmed by many others [5.11].

5.2. Principle of operation

The thin-film electroluminescent display is, as described above, made-up of a sandwich structure of conductors and dielectrics with a luminescent phosphor in the center. The thin-films are deposited on a glass substrate, often starting with an Indium Tin Oxide transparent conductor followed by a dielectric of high electrical break-down strength, followed by the manganese-doped zinc sulfide phosphor, the second dielectric, and finally, the rear conductor. As stated earlier, a black thin-film layer may be incorporated to the rear of the zinc sulfide phosphor before the rear electrodes as a means of providing contrast enhancement. The basic principles of operation are as follows. While the mechanism is not totally understood, it is generally agreed that when a high electric field (up to $1-2 \times 10^5$ V cm^{-1}) is applied to the sandwich structure, the ZnS layer, also a dielectric, breaks-down into avalanche conduction and current flows through it to the encapsulating dielectric interface. As the charge builds-up on the dielectrics, the internal field in the ZnS is reduced and conduction ceases until the field is reversed. Thus, one has a pulsed avalanche conduction of the ZnS for each field reversal. The break-down in the ZnS is apparently initiated by electrons tunneling out of the interface sites into the conduction states of the ZnS.

Another way of looking at the conduction mechanisms described above, which may be clearer to some readers, incolves the concept of the band gap. Figure 5.3 shows the energy level diagrams for a typical insulator, phosphor, insulator stack. The band gap is a region in the energy level structure of a material in which there are no available electron energy states. Below the bottom of the gap is the valence band in which the many available states are filled with electrons. Above the top of the gap, the conduction band has many states but they are generally empty of electrons. An insulator is a material which has a band gap so wide that very few electrons acquire enough energy to be boosted from the valence band across the forbidden zone of the band gap up to the conduction band where they would be free to move throughout the material. The electrons in the valence band cannot move about because the surrounding energy levels are all filled eliminating any place for an electron in any particular site to move to. Conductors have very narrow band gaps and thermal energy keeps the conduction band full of freely mobile electrons. Materials with band gap widths that fall in the range between conductors and insulators are semi-conductors of varying degreee. The figure shows that the semi-insulating EL phosphor has a narrower band gap than its surrounding insulator layers. Applying a voltage across the device raises the energy of electrons on the negative side and reduces the electron energy on the positive side causing a tilt in the whole band gap structure. Any electrons in the conduction band would be accelerated downhill resulting in current flow but there are too few electrons in the conduction band to result in a significant current. The hypothesis is that there are electrons in the region of the interface between the insulator and phosphor that lie in so called interface states some distance below the conduction band. As increasing voltage steepens the tilt of the bands, the barrier between these interface states and the conduction band becomes thin and eventually these electrons can tunnel through this barrier into the conduction band. They enter the conduction band with high energy and are further accelerated by the field (tilt) until they are sufficiently energetic to impact excite the manganese atoms in their path. The build-up of

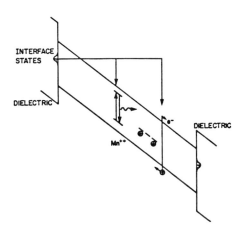

Fig. 5.3 Energy level diagram for a typical insulator-phosphor-insulator stack.

electrons at the positive interface reduces the internal field bringing further tunneling and conduction to a halt until the voltage is reversed.

At the very high fields involved, the electrons excite the manganese atoms, and photons are emitted, i.e., luminesce with a characteristic yellow-orange (585 nm) emission. To prevent destructive break-down under normal operation, the encapsulating dielectrics should not conduct until the field reaches levels of $10\text{-}14\text{x }10^4$ V mm^{-1}. When break-down occurs in the encapsulating dielectric layers, a destructive runaway current flows through the entire structure and the stored energy from the surrounding capacitance can discharge through the break-down area causing failure of the layers to occur (see section 5.3.3). Since it is beyond the capabilities of technology to produce perfect, defect free thin-films over large areas, the successful TFEL device structure is one which can survive its defects. As voltage is gradually applied to a newly fabricated display panel, minute defect areas burn-out and clear themselves leaving barely visible dark spots in some of the display element areas. In properly designed and fabricated devices, these cleared defects are stable and do not contribute to further degradation of the panel. In improperly built film structures, such defects have been known to propagate across the electrodes areas (called "Pacman" syndrome because of the similarity in appearance to the popular game) eventually destroying much of the display.

For some time, the EL phenomenon was considered to be due to ionisation. Vlasenko and Yaremko [5.12] investigated AC electroluminescence of ZnS:Mn films as a function of thickness between 0.04 and 2 μm. Below 0.1 μm, the emission dropped very rapidly. Thus, they concluded that the excitation mechanism was impact ionisation by accelerated electrons in the conduction band. There is now general agreement [5.13] that light emission from ZnS thin-films is due to electron impact excitation of the Mn activator. The processes of tunneling and impact excitation are both very non-linear with respect to the applied field. These help give the luminance vs voltage variation an unusually strong

steepness and   threshold behaviour, greatly facilitating multiplexing (figure
5.4).

The role of the phosphor/dielectric interface in providing   the   electrons   for
excitation is clearly demonstrated by observations on devices which were made
with different materials and process conditions for the   two dielectric
depositions. In some samples, light was emitted only on one polarity of the
drive voltage and none was measured on the alternate half-cycles when the
polarity was reversed. In other words, when the normal interface was driven
negative, electrons were driven into the phosphor in the usual manner and light
was produced. When the other interface was driven negative, current still
flowed back through the phosphor but apparently, due to the different interface
conditions, the electrons did not reach sufficient energy to excite the Mn
atoms. The obvious conclusion from this is that the excitation process does not

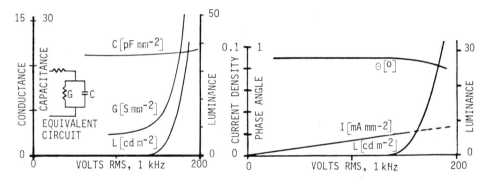

Fig. 5.4 Electrical circuit characteristics of an AC EL display element.

occur solely within the ZnS layer but depends strongly on the interface
conditions.

5.3. Physical characteristics

5.3.1. Size

The actual panel size of an AC TFEL is constrained by the need to provide a
hermetic seal and electrical connections around the periphery. At least 10 mm
must be provided to the length and width of the active area. The depth of
panels is usually less than 3 mm. The thickness of a display head incorporating
such a panel depends on the method of interconnection, the protective frame and

the drive circuitry included in the head. Typical displays are 20 to 30 mm thick regardless of the display area dimensions.

5.3.2. Life expectancy

Although "half-life" has often been used to describe the expected usefulness of flat panel displays, TFELs evidence a unique characteristic, due to "burn-in" or annealing. The annealing process probably results from a redistribution of luminescent centers. The burn-in establishes and maintains a higher luminance level than when first activated (figure 5.5) [5.14]. Once established, this level can be maintained as a linear function by minor voltage adjustments, sometimes called maintenance or flat maintenance. Life-tests have demonstrated that an MTBF (mean time between failures) of greater than 10,000 hours is readily achievable.

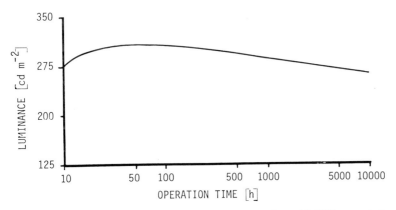

Fig. 5.5 Improved luminance effect of burn-in of an AC TFEL resulting in high maintenance profile.

5.3.3. Reliability

The reliability of AC TFEL devices not only depends on the quality of the polycrystalline manganese-doped ZnS layer, but on the quality of the dielectric layers as well. Since the light output of the device is directly proportional to the charge flowing through the capacitive layers per pulse, the number of pulses per second, and the voltage across the ZnS film, higher reliabilities can accrue if high dielectric constant and break-down strength insulator films are used [5.15]. Caution must be considered in the sealing of the sandwich, since the TFEL structure demonstrates considerable sensitivity to moisture. The devices must be passivated in some way to avoid progressive destruction of the device through a process of area break-down delamination, or peeling, of the film. On the other hand, work by Suntola, using atomic epitaxy deposited thin-

films, evidenced structures so dense and uniform as to be be seemingly
insensitive to moisture and, except at the edges, to require minimal sealing
[5.16]. Small burn-outs can occur at pinholes in the dielectrics because of
high current conduction. When the high current conduction occurs, the conductor
acts as a "fuse" and the circuit burns-open, thus limiting the degree of
destruction. This fusing action is very helpful in the overall performance of
the display since it blocks the possibility of catastrophic failure. The atomic
epitaxy deposition appears to reduce, although not eliminate, this problem.

Another characteristic of thin-film EL devices is the relationship between
localised failure mode and active device area, a relationship which is
critically dependent upon the electrode materials and thickness, as well as
upon the choice and perfection of the dielectric layers. Break-down is a non-
linear phenomenon. Once initiated at some defect in the dielectric layer, the
stored energy available to produce destructive local heating is proportional to
the contiguous active device area. Unless the electrode restricts or constrains
the lateral flow of energy, catastrophic failure may result. This is one of the
major problems in this technology, because, it is not yet feasible to make very
large areas totally free of local break-downs. Suntola slotted the electrodes
at the line intersections as a means to control the lateral destruction [5.17].
Another important factor is the possible contamination of the materials.
Efficient luminescence with an Mn^{+2} activator employs a ten times higher
concentration than with Cu or Cl activators. The practical significance is that
the Mn activated phosphor system is less affected by impurity contamination;
i.e., given the same order of contamination.

5.3.4. Memory

A potentially useful memory effect has been observed in AC TFEL devices which
have an appropriate thickness and manganese doping. When the applied voltage is
first increased, the light output in a memory device increases non-linearly as
a non-memory device would [5.18]; however, as the applied voltage is reduced,
the light output tends to remain at a high level (which depends on the
magnitude of the highest prior applied voltage). A memory display panel of 1248
characters (5x7) has been demonstrated [5.19]. Using this effect, one can also
obtain grey scale with memory by using (switching) voltage amplitude modulation
in the matrix address mode. This hysteresis behaviour can also be triggered by
light as demonstrated by Suzuki et al [5.20]. Driving techniques have been
developed such that an AC TFEL memory panel, operated as a video display panel,
can have a stop motion or frame store capability [5.21]. Furthermore, in a
storage display mode, the information may be electrically read out by applying
a reading pulse to sense the state of each cell (del) [5.22]. The major
importance of the memory effect is that it could significantly reduce the

demand on the luminous efficiency of the AC TFEL device since it obviates the need for refresh. On the other hand, the long-life characteristics typical of the non-memory panel have not been demonstrated in the memory version. The stability of the memory operation is also one of the critical problems still to be solved. These issues will need to be resolved before commercialisation of TFEL memory panels can be expected. No significant work has been reported in this area for several years.

5.3.5. Efficiency

The efficiency of the EL is strongly dependent on the Mn concentration, stoichiometry and crystallinity of the phosphor film and the structure and interface with the dielectric films. These factors contribute to the internal efficiency of the device. The external efficiency of the drive technique and circuitry will be considered separately (see section 5.4.3). For a given flow of charge through the phosphor film, the light output depends directly on the concentration of Mn atoms available to be excited up to a point. Somewhere between 0.5 % and 1.0 % Mn concentration in the Zns, a phenomenon known as concentration quenching occurs causing reduced luminance and efficiency in the device.

The process of light emission from excited Mn atoms is always in competition with other energy transfer processes which do not produce light but which result in energy loss by infra-red, phonon or thermal processes. This has a high probability of happening when defect states at an intermediate energy level exist in proximity to an excited Mn atom. Another process which does not itself contribute to energy losses is the resonant transfer of energy from an excited Mn atom to a nearby unexcited one. After such a transfer, the energy is still available to produce light emission; however, this process can increase the volume of the crystal from which a defect state can remove energy without emitting light. The increasing probability of resonant transfers with increasing Mn concentration along with subsequent loss to defect states is the cause of concentration quenching. The question remains open as to whether higher crystal perfection in the polycrystalline films would allow higher Mn concentrations before quenching became a limitation, thereby increasing both the luminance and efficiency of the devices.

Typical luminous efficiencies of AC TFEL panels are in the range of 1 to 2 lumen per watt, with 5 and 9 lumen per watt having been reported [5.17,5.23]. At 2 lumen per watt, a 150 mm square panel, with typically 20 % of that area in actual use, will consume about 185 mW for a 32 cd m^{-2} luminance (sufficient for sunlight legible alpha-numerics and graphics, but not grey scale video). Since the line driver technology now being used can dissipate up to forty times that power, this can result in a total power consumption of as much as 7 W.

5.4. <u>Addressing/driving</u>

5.4.1. **Approaches**

The ZnS:Mn EL film requires a strong electric field to cause it to emit light.
The electric field must be sufficient to cause conduction in the phosphor but
not so high as to cause the encapsulating dielectrics to conduct. An AC field
is necessary for sustained operation. Beyond these requirements, the wave shape
of the excitation signal is not usually critical. There are several ways to
implement the required voltage drive:

a) One method is active matrix addressing in which a thin-film transistor (TFT)
   circuit is integrated at each pixel in an array using thin-film fabrication
   techniques. As the TFT array is scanned, the individual dels are turned to
   an on, off or intermediate state which is then maintained until the next
   frame scan. Driving the TFEL dels is done continuously and independently
   from the scanning. The very significant advantages of this approach are that
   each del can be turned on with a 100 % duty cycle and that resonant power
   supplies can be easily implemented to save the imaginary power and reduce
   the overall power consumption. The disadvantage of this approach is that, so
   far, the yield has been low due to the great number (approximately 17) of
   thin-film layers required. As thin-films are deposited, one on top of the
   other, surface features that appear randomly tend to be propagated through
   subsequent layers; as the layers are built up, the surface eventually
   becomes too rough for the deposition of any more high quality thin-films.

b) The most expeditious addressing and driving method uses direct X-Y matrix
   addressing which takes advantage of the extremely steep threshold
   characteristic of the TFEL display medium.

5.4.2. **Matrix addressing**

The display matrix consists of line and column electrodes with each
intersection of the electrodes defining one del of the display (see section
I.1.5). The display is scanned, one-line-at-a-time, with modulation for all of
the dels in that line applied to the columns. Coupling between the lines and
columns at every del distributes the applied voltages throughout the entire
display. Because of this largely capacitive coupling, it is not possible to
apply drive voltages to selected dels without also having partial voltages
appear on all of the other elements of the display. Unless there exists a
strong difference in output characteristics at the drive voltage compared to
that at the partial voltage (known as discrimination ratio) there will be very

little difference in output between selected and unselected dels and the display contrast will be very low.

One of the most attractive features of TFEL technology lies in the fact that this material has a discrimination ratio that is extremely high and has been measured in the range of about 100. Because of this steep threshold characteristic, direct matrix addressing of 512 lines has been demonstrated and calculations indicate that as many as 1000 lines can be operated in this way.

The image on an EL display is generated by programming the luminance of the individual dels. The method associated with the programming of these dels is commonly called the drive scheme, while the electronics needed to incorporate this drive scheme is the exerciser or generator. The exerciser electronics is comprised of the logic and the high voltage drivers. The design of the logic is relatively straightforward given the vast source of micro-processors and integrated logic circuits available. The primary concerns for incorporating a drive scheme are the high voltage drivers and the interface of these drivers to the controller logic and the electrodes of the display.

Many drive schemes have been proposed for AC TFEL displays. All of them utilise the voltage-luminance characteristics of the thin-film structure. Typical curves are shown in figure 5.6. Most EL displays operate in a binary mode where a del is either on or off. Grey shading of the dels can be done by operating at points along the voltage luminance curve [5.24]. A characteristic of the thin-film electroluminescent material is that it emits bursts of light, with each burst having an exponential persistence decay (i.e., $L = L_0 \exp - t/t_0$) with a decay time to half-brightness about 1.25 ms.

Before a second burst of light can be activated, the AC TFEL material must be refreshed with a voltage pulse of the opposite polarity. Hence the drive of an AC TFEL display essentially requires an AC

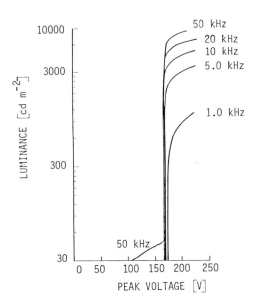

Fig. 5.6 Typical steep luminance versus voltage curves.

waveform. Also, because of this characteristic, the apparent luminance of the image on the display is dependent on the frequency at which each del emits the bursts of light.

Figure 5.7 shows an electrical model for a matrix EL display. The line and column electrodes are modelled as distributed resistances between each del. Each del can be modelled as an ideal capacitor in parallel with a non-linear resistor as shown in figure 5.4. The capacitance models the thin-film structure and the non-linear resistor models the electrical effect of light being emitted by the EL phosphor. The effect of the non-linear resistor has a negligible

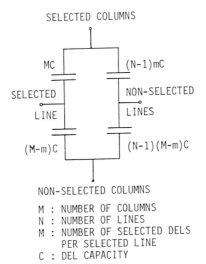

M : NUMBER OF COLUMNS
N : NUMBER OF LINES
M : NUMBER OF SELECTED DELS
    PER SELECTED LINE
C : DEL CAPACITY

Fig. 5.7 Electrical model of a matrix AC TFEL display.

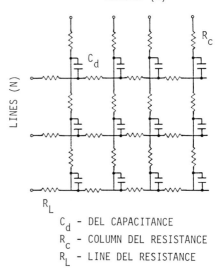

C_d - DEL CAPACITANCE
R_c - COLUMN DEL RESISTANCE
R_L - LINE DEL RESISTANCE

Fig. 5.8 Capacitive model of a matrix AC TFEL display.

impact on the entire circuit and is usually ignored for circuit analysis. Each del then, is represented by a single capacitance Cd.

For each selected line, the drive scheme applies a modulation voltage to each "on" del along that line through the corresponding column. Because of the matrix design of the EL display, all dels are capacitively linked therefore voltage applied to the selected line and columns is coupled to some degree to all of the unselected dels. When these non-selected dels are charged to a sufficient level to emit light, this is described as crosstalk. Eliminating this crosstalk is a principal requirement for the display device.

If there are N lines in the display, the line-at-a-time addressing technique allows the voltages to be applied to selected dels for, at most, a fraction equal to 1/N of the time. On the other hand, a partial voltage coupled through the matrix capacitance is applied to the unselected dels just about all of the time. In order to have good contrast and no crosstalk, the "on" luminance, even under drive conditions with 1/N duty cycle, must be orders of magnitude higher than any luminance generated at the "off" voltage levels. This requires that the display material have a very steep threshold characteristic in its luminance as a function of voltage as shown in figure 5.6.

A TFEL matrix display panel represents a very complex electrical circuit to analyse but there are valid simplifying assumptions which can be made to allow calculation of display parameters to a very high degree of accuracy. The non-linear impedance voltage characteristic of the phosphor film can be modeled as in figure 5.4 but since the effects of this characteristic are small when compared to the overall panel impedance, it can be ignored in calculating voltages and currents in the panel. The panel electrical model shown in figure 5.7 is fairly accurate but still too complex because for realistic panel sizes it results in thousands of simultaneous loop equations. Several more simplifying assumptions are needed. First, we assume a condition which is in any case necessary for panel uniformity. That is, that the driving pulse widths are wide enough such that the capacitance becomes fully charged and a condition of stability is achieved (no current flowing). Under this condition, the line and column resistances can be ignored and the del voltages can be calculated by considering only the capacitances. Next we must assume that the del capacitances are all equal, which is also a necessary condition for a uniform panel. This allows us to lump together those dels which experience the same drive condition, making use of the old electrical engineering trick of tying together (in the diagrams or equations) points which are at the same potential. This may be done because it will not cause any additional current flow and therefore will not change any results.

To analyse a panel which has M columns and N lines we need to examine what happens when a drive voltage is applied to one of the N lines and a modulation voltage Vm is applied to m of the M columns. Under a typical set of drive conditions, the N-1 off lines are open circuited but can be considered as a single point electrically because they will all experience the same condition and no current would flow among them even if they were connected together. All of the dels on the panel now fall into one of four conditions. These are:

a. the m driven dels between the selected line and selected columns

b. the M-m non-driven dels in the selected line

c. the (N-1).m dels between the selected columns and non-selected lines

d. the (N-1).(M-m) dels between the non-selected columns and non-selected lines.

This accounts for all of the pixels in the panel since:

m + (M-m) + (N-1).m + (N-1).(M-m) = M.N.

Multiplying each group of dels by the capacitance Cd of each element, results in the equivalent circuit of figure 5.8 which can be used to calculate the distribution of voltage on dels or the power consumption for various drive conditions.

For example, when -Vd is applied to one line, Vm is applied to m columns, while non-selected columns are grounded and non-selected lines are floating: the selected dels see:

Vm - (-Vd) = Vm + Vd volts

non-selected dels in the selected line see

0 - (-Vd) = Vd volts

and the dels in the non-selected lines divide the modulation voltage as:

Vm . (M-m)/M   for dels in selected columns and
Vm . m/M   for dels in non-selected columns.

The resistance in the lines and columns must be considered in calculating the rise time of a voltage pulse but this can be done for a line-at-a-time and a transmission line type of calculation provides accurate results.

The effect of line resistance on power consumption depends on the waveform of the driving pulse. For a steep risetime approximating a step function, the resistance does not matter. When a step voltage is applied through a resistor to a capacitor long enough to fully charge the capacitor, 1/2 of the energy consumed is stored on the capacitor and the other 1/2 is dissipated in the resistor regardless of the resistance: or Er/Ec = 1. The resistance only determines how long it will take to charge the capacitor. Minimum energy dissipation in the resistor will occur when the capacitor is charged with a constant current over a time t > 2RC. In this case, Er/Ec = 2RC/t which becomes much less than 1 for large t.

Measurements have confirmed that in a typical TFEL device, the energy consumed in generating light is approximately 5 % of the total energy deposited on the del. The remaining 95 % of the energy, being stored on the capacitor, is available to be recovered by a drive scheme designed for that purpose. A resonant drive scheme was reported by Miller and Tuttle [5.25] in 1981 and various other energy recovery drive schemes have been implemented since that time with power savings greater than 50 %.

5.4.3. Drivers

The production of cost effective TFEL displays depends very heavily on the existence of integrated line and column drive circuits combining logic functions with multiple high voltage outputs. Such drivers are available in on/ off configurations for text and graphics displays from several manufacturers. Integrated drivers with from 16 to 64 outputs are being produced. In addition there is a set of 16 output grey scale drivers available for use with video and other imagery applications. The column drivers in this set include 4-bit wide shift registers to handle 16 grey shade levels. The grey shade information for a line is shifted into the drivers and latched. The drivers then convert the 4-bit code to one of 16 voltage levels between 0 and 60 volts to be applied to the columns during the line pulse. These have been used to produce live TV displays on a 512 line by 640 column panel at a 30 Hz interlaced frame rate.

5.5. System interface

Although the AC TFEL is digital in nature, one current approach to the implementation of AC TFELs is to consider the display simply as a display monitor, i.e., the signal input would be compatible with, for example, a U.S. standard television RS-170 composite video input; all conventional signal formatting, symbol generation, and conditioning, etc., to be handled externally to the display monitor. The display monitor will provide the proper decoding (timing, x and y sync strip-off, etc.) to match to the internal refresh and matrix addressing necessary for the display itself. In this manner, the display monitor can be used as a direct replacement for a CRT. The other system interface considerations appropriate to the CRT as enunciated in chapter II.1 generally hold for the AC TFEL display monitor. The addition of a bit map memory (storage per display element address of the number of grey shade bit planes) greatly increases the flexibility of the display by removing the constraint that the information from the data source be at the same rate as the display refresh. This allows higher luminance to be obtained by operating at a higher refresh rate and also simplifies interfacing computer generated images.

5.6. <u>Visual characteristics</u>

5.6.1. **Reflectivity**

Specular reflectivity associated with a front glass surface is essentially independent of the technology. The now standard treatment is to use a high efficiency reflection reducing coating. Since the front surface of the EL display is glass, the same coating requirements would prevail.

In general, for an excited TFEL element to be directly visible against its background, the reflection of ambient light must be minimised. Most conventional emitting display schemes employ filters to reduce the internal specular and diffuse reflectance. This, in turn, also interferes with the emission from the display by introducing optical attenuation. The high contrast dark field TFEL requires no filters, thus all the light emitted forward from the TFEL layer is transmitted without attenuation. Because of the high index of refraction the ZnS layer, about 2.3, only a small cone with an angle of rotation of about 18 degrees is not totally reflected.

5.6.2. **Contrast**

With a vacuum deposited thin-film phosphor structure, ambient light is scattered, i.e., diffusely reflected, only at the rear surface of the phosphor, in the region designated by dielectric film 2 (see figure 5.2). Ambient light arriving through the display structure can be absorbed here if dielectric film 2 matches the index of refraction of the TFEL film and is opaque. Taking advantage of this situation, high contrast display structures with a diffuse reflectivity less than 0.25 % [5.26] were achieved. More recently, black layers measuring between 0.1 % and 0.14 % reflectivity have been demonstrated. Because very high contrast ratios can be obtained with a black layer, TFELs are readable in sunlight levels beyond 10^5 lux. As a specific example, a black layer TFEL display having a diffuse reflectance of 0.1 % would reflect a 32 cd m^{-2} in a 10^5 lux ambient. If the display element itself emits about 64 cd m^{-2} the contrast ratio would be the 64 emitted, plus the 32 reflected, divided by the 32 reflected off the black layer or equal to 3. Obviously, higher outputs will provide more than adequate contrast.

Thin-film EL alpha-numerics have been demonstrated as more legible than comparable dot matrix CRT formats because of the sharp, square dot pattern, though this has not been extensively tested.
The practical grey scale range currently depends on the number of lines multiplexed and the ambient illumination. The inherent non-linearity and the

peak luminance allows refresh matrix operation up to 500 lines with about 10-15 cd m^{-2} luminance and negligible loss of contrast due to crosstalk. However, for large area displays the peak currents will be large, and beyond 3000 mm^2 one must look closely at the properties of the transparent electrode material to avoid waveform distortion in the interior of the panel. With the TFT approach or an equivalent process not only this problem is obviated, but luminance levels of up to 200 cd m^{-2} can be easily obtained.

5.6.3. Flicker

Flicker is easily detectable when a TFEL pixel is refreshed at below 50 Hz, although flicker or "strobing" becomes apparent under high vibration at higher refresh frequencies. The persistence of the phosphor is relatively independent of the current. Normally operated AC TFELs otherwise exhibit no flicker.

5.6.4. Resolution

Resolution, once a serious problem, is no longer a limiting factor. A panel with a density of 20 lines per mm across a linear dimension of 25.4 mm was demonstrated [5.27] for use in a helmet-mounted display as well as a 512 by 640 line display at a density of 4 dels per mm (100 dels per inch) [5.28]. The latter would appear to satisfy most graphics and video requirements. Line and column drivers were developed to utilise the capability of these panels.
The ultimate resolution is probably limited by the size of the pinhole destruction phenomenon [5.29].

5.6.5. Colour

Work on colour TFEL, that is, colours other than yellow-orange Zn:Mn emission, is expanding rapidly. For the first time in 1986, the International Symposium of the Society for Information display included a separate session on colour electroluminescence as well as the general electroluminescence. Corporate and university research world wide in the area of colour TFEL is extensive.

The usual TFEL colour one thinks of is the yellow-orange ZnS:Mn emission. This is a broad band emission peaked at 585 nm with a width at half maximum of about 50 nm. Typically a commercial panel will have a luminance of 65 to 160 cd m^{-2} at 60 hz drive and an efficiency of 2 lumen per watt. For colour phosphors to be useful they must have characteristics similar to the ZnS:Mn. Note: when referring to the frequency, one period is the time from the application of a positive voltage pulse to the application of the next positive voltage pulse.

During this period a negative voltage pulse will be applied. Therefore a TFEL display running at 60 hz will have voltage pulses applied at an 120 hz rate.

The history of colour TFEL goes back to the original work with powder EL. Large numbers of colours were found to be available using the rare-earth fluorides. A survey of these phosphors was published in 1969 [5.30]; a recent survey is the Okamoto thesis published in 1981 [5.31]. More readily available is the tabulation found in the conference record of an invited address to the International Display Research Conference [5.32].

5.6.5.1. Phosphors.

Green: The green phosphor which has had the most development and research is ZnS:TbF3. This phosphor has an advantage in that the host material is ZnS which has been studied extensively, so the technology for making good TFEL layers is well known. This green is typical of the rare-earth phosphors in that it has besides the main peak several lower intensity emission peaks. For ZnS:TbF3 the approximate location of the peaks are 490 nm, 590 nm, 625 nm, with the main peak at 550 nm. The efficiency and luminance obtained with this phosphor has shown a dramatic increase in the last few years. The results from a 1979 paper gave a luminance of 20 cd m^{-2} and efficiency of 0.25 lumen per watt [5.33]. In 1985, the best reported results were 185 cd m^{-2} at 1.5 ℓm W^{-1} (this luminance is linearly extrapolated from the 310 cd m^{-2} figure at 100 hz) [5.34].
The continuous improvement of the luminance and efficiency of ZnS:TbF3 is the result of gradual improvement, by various researchers modifying fabrication techniques and materials. Theory predicted that the material should have similar characteristics as Mn in ZnS [5.35]. The combination of basic analytical tools coupled with theory and experiment have steadily increased the luminance and efficiency of the green phosphor until it is at the level where it may be used effectively in displays.

Red: The red phosphors which have been reported to date have been well below the 'yellow' standard. Various phosphors have been tried with ZnS:Mn,TbF3 giving the best results. Luminances of 14 cd m^{-2} with .1 ℓm W^{-1} have been obtained at 60 hz [5.36]. While this is less than an order of magnitude below the luminance obtained from ZnS:Mn, it is not a well saturated red. It has a very broad emission starting about 520 nm and extending out into the 700 nm region. The rare-earth phosphors which have been tried in ZnS have not shown the steady improvement like the Tb. This may only be the result of less research on these materials or it may be due to an incompatibility between the host and the dopant. New host materials such as CaS and SrS are being explored with encouraging results. SrS:Eu TFEL samples have been made with a luminance of 1000 cd m^{-2} at 5 khz [5.37]. Linearly extrapolated to 60 hz this figure becomes 12.5 cd m^{-2}. While still less than the 'yellow' this order of magnitude increase over the previous best of 100 cd m^{-2} at 5 khz reported only six months earlier demonstrates the strides which can be made with improved processing and fabrication [5.38].

Blue: The blue phosphors generally have yielded lower luminance and efficiency than other colours in TFEL devices. This may be due to the higher energy needed for blue emission, which means that the excitation mechanism must have a higher energy distribution. In 1984, a new host for CeF3 was reported, SrS [5.39]. This change from the traditional ZnS host material to SrS has resulted in a 2 order of magnitude improvement in luminance. These samples exhibit 155 cd m^{-2} at 1 khz, which extrapolates to 10 cd m^{-2} at 60 hz. This phosphor may yield further improvement with more work on the processing of these materials.

5.6.5.2. Panel structure.

The panel structure for any monochrome matrix TFEL will be the same as the design used for the ZnS:Mn except for changes in materials. To make multi-colour TFEL displays, changes have to be made. The most obvious structure is the stacked film structure which means that one TFEL stack is fabricated on top of the preceeding stack. Each stack could have its own electrodes or they could share one common electrode between stacks. For this structure, there is no loss in the active area of the dels, since the light from the stack in the back would pass through the front stack. Simple demonstration devices of this type have been reported [5.36]. In this structure all the electrodes, except for the last one deposited, would have to be transparent to allow the light to get out to the viewer. A critical problem is that of yield and reliability in making the devices. Most commercial TFEL devices have the characteristic of self healing break-down, i.e. when a weak portion of a del breaks-down it clears itself and will not open up the line. Using ITO (Indium Tin Oxide) as the second electrode in a TFEL stack usually reduces this effect although thin metallic electrodes are a possibility. Further, most devices do not have solid seal in contact with the films, since this also reduces the effect of self healing. To date, no reliable samples of this type have been reported.

A structure which is similar to the stacked film structure is the stacked substrate. For a two colour display, each TFEL colour is made as an independent panel. The two panels are then mounted together with a small gap between the stacks. This structure has the advantage of improved yield because the two stacks are independently made and only good monochrome panels will be mounted together. As with the stacked film structure, the del areas need not be reduced, and all the electrodes except the back (with respect to the viewer) electrode will have to be transparent. This type of display has been made in a 240 line by 320 column format [5.40]. To carry this type of structure to a three colour display would require that the middle colour be fabricated on an a very thin substrate, which implies handling and processing problems. The reason for the thin substrate is that the there would severe parallax problems due to the separation of the emitting layers if a normal thickness substrate were used. Normally the glass used in TFEL displays is between 1 mm and 1.25 mm thick. With the two colour displays the spacing between the TFEL stacks is a tenth of a mm which does not seem to cause problems.

A third way of obtaining multi-colour TFEL panels is the patterned phosphor method. As the name implies, the phosphor is pattered so that all the colours are on one substrate. The simplest configuration is stripes of alternating colours. An advantage of this method is that normal ITO front electrodes and Al back electrodes may be used. The electrodes may be independent or one may be common to all colours. Since all the colours are on the same substrate, the effective fill of each colour is reduced. To achieve this structure with different host materials, each of the phosphors must be deposited and patterned. The processes for depositing and patterning each of the phosphors must be compatible. An alternative method for phosphors with the same host material is to deposit one layer of the host material and then spatially introduce the dopants to the host material. This was demonstrated with Mn in a ZnS:TbF3 film using photochemical vapor deposition to define the lines [5.41]. Since this structure requires more processing steps to deposit and pattern the different colours, the production yield will be reduced.

5.6.5.3. Electrodes.

Most TFEL displays have from 2 to 4 lines per mm [ℓ mm^{-1}]. This does not represent an upper limit on what can be achieved. In most displays the electrodes are interdigitated so that the electrodes must be accessed at only half the display spatial frequency. On a 4 ℓ mm^{-1} panel the electrodes would be accessed at 2 ℓ mm^{-1} on each edge. Current methods for making connection between the electrodes and the drive chips and electronics are: wire on kapton, elastomeric, and soldered wire waterfall. These methods for making connection are adequate for present displays; current elastomeric connectors should work up to 8 ℓ mm^{-1}. In a 3-colour display the electrode density would be three times as high as in a monochrome display if the electrodes for each colour were independent. In this case a 4 ℓ mm^{-1} display would have electrodes at 12 ℓ mm^{-1} and if it was an interdigitated sample it would be accessed at 6 ℓ mm^{-1}. To overcome this connection problem the electrodes can be fanned out to reduce the

line density. Although this takes up more space on the periphery of the panel
it is a workable solution. In the future alternative connection methods may be
devised or present technology be improved. Chips on glass is a way to reduce
the number of connections to the panel to the data, DC voltages, and timing
signals. In this scheme the driver chips are bonded directly to an electrode
pattern on the glass.

5.6.6. Viewing angle

Viewing angle, often a critical factor such as in liquid crystal displays, is
not a concern with TFELs. The image can be seen almost to ± 90 degrees,
although the practical limit for alpha-numerics and complex graphics is
probably about ± 70 degrees.

5.7. Other EL devices

5.7.1. AC powder EL

AC powder EL, the variety associated with Destriau, was the object of an
extensive research effort in the 1950s and early 1960s aimed (unsuccessfully)
at making it an efficient source of illumination. This was described by Ivey
[5.42].

The active layer, which consists of a suitably doped ZnS powder (5-20 μm
grains) suspended in a dielectric, is sandwiched between two electrodes, one of
which is transparent, and is supported by a substrate. The substrate may be
glass, flexible plastic, or may be a metal such as aluminum. In the latter
case, the top electrode must be transparent. The dielectric may be organic
(e.g. cyanoethyl cellulose) or a low melting point glass. A second dielectric
is often used as extra protection against catastrophic dielectric break-down.
Applied fields of 10^3-10^4 V mm^{-1} are sufficient to produce luminance as high as
340 cd m^{-2} at efficiencies which are reasonable by display standards (1-5 lumen
per watt). Luminance increases with frequency at moderate frequencies (10^2-10^4
Hz), but device life is usually decreased in the same proportion.

One has a choice of colours depending upon the activation of the ZnS powder.
Copper is always used in these EL powders. The combination of Cu and Cl can
give either blue (460 nm) or green (510 nm) emission, depending upon the
relative amount of Cl, while the combination Cu, Cl and Mn yields yellow
(585 nm). A key parameter is particle size, which is involved in some critical

trade-offs. Small particle size increases efficiency and non-linearity, but decreases life.

Efficiency generally has the functional form $L^{0.5}V^2$ (where L is luminance) and usually peaks well below the highest luminance levels; this ultimately relates back to particle size which is increased to improve life, yet decreased to improve luminance, non-linearity and efficiency. Much of the behaviour can be understood if one looks at the microscopic nature of the Destriau phenomenon. The best synthesis of observations and ideas on this subject is the work of Fischer [5.43]. EL powders are typically fired at high temperatures where the hexagonal phase dominates. When they are cooled, there is a transformation to a cubic zinc-blend structure, and the copper (previously mentioned as essential) which exceeds the solubility limit, precipitates on defects resulting from the hexagonal to cubic transformation. The result is imbedded Cu_xS conducting needles which act to concentrate an applied electric field at their tips. Thus an applied field of 10^3–10^4 V mm^{-1} can induce a local field of 10^5 V mm^{-1} or more. This is enough to induce tunneling of holes from one end of the needle and electrons from the other. The holes are trapped on Cu recombination centers, and upon reversal of the fields the emitted electrons recombine with the trapped holes to produce light. Larger particle sizes lead to longer needles and greater field enhancement, but as the shunted material still contributes to the losses, efficiency is reduced. In addition, the localised current flow results in localised heating. This, together with the high fields, can result in diffusion and electromigration of active species such as Cu^{+2} or harmful sulfur vacancies which interfere with the emission process. The Lehmann [5.44] hypermaintenance process was, in fact, an attempt to reduce and control sulfur vacancies, which also affect the mobility of Cu ions. Since moisture is thought to be a source of sulfur vacancies at the particle surfaces, Fischer et al [5.45] proposed that phosphor particles should be microencapsulated with a phosphosilicate after Lehmann-type processing, to yield even longer life. The longer life associated with larger particle diameter is related to a longer diffusion time for defects generated at the surface of the grains. Life is still a key problem for the application of this technology, but there is indeed some expectation for more than 10,000 hour life at moderate luminance, especially for green-emitting devices. The problem is that peak luminance is a serious limitation, and low duty cycle operation is not acceptable. This is especially important because the powder reflects ambient light, and good contrast is achieved by the use of an absorbing filter on the front of the display, as in a CRT display. These absorbing filters, however, seriously reduce the emitted light.

The most promising approach to using AC powder EL in complex displays is in conjunction with arrays of thin-film transistor (TFT) circuits. Several groups pursued this approach [5.46,5.47] with the most impressive results obtained by Brody et al [5.48]. Currently, AC powder ELs are used in civil aircraft control panels for night illumination and are being incorporated in military aircraft as well, particularly because of the need for compatibility with night vision goggles. These displays, however, cannot satisfy high luminance environment like outdoor uses and in aircraft, marine and landvehicle applications. AC powder ELs are used extensively in backlit (transflexive) liquid crystal displays especially now in cars and in "laptop" portable computers.

5.7.2. DC powder EL

There is another type of powder EL device which is associated primarily with A. Vecht and his collaborators [5.49,5.50]. In the AC powder case, the high temperature firing and cooling results in a conducting Cu_xS surface layer on the particles, which is removed chemically. In the DC powder case, a fine grain (0.5-1 μm) Mn-doped ZnS powder is prepared with Cu_xS coating, and a layer of the powder (with a small amount of binder) is formed into the device. If a matrix display is being fabricated, the powder layer, which is 40-50 μm thick and is conducting, must be patterned into strips so that there is no shorting from line to line. Such a device must be "formed". When a voltage is first applied, a large current flows, the layer heats up and gradually a narrow region about 1 μm thick, adjacent to the S_nO_2 anode, begins to luminesce. In the standard ZnS:Mn device, the light is yellow (585 nm), characteristic of Mn emission in ZnS.

There are three points to be made regarding this structure. First, the general requirement that the thick powder layer be patterned, limits the resolution which one can easily achieve relative to thin-film structures and even relative to AC powder structures.
Second, this structure, by itself, provides no protection for the phosphor powder; thus it must be hermetically sealed. Since one is dealing with a white powder device viewed through the glass substrate, good contrast in high ambient illumination is at the expense of device luminance, i.e., system efficiency. One valuable characteristic of DC powder devices is that they can be operated in a pulse mode beyond the voltage and luminance of the forming condition, and the net effect is a better discrimination ratio and average luminance than one might have expected from the DC characteristic [5.51].

Luminance maintenance seems to be a somewhat complex question for these devices. There is no simple functional form for the decay; although pulsed operation, in particular, and low luminance operation, in general, lead to longer life. Devices operated at 8 cd m^{-2} (pulsed) can be expected to retain more than half their initial luminance for more than 10,000 hours in the best cases. Recently Alder et al, [5.52] reported good quality panels with mean luminances of between 175 and 350 cd m^{-2} and a half-life luminance of 3000 hours.

An interesting application of DC powder is the alpha-numeric matrix display development first reported by Mears et al [5.51], taking advantage of the unique pulse response characteristics. While the initial report described a 36 character display, they also discussed the capability for increasing the size to a 200 line display of 1250 characters, which they subsequently achieved [5.53]. Other mechanisations and interface electronics were discussed by Smith and Werring [5.54].

Much of the recent work in this technology has been aimed at extending the available colour range by using different powder materials and different activators. DC electroluminescence has been demonstrated for various rare-earths in ZnS and for Mn in Ba2ZnS3 [5.55] as well as for a number of alkaline-earth sulfides doped with rare-earths [5.56,5.57]. Most of these alternatives have been much less efficient than ZnS:Mn,Cu, but recent results are encouraging; for example, ZnCaS:Ag produces green (70:30) and red (30:70) emissions which are almost one third as efficient as the standard yellow phosphor. There is also a fairly good blue-green (SrS:Ce,Cl) phosphor. Most important is the steady rate of progress in these new EL materials, which may stimulate renewed interest in this technology, especially where a range of colours is important. DC powder ELs are being used in several applications where the luminance environment is controlled. However, is in the case with the AC powders, these displays do not meet the high illumination outdoor and vehicle environment.

5.7.3. DC thin-film EL

The simplest electroluminescent structure one can conceive of is a thin-film of phosphor with electrodes on both sides driven by a DC source. This is probably the reason why the intuitive urge to exploit the phenomenon has so often led people in this direction. Attempts to make devices of this sort go back many years, and the materials most often used are the familiar ones, ZnS:Mn, ZnS:Cu,Cl, and ZnS:Mn,Cu [5.58-5.62]. The basic problem with thin-film devices excited by DC voltages has been a tendency toward catastrophic failure, especially in the case of films without copper. For devices employing ZnS:Cu,Mn, there is at least one layer of Cu$_x$S adjacent to the phosphor film.

It appears that in the absence of copper, or for low concentrations of copper, the fields required to initiate luminescence are so high that avalanching takes place, leading to negative resistance and a runaway current. A resistive layer stabilises the current distribution. When Cu_xS is present, not only can lower fields be used but very likely, the recombination of Cu sites suppresses space charge formation and negative resistance effects (films with copper can under AC excitation, behave analogously to AC powders [5.63] and probably involve alternate injection of holes and electrons from Cu_xS).

A variety of luminance voltage dependencies have been reported for DC thin-film devices, but the common characteristic is a greater non-linearity than is typical for AC powders. The most interesting work in the AC film area was that of Abdalla and Thomas [5.64]. This device seems to be a direct film analog of the DC powder device of Vecht et al [5.50], with a Cu_xS injecting layer and a copper-free active layer being created during the forming process. As with DC powder devices it works best in the pulse mode, which apparently allows the high fields necessary for impact excitation of manganese without excessive heating. Under these conditions, they reported good maintenance.

Further performance data was provided by Abdalla et al, at the 1980 IEEE-SID-AGED Biennial Display Conference [5.65,5.66]. Similar structures were reported by Vlasenko and Gergell [5.67], but no indication of device life was given. Finally, there are reports of DC electroluminescence of reasonable efficiency in films of other materials, such as CdF_2, and La_2O_3, but neither of these appears to be suitable for practical use at this stage; CdF_2 films are not sufficiently stable, and La_2O_3 has low luminance output.

5.8. State of development

While it can be concluded that each of the four types of electroluminescent displays may find useful commercial and industrial application, it is the AC-ELs which are currently popular in the marketplace. Military requirements, however, are stringent. In the case of AC powder devices, if the applications are to go beyond the simple direct address displays (or lamps, as they are commonly called), it is important that good TFT arrays be applied. AC powder devices have been proposed for night lighting in military vehicles where night-vision goggles are to be used. Although the DC powders now appear to have enough life to be useful and are being exploited for several applications in the U.K., they have the drawback of all matrix displays - high drive cost. Like the AC powders, they can be used in limited outdoor and airborne applications, such as for night lighting at low illumination levels. The principal advantage presently seems to be the variety of colours which can be provided. Although there appears to be little or no current work being done, DC films may also be achievable; they can be driven with a relatively low voltage and used at room temperature. Nevertheless, considerably more development must take place before they will be competitive with the powders.

The AC thin-film EL devices are still the most interesting for outdoor professional matrix display applications because of their many attractive features: high peak luminance, long life, non-linearity, contrast, memory, and currently, the increased capability to provide several colours. Nevertheless, matrix displays will not find wide application without reduction in drive costs. This is demonstrated by the fact that AC TFEL are being provided in some laptop portable computers at a premium over AC powder EL backlighted liquid crystal displays. Although the readibility of the AC TFEL is clearly better than the backlit TN liquid crystal display, the cost differential as well as the increased battery load is a mitigating factor.

Not included under reliability is surviveability. Because the structure is basically solid-state, vibration and shock are not critical concerns. Since the substrate is a glass plate, the usual precautions for glass must be taken. There have been instances where these displays have continued to operate, although with reduced resolution, with a crack in the display glass entirely across the panel. This kind of "soft failure" is an important consideration in harsh environments. Finally, as indicated in 5.3.5 and 5.4.3, power dissipation is principally contributed by the drive system. Nevertheless, the estimated power requirement for a 512x640 display, including the drive system, is under 20 watts.

5.9. Acknowledgements

Special credit is due W.E. Howard [5.68] and I.F. Chang [5.69], both of IBM U.S. from whose works this chapter has liberally borrowed. In addition, special thanks are due L. Thannas, Consultant, U.S., M.J. Abdalla, Sylvania, U.S., E. Schlam of ETDL, U.S. Army for their helpful contributions and criticism; and to M. Robert Miller and David C. Morton who assisted in developing this text from the original AGARD contribution.

REFERENCES

[5.1] Destriau, G., Research into the scintillations of zinc sulfides to alpha rays, J. Chim. Phys., <u>33</u>, 587, (1936).

[5.2] DeBoer, J.H. and Dippel, C.J., X-ray intensifying screen, U.S. Patent 1,954,691, (1934).

[5.3] Williams, F.E., Some new aspects of germanate and flouride phosphors, J. Opt. Soc. Am., <u>37</u>, 302, (1947).

[5.4] Feldman, C. and O'Hara, M., Formation of luminescent films by evaporation, J. Opt. Soc. Am., <u>47</u>, 300, (1957).

[5.5] Thornton, W.A., Electroluminescent thin-films, J. Appl. Phys. <u>30</u>, 123, (1959).

[5.6] Piper, W.W., Phosphor screen, U.S. Patent 2,658,915, (1955).

[5.7] Soxman, E.J. and Ketchpel, R.D., Electroluminescent thin-film research, Final Report JANAIR no. 720903, Jul. (1972). See also JANAIR Reports: EL-1, AD 475-700L, Aug. (1965), EL-2, AD 800-992L, Aug. (1966), EL-3, AD 815-950L, Jan. (1967), EL-4, AD 682-547, Jul. (1967), EL-5, AD 704-536, Apr. (1969), EL-6, AD 704-537, May (1969).

[5.8] Steele, G. and Soxman, E., Dark field, high contrast light emitting display, U.S. Patent 3,560,784, Febr. (1971).

[5.9] Inoguchi, T., Tekeda, M., Kakihara, Y., Nakata, Y. and Yoshida, M., Stable high brightness thin-film electroluminescent panels, SID Digest, <u>V</u>, 84, (1974).

[5.10] Mito, S., Suzuki, C., Kanatani, Y. and Ise, M., T.V. Imaging system using electroluminescent panels, SID Digest, <u>V</u>, 86, (1974).

[5.11] Hurd, J.M. and King, C.N., Physical & electrical characteristics of co-deposited ZnS:Mn EL thin-film structure, J. Electr. Mat., <u>8</u>, 879, (1979).

[5.12] Vlasenko, N.A. and Yaremko, A.M., On the mechanism of the excitation of electroluminescence in ZnS:Mn films, Optics and Spectroscopy, <u>18</u>, 263, (1965). Study of the simultaneous action of electric field and ultra radiation on the luminescence of a sublimed ZnS:Mn phase, optics and spectroscopy, <u>18</u>, 461, (1965).

[5.13] Tanaka, S., Evidence for the direct impact excitation of Mn centers in EL ZnS:Mn films, J. Appl. Phys., <u>47</u>, 12, (1976).

[5.14] Fugate, K.O., High display viewability provided by thin-film EL, black layer and TFT drive, SID Proc, 18, 2, (1977), pp. 125-133.

[5.15] Howard, W.E., The importance of insulator properties in a thin-film electroluminescent device, SID Proc., <u>18</u>-2, 119, (1977) and IEEE Transactions Electron Devices, vol. <u>ED-24</u>, 909, (1977).

[5.16] Suntola, T., Antson, J., Pakkala, A. and Lindfors, S., Atomic layer epitaxy for producing EL thin-films, SID Digest, 11, 108, (1980) and private communications.

[5.17] Suntola, T., Performance of atomic layer epitaxy devices, SID Digest, 12, 20, (1981) and private communications.

[5.18] Yamauchi, Y., Takeda, M., Kakihara, Y., Yoshida, M., Kawaguchi, J., Kishihita, H., Nakata, Y., Inoguchi, T. and Mito, S., Inherent memory effects in ZnS:Mn thin-film EL devices, IEDM Digest, (1974), pp. 348-351.

[5.19] Marrello, V. and Anton, A., The dependence of the memory effect in ZnS:Mn AC TFEL on Mn distribution, Appl. Phys. Lett., 31, 7, (1977).

[5.20] Suzuki, C., Kanatani, Y., Ise, M., Misukami, E., Imazaki, K. and Mito, S., Character display using thin-film EL panel with inherent memory, SID Digest, 7, 50, (1976).

[5.21] Ibid., Optical writing on a thin-Film EL panel with inherent Memory, SID Digest, 7, 52, (1976).

[5.22] Kako, N., Yamane, Y. and Suzuki, C., EL TV display with stop motion, SID Digest, 9, 134, (1978).

[5.23] Ketchpel, R., Efficiency of thin-film AC EL emitter. IEDM, 685, (1979).

[5.24] Gielow, T., Holley, R., Lanzinger, D. and Tuttle, R.P., Monolithic driver chips for matrixed grey shaded TFEL displays, SID Digest, 12, 24, (1981).

[5.25] Miller, M.R. and Tuttle, R.P., A drive method for electroluminescent matrix displays, SID Digest, 12, 26, (1981).

[5.26] Soxman, E.J., Electroluminescent thin-film research, JANAIR Rept. EL-2, AD 800-992L, (1966).

[5.27] Ketchpel, R.D., Santha, I.S., Hale, L.G. and Lim, T.C., High resolution, thin-film matrix display device, SID Digest, 9, 138, (1978).

[5.28] Gielow, T.A., Holley, R.H., Shaikh, S. and Lanzinger, D., Tactical video display. Hycom, Inc. Tech Rept., Contr. DELET-TR-79-0251-2, Dec. (1979) – Jul. (1980).

[5.29] Tannas, L.E. Jr., Thin-film electroluminescent emitter, SID Digest, 12, 22, (1981).

[5.30] Chase, E.W., Hepplewhite, R.T., Krupka, D.C. and Kahng, D., Electroluminescence of ZnS lumocen devices containing rare-earth and transition-metal flourides, J. Appl. Phys., 40, (1969), p. 2512.

[5.31] Okamoto, K., Thesis, Osaka University, (1981).

[5.32] King, C.H., Progress in achieving full-colour EL displays, Conference. Record 1985 International Display Research Conf., San Diego, CA, (1985).

[5.33] Okamoto, K. and Hamakawa, Y., Bright green electroluminescence in thin-film ZnS:TbF3, Appl. Phys. Lett., 35, (1979), p. 508.

[5.34] Ohnishi, H., Yamamoto, K. and Katayama, Y., Improved efficiency of green colour ACTFEL devices grown by rf-sputtering, Conf. Record 1985 International Display Research Conference, San Diego, CA (1985).

[5.35] Bernard, J.E., Martens, M.F., Morton, D.C. and Williams, F., Mechanisms of thin-film electroluminescence, Conf. Record 1982 International Display Research Conference, Cherry Hill, NJ, 20, (1982).

[5.36] Coovert, R.E., Christopher, N. and King, C.N., A dual colour AC thin-film electroluminescent (TFEL) display, Proceedings of the SPIE, Advances in Display Technology III, Vol. 386, Los Angeles, CA, Jan. (1983).

[5.37] Kane, J., Harty, W.E., Ling, M. and Yocom, P.N., New electroluminescent phosphors based on strontium sulfide, Conf. Record 1985 International Display Research Conference, San Diego, CA (1985).

[5.38] Tanaka, S., Shanker, B., Shiiki, M., Degchi, H. and Kobayashi, H., Multi-colour electroluminescence in alkaline-earth-sulfide thin-film devices, Society for Information Display, International Symposium Digest of Technical Papers, Vol. 16, Orlando, FL (1985).

[5.39] Barrow, S.A., Coovert, R.E. and King, C.N., Strontium sulphide: the host for a hew high efficiency thin-film EL blue phosphor, Society for Information Display, International Symposium Digest of Technical Papers, Vol. 15, San Fransisco, CA (1984).

[5.40] Barrow, W.A., Tuenge, R.T. and Ziuchkovski, M.J., Multi-colour TFEL display and exerciser, Society for Information Display, International Symposium Digest of Technical Papers, Vol. 17, San Diego, CA (1986).

[5.41] Kitai, A.H. and Wolga, G.J., Two-colour thin-film electroluminescence with spatially-selective activator doping, International Symposium Digest of Technical Papers, Vol. 14, Philadelphia, PA (1983).

[5.42] Ivey, H.F., Adv. In Electronics & Electron. Phys. Supplement, Academic Press, New York (1963).

[5.43] Fischer, A.G., Electroluminescent lines in ZnS powder particles, J. electrochem. Soc., 109, 1043, (1962) and J. Electrochem. Soc., 110, 733, (1963).

[5.44] Lehmann, W., Hyper-maintenance of electroluminescence J. Electrochem, Soc., 113, 40, (1966).

[5.45] Fischer, A.G., Koger, K., Herbst, D. and Knufer, J., Advances in AC electroluminescent powder layers, SPIE, 99, 202, (1977).

[5.46] Kramer, G., Thin-film transistor switching matrix for flat panel displays, IEEE Transactions Electron Devices, ED–22, 733, (1975).

[5.47] Fischer, A.G., White-emitting AC electroluminescent powder layers for flat panel television, Electr. Lett., 12, 30, (1976).

[5.48] Brody, T.P., Luo, F.C., Szepest, Z.P. and Davies, D.H.,
Electroluminescent display panel, IEEE Transactions Electron Devices,
ED–22, 739, (1975).

[5.49] Vecht, A., Werring, N.J. and Smith, P.J.F., High efficiency DC
Electroluminescence in ZnS(Ma,Cu), Brit. J. Appl. Phys. (J. Phys. D), 1,
134, (1968).

[5.50] Vecht, A., Werring, N.J., Ellis, R. and Smith, P.J.F., Materials control
and DC electroluminescence in ZnS:Mn,Cu,Cl powder phosphors, Brit. J.
Appl. Phys. (J. Phys. D), 2, 953, (1969).

[5.51] Mears, A.L., Parker, J., Sarginson, R.W. and Ellis, R., An operating 36
character DC EL –n display, SID Digest, 4, 30, (1973).

[5.52] Alder, C.J., Cattell, A.F., Dexter, K., Kirton, J. and Skolnick, M.S.,
Forming and failure mode studies of DC electroluminescent display, Conf.
Rec. IEEE, SID, AGED; IEEE, ED–28, 6, (1980).

[5.53] Mears, A.L. and Sarginson, R.W., Proc. int. conf. displ. for man–machine
systems, IEEE, 10, London (1977).

[5.54] Smith, P.J.F. and Werring, N.J., Progress in DC electroluminescent
displays and systems, Proc. SID/NTG-Eurodisplay, Munich (1981), pp.
149–151.

[5.55] Vecht, A., Electroluminescent displays, J. Vac, Sci, Technol., 10, 789,
(1973).

[5.56] Vecht, A., Mayo, J. and Higton, M., Blue, green and red DC EL CaS and
SrS displays, SID Digest, 8, 88, (1977).

[5.57] Higton, M., Vecht, A. and Mayo, J., Blue, green and red DC EL display
development, SID Digest, 9, 136, (1978).

[5.58] Halsted, R.E. and Koller, L.R., Electroluminescence in thin–films of
ZnS:Mn., Phys. Rev., 93, 349, (1954).

[5.59] Goldberg, P. and Nickerson, J.W., DC electroluminescence in thin–film of
ZnS, J. Appl. Phys., 34, 1601, (1963).

[5.60] Vlasenko, N.A. and Popkov, I.A., Study of the Electroluminescence of a
sublimed ZnS:Mn phosphor, Optics & Spectroscopy, 8, 39, (1960).

[5.61] Thornton, W.A., Electroluminescent thin–film, J. Appl. Phys., 30, 123,
(1959).

[5.62] Thornton, W.A., DC electroluminescence in Zn Sulfide Films, J. Appl.
Phys., 33, 3045, (1962).

[5.63] Plumb, J.L., DC characteristics of electroluminescence in evaporated
Zns:mn,Cu,Cl films, Japan, J. Appl. Phys., 10, 326, (1971).

[5.64] Abdalla, M.J. and Thomas, J.A., Low voltage DC electroluminscence in
ZnS(Mn,Cu) thin–film phosphors, SID Digest, 9, 130, (1978).

[5.65] Abdalla, M.J., Thomas, J., Brenac, A. and Noblanc, J.P., Performance of
DC EL coevaporated ZnS:Mn,Cu low voltage devices, Conf. Rec. IEEE, SID,
AGED, 165, (1980).

[5.66] Abdalla, M.J., Godin, A., Brenac, A. and Boblanc, J.P., Electrical conduction and degradation mechanisms in powder ZnS:Mn,Cu direct current electroluminescent devices, Conf. Rec. IEEE, SID, AGED, 174, (1980), IEEE, ED-28, 689, (1981).

[5.67] Vlasenko, N.A. and Gergell, A.H., Phys. Stat. Sol., 26, K77, (1968).

[5.68] Howard, W.E., Electroluminescent display technologies and their characteristics, Proceedings SID, 22-1, 47, (1980).

[5.69] Chang, I.F., Recent advances in display technologies, Proceedings SID, 22-2, 45, (1980).

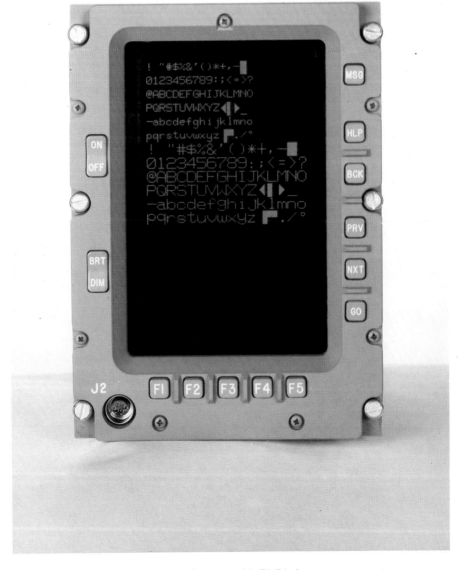

A Typical Programmable EL Display.

DISPLAY ENGINEERING: D. Bosman (Editor)
© Elsevier Science Publishers B.V. (North-Holland), 1989

IL 6: LIGHT EMITTING DIODES

D. PRICE
General Electric Aviation
FARL, United Kingdom

K.T. BURNETTE
Consultant
U.S.A.

6.1. Introduction

The evolution of Light Emitting Diodes (LEDs) started in the 1950s as a result of the initial search for materials suitable for the formation of improved quality diodes and transistors. A low efficiency yellow emitting silicon carbide device using an insulated metal contact to inject electrons has been demonstrated in the 1920s but its development was not pursued. The end of the 1960s marked the first real attempts to apply the research which had been conducted to the development of practical LED display devices; at first for watches and calculators, soon followed by road vehicle and airborne applications.

Any semi-conductor having an energy band gap wide enough to support a visible radiative recombination process is a potential candidate for the fabrication of LEDs. Materials used successfully for the formation of light emitting diodes include SiC, a compound of Chemical Group IV elements, several Group III-V compounds and several Group II-VI compounds. Forming pn diode junctions within Group II-VI compounds experienced extreme difficulties. The development resulted in a metal-insulator semi-conductor electron injection structure that to date has in general been characterised by relatively low light emission efficiencies.

Group III-V compounds turned out to be suitable for the formation of relatively efficient p-n junction type LEDs. Commercial LED development was initially restricted almost exclusively to the high market volume red LEDs, with a more recent expansion into the orange, yellow and green colours, which can also be made using the GaAs/GaP system. The potential for efficiency improvements in most types of III-V compounds LEDs still remains relatively good, however, a commercial product incentive for making these improvements is lacking because of availability of good, low voltage compatible devices (LCD), except for the aircraft and military market. LEDs have reached the point where they are being successfully applied to airborne numeric and alpha-numeric display tasks with more sophisticated graphics displays designed for aircraft installation nearing completion.

6.2. Principles of operation

The physics of LEDs is based on photons generated in a semi-conductor when an electron associated with the conduction band recombines with a hole associated with the valence band. This is a non-equilibrium process and is most easily effected by injecting minority carriers across a forward biased p-n junction. To a first approximation the energy of the emitted photon is equal to the energy difference between the initial electron and hole states and must lie between 1.9 and 2.7 eV for the photon to be in the visible region of the spectrum.

The efficiency of the radiative recombination process depends on various parameters, the most important of which are detailed band structure and crystal perfection. Solid state theory predicts radiances much greater for direct gap semi-conductors than for indirect gap semi-conductors. GaAs has the required direct gap band structure but its energy gap of 1.44 eV corresponds to near infra-red radiation. GaP has the energy gap of 1.26 eV for green radiation but has an indirect gap. However, these two compounds are completely miscible throughout the ternary composition to form a range of alloys $GaAs_xP_{1-x}$. The different types of energy gap band structure and alloys with larger phosphorous content have an indirect band structure. Diodes made from an alloy composition close to the cross-over point exhibit deleterious inter-band carrier transfer effects; optimum radiant efficiencies are obtained at x = 0.4, which corresponds to the red displays once familiar in pocket calculators and digital watches.

The argument outlined above on radiant efficiencies relates to recombination processes associated with free carriers or shallow donor and acceptor levels. In indirect gap alloys in the $GaAs_xP_{1-x}$ system the efficiencies may be increased by the use of nitrogen as an iso-electronic trap. Nitrogen, being a Group V atom, should not be electrically active but the large difference in its electro negativity and convalent radius over those of the arsenic or phosphorous atom it replaces in the lattice, creates a shallow trap close to the conduction band edge. Once an electron is bound in this trap the centre becomes negatively charged and can readily capture a hole in the long range coulomb potential to form a bound exciton, which decays with the emission of a photon. This process results in more than two orders of magnitude increase in efficiency for green GaP radiation at 565 nm than when nitrogen is not present. If nitrogen is in the lattice in excess, excitonic recombination at N-N pairs shifts the peak wavelength into the yellow region of the spectrum at 575 nm. A similar process, involving the "molecular" Zn,O iso-electronic trap is responsible for the extremely efficient red emitting devices made in GaP.

It is difficult to prepare large bulk single crystals of the ternary alloys, and devices are made in epitaxial layers grown on either GaAs or GaP substrates, depending on the precise alloy composition required. Since the lattice constants of these two compounds are significantly different, it is first necessary to grow a graded layer in which the initial composition is that of the substrate, and in which the phosphorous content is gradually increased until the desired alloy composition is reached. A layer of constant composition is then grown. The crystal perfection, and hence device luminescence efficiency, is a function of both the graded and constant composition layers. This explains the large differences in efficiencies reported for commercial and laboratory performance of devices, where the latter use more careful prepared material. A summary of the range of efficiencies is given in table 6.1. When corrected for the response of the human eye, the green 565 nm (0.3 % efficiency) commercial LED has a luminous efficiency of about 2 ℓm W^{-1} as compared to 0.5 ℓm W^{-1} for the 640 nm red device. Efficiencies of blue LEDs are an order of magnitude lower than for red and green.

Table 6.1

Efficiences of various LEDs

Colour	Wavelength (nm)	Material	Commercial (%)	Laboratory (%)
Green	565	GaP:N	0.3	0.7
Yellow	585	$GaAs_{.15}P_{.85}$:N	0.05	0.15
Red	640	$GaAs_{.35}P_{.65}$:N	0.3	0.5
Red	660	$GaAs_{.60}P_{.40}$	0.2	1.0
Red	698	GaP:Zn,O	3.0	15.0
Red	660–670	GaAlAs		1.0

Junction depths vary in the range 1–3 μm for direct gap alloys and in the range 5–25 μm for the indirect gap material. Slice processing is completed by depositing, delineating and alloying suitable ohmic contacts to the p- and n-regions of the devices.

The desired display format may be achieved with LEDs either by monolithic or hybrid construction. In the former case the format is obtained by diffusing the diodes into a single n-type substrate, which acts as the common cathode, and bringing out contacts for connecting the individual anodes to the address circuitry. In the latter case slices are diced into discrete diodes which are mounted and bonded in the required pattern onto a thin- or thick-film circuit board, which in most cases also contains the address and drive electronics.

In either type of display it is important to maximise the contrast ratio by eliminating visible optical crosstalk at distances of one LED or greater from the activated diode. Crosstalk out to the imaginary boundary separating adjacent LEDs is actually a benefit since it increases the area averaged (i.e. perceived) luminance of the picture element. Photons are generated iso-tropically across the p–n junction and, unless there is a high internal self absorption, those travelling through the bulk will be emitted through any convenient window for which the angle of incidence is less than the critical angle. In direct gap LEDs (e.g. $GaAs_{0.6}P_{0.4}$) the high absorption coefficient for the emitted radiation, particularly within the graded layer, eliminates virtually all optical coupling and only the surface of the addressed diode lights up.

Although high optical crosstalk might be considered a likely consequence of the emitted photon transparency of indirect band gap LEDs, the optics of the monolithic chip geometry instead predicts that virtually no optical crosstalk should occur. The large refractive index of indirect band gap LEDs causes the critical angle to define a very small cone (i.e. 34° included angle for unencapsulated GaP) through which light reaching a surface can be emitted. Due to this emission angle restriction a perfect monolithic chip with parallel surfaces could only exhibit optical crosstalk for multiple reflections of those photons which are initially emitted within emission cone windows (i.e. symmetric about normals to the display surface) which have apexes located within the LED junction area. All other emissions would experience total internal reflections. The Fresnel reflections within the emission window would result in a very rapid decay of luminance with distance from the LED junction edge and visible crosstalk on the chip's planar surface would be absent. The majority of photons are emitted outside the emission window and would eventually reach the monolithic chip edges where some would be emitted but most would again experience total internal reflection. LED emissions viewed microscopically through untreated LED chip edges do in fact allow the image of the LED junction to be observed.

In practice, well made indirect band gap LED monolithic chips perform as described above with one major exception. The bulk GaP substrates are far from being perfect crystals and as a result light scattering into the LED emission cone occurs both within the bulk crystal and at its unpolished surfaces. In a darkroom lighting one LED will cause the entire monolithic chip to illuminate at a low luminance level. Under normal cockpit display contrast, optical coupling 0.2 mm from the edge of an energised LED is no longer visible. The measured optical coupling distances of the indirect band gap LEDs would pose a

potential image quality problem for video displays having resolutions   much   in
excess of 5 LEDs per mm.

LED monolithic arrays (figure 6.1) intended for use in dedicated display

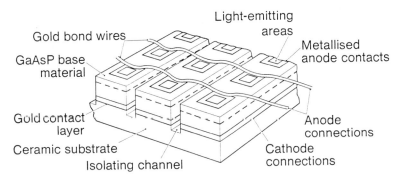

Fig. 6.1 Method of fabrication of a monolithic LED matrix.

formats can, at the design stage, be layed out in virtually any desired
geometric configuration. More general display surfaces are obtained using
matrix addressable arrays in which the anode and cathode connections are
orthogonal to one another.
The most difficult issue encountered in fabricating good monolithic LED arrays
is achieving diode electrical isolation. Anode isolation is readily achieved
since the diode junction of unaddressed LEDs are reverse biased when the
addressed LED is energised. Several approaches have been used to obtain the
more difficult cathode isolation. A truly monolithic process, analogous to that
used in silicon integrated circuit technology, has been demonstrated in
monolithic chip sizes of up to 125 mm square. The technique requires special
epitaxial material and a complex processing schedule involving both deep and
shallow diffusions. The objectives of the research, which centred on achieving
cathode isolation and a flip chip electrical connection structure, were fully
achieved. Further research on this LED structure is being conducted to improve
it from a 2.5 to 5 dels mm^{-1} resolution. The major advantage is an unobscured
viewing surface, since both the cathode and anode connections are made on the
rear side of the monolithic chips. The latter feature is highly desirable when
used for the construction of mosaic display surfaces based on the four edge
abuttable LED display module technique described in geometric configuration
(see below). An alternative display fabrication technique which has been in use
now for over ten years involves conductive epoxy bonding monolithic LED array
chips to a thermally conductive electrically insulating substrate. The
isolation is then achieved using a precision mechanical saw that cuts through
the complete semi-conductor slice and the gold surface layer on the ceramic

substrate carrier. Connections to the anode rows are made either by stitch bonding or beam leading. This technique has been used with success to fabricate small single and multi-colour non-abuttable arrays with resolutions up to 5 dels mm^{-1}. It has also been used for constructing large area LED displays at a 2.5 dels mm^{-1} relation.

A third fabrication technique involves the hybrid assembly of discrete LEDs using automatic die placement equipment. This approach has also resulted in succesful matrix arrays up to 2.5 dels mm^{-1} resolution. Luminance outputs of 17×10^3 cd m^{-2} are typical of recent GaP material at drive currents of 0.03 A mm^{-2} leading to contrast ratios over 4:1 in sunlight.

6.3. Physical characteristics

6.3.1. Luminance characteristics

Light emitting diodes produce a luminance output which is roughly proportional to the current density passing through the forward biased LED junction area. The current required to produce a desired luminance level is therefore directly proportional to the junction area of the particular LED being addressed and is the quantity which has to be controlled when using a LED as a display device. The forward biased voltage drop across a LED establishes the minimum supply voltage level needed to operate the LED (typically from 5 to 8 V with voltage drops up to 3.5 V) and also determines the power efficiency of the LED/driver circuit combination.

Non-linearity in the LED luminance versus current characteristic is introduced by two effects: junction temperature induced efficiency (and colour) changes and current density induced luminance saturation. Both of these effects can result in a drooping characteristic. The temperature rise produced by the current flow is determined by a large number of factors which can vary significantly based on the layout, materials and fabrication techniques used in the construction of a specific display. The thermal conductivity of the LED array substrate, of the electrical and mechanical bonding techniques used to affix the LEDs to it, and of the ceramic/heat-sink interface are particularly important if current magnitude induced non-linearities are to be minimised.

In the family of LED materials described by the chemical notation $GaAs_x P_{1-x}$, as the fractional composition (x) of arsenic increases, both the junction temperature and luminance saturation induced non-linearity effects in the luminance versus current characteristic become more pronounced. Green 565 nm GaP (i.e. with x = 0) typically exhibits a luminance characteristic which is linear in current from very near its maximum saturated luminance of 10^4 cd m^{-2}

to near luminance extinction (i.e. linearity has been tested down to 3.4×10^{-3} cd m^{-2} but based on LED theory should continue to decrease thereafter in a linear fashion). A junction temperature rise in a GaP LED causes a linear decrease in luminance in the -40 C to $+125$ C temperature range of nominally 0.8 % C^{-1} and a 0.12 nm C^{-1} shift in the emission spectrum toward longer (yellow) wavelengths. As the arsenic concentration in the $GaAs_xP_{1-x}$ compound LEDs is increased the non-linearity of the luminance versus current characteristic becomes more severe at high current levels with saturation occuring gradually over a larger range of currents as the maximum luminance is approached. Red (655 nm) $GaAs_xP_{0.4}$ exhibits a luminance decrease of greater than ten times that of GaP and a colour shift of 0.2 nm C^{-1} (nearly twice that of GaP) as the junction temperature increases, see figure 6.2. It should be

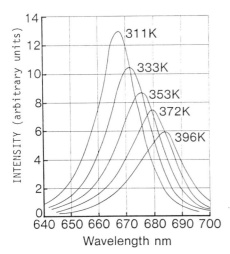

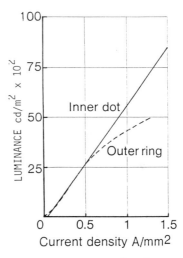

Fig. 6.2 Radiant intensity spectrum of red emitting LED as a function of heat sink temperature.

Fig. 6.3 Luminance of red emitting LED as a function of current density.

noted that all of the preceding temperature sensitivities are only nominal values and will vary somewhat depending on the fabrication techniques employed.

6.3.2. Luminance control options

The nearly linear luminance versus current characteristic of LEDs (figure 6.3), particularly at low luminance values, allows current amplitude dimming of LEDs to levels well below the 0.2 cd m^{-2} level desirable for night driving and flying or the 0.04 cd m^{-2} level needed for use with night vision goggles. The practical problems encountered with achieving these dimming levels in actual displays are those associated with implementing a current source control circuit design that is capable of handling a dynamic dimming control range of from 2,500 to 30,000 (depending on the type of display application) while continuing to look like a current source to the LEDs. A dimming control implemented to accept a digitally encoded input would for instance have to be capable of using a 16 bit binary luminance control word to avoid the perception of luminance changes as steps at the low end of the control range. Failure of the drive circuit to perform as a current source during dimming can result in luminance uniformity problems since the variations present in the knee region of the LED voltage-current characteristics can then influence the luminance produced by the LED.

The very short luminance response times of LEDs, in the medium to low nanosecond range, also allow either or both linear pulse duration and linear pulse frequency control of LED luminance to be achieved. The response times of typical silicon integrated circuit logic and driver circuits serve as design limitations on the linear dynamic dimming ranges that can be achieved using these LED luminance control techniques.

6.3.3. Geometric configuration: small area displays

In small area displays, there are virtually no restrictions on the geometric configurations in which LEDs may be employed. Numeric readouts, bargraphs, scales, reticles, special geometric figures and small dot-matrix arrays can be fabricated and packaged to meet almost any specification of environmental requirements.

At resolutions of up to 2.5 dels mm^{-1}, hybrid LED arrays have been successfully formed using both semi and fully automated single diode placement and bonding techniques.

An alternative technique which has proved to be satisfactory for display resolutions between 1.6 and 5 dels mm^{-1} is the manual placement of monolithic array LED chips. The chips, which typically have edge dimensions from 3 to 12.8 mm, are abutted to one another in a mosaic fashion on high thermal conductivity substrates so as to produce the desired uninterrupted resolution across the entire display surface area and are then saw cut isolated and bonded. Both construction approaches have been used to produce military specification qualified LED arrays, with the monolithic approach used at resolutions of up to 4.6 dels mm^{-1}, for film annotation and for helmet-mounted applications.

As is the case for virtually all electronic displays, the overall design of a

LED display system must be carefully optimised to produce a result that will be considered satisfactory by the observer. In the case of LED direct view displays used in non-controlled environments this requires that special attention be paid to the optical filtering (section I.1.4.2,[1.12]), to the electrical drive/address techniques employed and to the thermal design.

Overall display system design optimisation is equally critical when attempting to employ a LED display as an image source for a helmet-mounted display application (see section II.6.7.4). In this case heat dissipation, weight and volume impose design restrictions. In addition, the luminance required of the image source is critically dependent on a variety of design factors only indirectly related to the LED. These factors for instance include: the image combiner field of view, the exit pupil size, the quality of the image combiner narrow band interference filter (which reflects the LED light with minimal blockage of the background scene), the transmittance of the high ambient light visor, and the scene luminance level when the pilot changes to the clear visor.

6.3.4. Geometric configuration: large area displays

The low operating voltages associated with dot-matrix LED displays require that relatively large average drive currents be used if the displays are to be operated at the same power levels as other higher voltage displays. Short low resistance current leads are therefore necessary to minimise lead losses and to provide the near negligible voltage drop between the LED drivers and the addressed dot-matrix display element (del), that is required to produce displays which have a uniform luminance distribution across the entire surface. The method used to achieve this objective and still be able to have large-area LED head-down displays is to form the display surface using a mosaic of edge abuttable independent display modules, each having its own set of integrally mounted drive and address integrated circuits.

The first successful large-area LED graphics display to be constructed using the modular building block approach was completed in 1978 [6.2]. The display has active area dimensions of 104x78 mm and consists of four green emitting two-edge abuttable LED modules mounted side by side to form the display surface. Each module consists of a ceramic substrate of the same width as the LED surface but 133 mm long and 1 mm thick with a nominal 2.5 del mm^{-1} green emitting LED array of 26x78 mm dimensions centred on the ceramic and having its column driver hybrid integrated circuits mounted on the rear surface. The objective of this display was to allow evaluation of the feasibility of integrating the LED, optical filter, ceramic, metallurgy and silicon integrated circuit technologies required to build modular LED displays suitable for use in portraying real-time flight control information to a pilot. Based on the

success   of   this demonstration display, a programme was carried out to develop
flightworthy graphics displays utilising a full four-edge abuttable module.
Figure 6.4 is a photograph of a green emitting, nominal 2.5 del mm^{-1} display

Fig. 6.4 Green emitting 64x64   Fig. 6.5 High speed graphics display with
del, 26 mm square, 4-edge   20 module mosaic surface (Photograph
buttable LED display module   courtesy Litton Systems Canada).
(Photograph courtesy Litton
Systems Canada).

resolution, GaP LED display module. A heat sink serves as the basic structural
element for the module to which the LED display surface ceramic substrate,
driver substrates, connectors and a power line filter capacitor are solidly
attached.   The display module is a cube having about 26 mm edge dimensions. The
modules are designed to permit their individual removal or replacement in the
display mainframe. Figure 6.5 is a photograph of an operating 130x104 mm^{2}
active area LED display formed with a mosaic of 4x5 four-edge abuttable modules
of the type shown in figure 6.4.

The size of the standard module was selected to accommodate either a 64x64
dels, nominal 2.5 del mm^{-1} resolution graphics LED array; or a 128x128 dels,
nominal 5 del mm^{-2} resolution video LED array where the binary equivalents of
64 and of 128 are compatible with standard silicon digital drive/address
circuitry. The choice between different digitally compatible module sizes
represented a trade-off between minimising the total number of modules in large
displays, and achieving a module small enough to permit reasonable flexibility
in the dimensioning of displays built using the modular building block
approach. The ultimate maximum size of a module is limited by the current
sourcing and sinking capabilities of the module's silicon integrated circuit
drivers.

The primary performance advantage gained from using modular display construction stems from the fact that the image legibility of the complete display is the same independent of the number of modules used to construct it. This follows because the display modules are all updated with information in parallel (i.e. at up to their 500 Hertz refresh rate capability) with each module acting effectively as an independent display. As a result, display size does not influence display maximum image speed, image quality, image positioning accuracy, emitted luminance, contrast, or viewing angle; these quantities are known and fixed when the display module is constructed. An in-depth discussion of both the advantages and disadvantages of modular flat panel display construction is contained in [6.3]; see also the paragraph on addressing techniques.

6.3.5. LED failure mode

LEDs have two potential failure modes. Under the forward biased light emitting drive condition, excessive drive current densities can result in junction temperature rises which permanently alter the diode luminance/current density characteristics; extreme overdrive conditions destroy the device. Junction temperatures above approximately 200 C, whether induced through driving the LED or by elevated ambient temperatures, can cause degradation in green GaP LEDs. In general, increasing arsenic compositions in the $GaAs_xP_{1-1}$ tertiary compounds causes the temperature limit to be reduced. When adequately cooled the GaP LEDs can tolerate continuous current densities of up to about 2 A mm^{-2} without damage and higher levels in a pulsed drive mode. Comparing this drive limit with the 15 to 30 mA mm^{-2} maximum current densities needed to satisfy practical aircraft dot-matrix display applications, explains the large tolerance the displays exhibit in the presence of inadvertant short duration overcurrent or timing fault induced 100 % duty cycle drive conditions.

The second potential LED failure mode occurs when the reverse bias breakdown voltage limit of the LED is exceeded by a sufficient amount. Depending on the diode fabrication techniques used, the Zener or avalanche multiplication mechanisms (that are made use of in silicon Zener diode applications) will be operative. The excellent reliability record of LEDs in practical display applications is testimony to the ease of eliminating these failure modes by proper design.

6.3.6. Luminance degradation

Luminance degradation as a function of operating time cannot be avoided. All of the present light emitting display technologies including CRTs, plasma panels,

EL panels and LED panels exhibit some degree of luminance degradation as a function of operating time. For LEDs operated near their density/junction temperature limits, the half-life (time to degrade to half the initial luminance value) of green LEDs is about 15,000 hours to 25,000 hours. Red LEDs typically exhibit somewhat shorter half-lives under the same operating conditions.

The half-life of LEDs increases significantly with reduced junction operating temperatures and current densities, with the temperature of the diode junction playing the most important role. As an illustration of this, half-lifetests run on continuously scanned xy matrix arrays of green LEDs for a period of five years (i.e. 43,800 hours) predict luminous half-lives of at minimum 500,000 hours. The 2.5 del mm^{-1} LED arrays used for these tests were operated without heat sinks and using drive conditions consisting of 500 Hz refresh rate current pulses of 100 mA amplitude and 10 % duty cycle in a 25 C ambient temperature environment. This drive condition is equivalent to that used for the graphics portrayal LEDs built to date with the exception that the test uses a factor of 20 higher pulse duty cycle in order to stress the array.

6.4. Addressing/driving

6.4.1. **Addressing techniques**

Lacking either inherent memory or a convenient method of either forming or incorporating an active control drive at each LED location, the most effective method available for activating LED displays is through one line-at-a-time (section I.1.5) scanning of data onto the display. When displaying segmented characters (such as numeric readouts) the readouts are typically scanned so that all of the data needed to describe each character to be displayed is provided in parallel as each readout location is sequentially scanned.

When dot-matrix display arrays of greater than 2x2 dels are to be addressed, a saving in the number of drivers and a reduction in the number of address lines is achieved when matrix rather than direct drive addressing is used.

The total time averaged current, I, required to drive an individual LED at the specified luminance level, is the same independent of the addressing/driving technique employed to operate it. This implies that in a dot-matrix array of M columns and N lines the peak line driver current is N x M times the average del "on" current; this peak current pulse requirement must be met by the array line scanning drivers. These requirements can act as practical constraints on how large a single LED display panel can be made, as can the electromagnetic interference (EMI) induced by high pulsed currents.

Another potential constraint on the display element dimensions of a single LED display panel is the saturation of the LED luminance. To hold the perceived (i.e. time averaged) luminance of a LED display constant while scanning larger and larger array sizes requires that the LED instantaneous pulse current be increased so as to maintain the same average drive current and hence luminance (section I.1.5). Eventually this process results in increasing the LED current density to the point where luminance saturation occurs.

The actual magnitude of the current density induced luminance saturation limit varies with the LED materials used, the techniques used to fabricate the LED junctions and the metalisation structure/metallurgy employed. Red emitting LEDs in general have lower pulsed current density saturation limits than do green GaP LEDs, however neither would be adequate for use in fabricating for instance single panel, 1024 line, high luminance video displays. To avoid the ultimate size limitation on sequentially scanned matrix addressed displays, the objective of LED display development has been the development of mosaic displays that use independently refreshed display modules (figure 6.4) to limit to a small manageable quantity the number of dels that have to be sequentially addressed using the line-at-a-time addresss technique. Using this approach, drive current and saturation problems have been successfully avoided for pulse duty cycles of as low as 0.5 % or 200 sequentially addressed LEDs per module. It should be noted that in general driver circuit capacity problems are encountered before saturation limits are reached.

6.4.2. LED drivers

The low voltage drive requirements of LEDs can usually be satisfied using commercially available integrated circuits selected to meet the necessary specification. The choice between common anode and common cathode drive circuit configurations is usually based on the electrical lead current handling capacity of the LED array design employed, which in turn is based on minimising LED junction obscurations due to wires or surface metalisation layers. The common terminal driver, whether it is associated with the common cathode or the common anode configuration, is the one which is scanned and which must handle the high instantaneous current loads. Its primary function is that of a switch having a virtually constant saturation voltage independent of whether it is passing the current of one or all of the LEDs in a data addressed line or column. Whether the inputs to the data line drivers are provided by serial shift registers or by parallel multiplex circuits, the LED address circuitry used with the driver generally must be able to store/latch the transient input data until new data is entered. The essential feature of the data driver is that it functions as a current control circuit. Enabling this circuit using a pulse duration control signal fed simultaneously to all data drivers has provided one straightforward method of controlling the overall luminance of the display under either automatic or manual operator control without influencing the relative luminances of the data displayed. Circuit designs for use with LED displays are still evolving, particularly for video drivers where digital grey shade storage and control are necessary.

For applications such as the graphic display module shown in figure 6.4 it was found necessary to have custom made silicon integrated circuits built, in order to make the combined line and column address/driver circuits small enough for mounting on the sides of the module. The advantage of this construction approach is the reduction in driver line power losses due to the shorter lead lengths, and the reduction in external module connections (i.e. reduced from 128 to 16 parallel data lines/module plus power, ground, clock, control and status sensing for the graphics display module).

6.5. System interface

The low signal level requirements present at the input to LED display address lines make these displays compatible with virtually all available types of logic, multiplex and memory interface circuitry. The type and complexity of the LED display system interface is therefore determined almost exclusively by the intended display application and in particular by the complexity and rate of change desired in the information to be displayed, not the display technology to be employed.

6.6. Visual characteristics

6.6.1. Optical properties

The legibility and efficiency achievable with a LED display depends rather critically on the effectiveness of the optical design which is implemented. In general the design of a LED display should treat the LED material, the substrate to which it is affixed, the visible electrical connections to it, any optical coatings applied to the display surfaces, and the optical filter or cover glass employed, as integral parts of a single composite optical system. Typical LED processing results in LED wafers and dice having flat smooth reflecting surfaces of between 30 and 38 % specular reflectance prior to encapsulation and 18 to 23 % afterwards. The diffuse reflectance associated with the LEDs varies from much less than 1 % up to 2 %.

The objective of optical filtering in any type of direct view display is to reduce the maximum combined display reflectance. In severe environments such as aircraft the allowable figures are less than approximately 0.5 % or a reflected maximum background luminance of approximately 170 cd m^{-2}. This guideline also would satisfy the requirement that foreground luminance which is legible against the 170 cd m^{-2} background luminance should be adequate in the glare induced veiling luminance environment created when the sun is within the pilot's instantaneous field of view.

A variety of optical designs capable of meeting the foregoing requirements have been developed for LEDs. The most cost effective of these designs utilise colour transmission selective filters (example I.1.7). Since the inherent diffuse reflectance of the LED is low, the primary objective of the filters is to maximise the filter transmittance to the LED light while significantly reducing specular reflectance and making a slight reduction in the diffuse component. A filter developed for use with green GaP LEDs, and which meets military specification environmental requirements, may be used as an example of a filter requiring little in the way of special display surface treatment. The filter utilises a laminated sandwich structure consisting of an anti-reflection coated UV attenuating filter, a neutral density circular polariser, and an IR absorbing filter with an EMI attenuating anti-reflection coated rear surface. The circular polariser effectively attenuates the LED specular reflectance component, the colour and neutral density filtering attenuate the diffuse component and the anti-reflection coatings minimise the external filter surface Fresnel reflections. A combined reflectance of 0.3 % has been achieved using this filtering method. Such techniques are expected to permit the display of up to 8 shades of 'grey' (section I.1.4.4). While the foregoing approach accomplishes the filtering objective, it also restricts the display emitted luminance to between 27 and 35 % of that emitted by the LED. As a result more efficient filtering techniques have been developed which concentrate on improving the display surface optical design. One method is essentially equivalent to the black matrix shadow-mask approach used on colour CRTs. The other involves the application of refractive index matching anti-reflection coatings to the LED arrays. Combined reflectances of well below 0.5 % are obtained, in combination with a circular polariser which at best transmits only about 40 % of the LED emitted light.

6.6.2. Luminance/contrast

The present performance of green 2.5 del mm^{-1} GaP LED arrays operated at a 1 mA/LED time averaged drive current (i.e. 64 mA, 31.25 µs duration pulses every 2 ms) is 7.54x10^3 cd m^{-2} of luminance spatially averaged over a 0.2 mm diameter LED measurement area (i.e. 15.4x10^3 cd m^{-2} for laboratory devices).

When this luminance is area averaged over the surface of a dot-matrix display character, symbol, or image so as to include both emitting and non-emitting areas (i.e. assuming no emission between LEDs) an area averaged luminance of 1.5×10^3 cd m^{-2} is obtained. Roughly equivalent performance can be achieved with red LEDs which, while more efficient radiometrically, lose this advantage due to the spectral sensitivity of the eye (figure I.1.15).

Contrast ratios achievable using LED displays vary depending on the filtering techniques employed. A low altitude (green LED numeric readout) radar altimeter in the F-111/FB-111 and EF-111 series aircraft produces an area averaged emitting to reflected luminance contrast ratio of one in a 10^5 lux illuminance environment. A green 78x26 mm active area 2.5 del mm^{-1} resolution LED alpha-numeric readout dot-matrix display being produced for a F-16 data entry display application has been characterised as having an area averaged emitted to reflected luminance contrast ratio of between two and three.

Radiant intensity spectra of a red LED standby sight plotted as a function of heat sink temperature are shown in figure 6.1. The measured intensity distributions show that as the temperature increases, the peak intensity decreases and moves to longer wavelengths. The effect on perceived brightness is even more pronounced since the eye's sensitivity drops rapidly with wavelength in this region from 72.8 ℓm W^{-1} at 650 nm to 11.6 ℓm W^{-1} at 680 nm. Colour uniformity variation either due to process control constraints or temperature variations have proved insignificant for green or red LEDs.

6.6.3. Image quality

Like other dot-matrix display techniques, the LED display exhibits excellent image edge definition. Special design eliminates visible indications of electrical crosstalk. Optical crosstalk between "on" and "off" elements is most severe in monolithic flip chip GaP LED arrays where the material is nearly transparent to the light it generates and no physical boundaries exist between adjacent LEDs. Even in this type of array the luminance measured at an "off" LED immediately adjacent to an "on" LED is typically less than 2.5 % of the "on" LED luminance. A greater than 5 % optical crosstalk level is capable of producing visually noticeable effects for displays adjusted to provide a peak image contrast ratio of ten to one. Increasing the contrast by either increasing luminance or reducing the ambient will of course eventually make the crosstalk noticeable.

Dot-matrix displays of all types exhibit spatial variations in del luminance and potentially also in del colour. These variations can stem both from variations in the drivers used to apply signals to the del and from the differences in the electro-optical characteristics of the individual elements.

LED displays having normalised standard deviations in luminance of 12 % of the mean value or less have been found to provide satisfactory graphics display image luminance uniformity provided that the luminance variations are spatially randomised across the display surface. By comparison, variations perceptable as line or column luminance variations can, at the same variation magnitude, be very obtrusive due to the larger size of the line image.

Luminance variations permissible on displays intended for professional video applications are determined by the need to avoid overlap between adjacent grey shades rather than just the issue of whether the variations are noticeable. Using the overlap criteria, a maximum variation of up to 17.2 % from the mean grey shade operating level would be permissible for an eight grey shade, $\sqrt{2}$ grey scale ratio video display. Luminance uniformity distributions having a normalised standard deviation of between 8 and 10 % from the mean would therefore be capable of satisfying this video LED display luminance uniformity criterion, which can be achieved using developed 5 del mm^{-1} monolithic chips. However, further processing improvements are desirable to achieve higher chip yields.

6.6.4. Flicker/dynamic visual effects

Current production, prototype and development aircraft cockpit LED displays employ 500 Hz or higher refresh rate in order to avoid in the vibrating environment both static and dynamic display image flicker effects. Flicker associated with display imagery which is static or nearly static on an observer's retina may be overcome at a refresh rate of 60 Hz or higher; the higher rates being required for larger LED displays (section I.2.4). Dynamic flicker or more accurately the appearance of multiple spatially separated images occurs at these low refresh rates as a result of motion of the display image on the retina of the observer's eye. The effect is observed under vibration and also when the observer averts his eyes to look at display information located elsewhere within the cockpit. This dynamic image phenomenon is experienced for all periodically refreshed display media having image persistence durations less than the refresh period (i.e. P-43 phosphor CRTs, plasma panels, TFEL and LED are examples). The effect is emphasised in dot-matrix displays due to the sharpness of the imagery portrayed. Refresh rates of 450 Hertz and greater effectively eliminate the phenomenon under practical aircraft cockpit illumination conditions [6.4].

6.6.5. **High resolution graphics/video**

LED arrays with resolutions of up to 5 dels mm^{-1} have been successfully demonstrated in both green and red monolithic chips (green in sizes of up to 26x26 mm). Military specification qualified devices being applied as film annotation arrays in airborne reconnaissance cameras utilise resolutions of up to 4.6 dels mm^{-1}. A number of arrays at resolution between 4 and 5 dels mm^{-1} have been applied to head-up and helmet-mounted displays and to date have been found to provide satisfactory performance when displaying alpha-numerics and graphic symbology.

The very high resolutions being sought for use in helmet-mounted displays video image sources (i.e. 20 dels mm^{-1} or greater) have not been attempted using LEDs. Diode junction formation at these resolutions would be feasible; however, a significant advancement in present LED electrical connection fabrication processing techniques would be required to achieve reliable devices. Purer LED substrates would probably also be required to avoid problems in achieving adequate luminance uniformity. No research to develop very high resolution displays has been reported.

6.6.6. **Colour**

Very little colour LED research has been conducted. Red, orange, yellow and green single colour LEDs, due to their importance in consumer and commercial products, have received the greatest research and development emphasis. However, this has been restricted largely to improvements in existing products. Blue and dark green emission colours, which have been demonstrated in large energy gap materials such as gallium nitride and silicon carbide, have received scarce attention from a research and development standpoint. Difficult processing, very low initial emission efficiencies, and a generally conceded low prospect for high volume device sales in green and blue displays has been responsible for this situation. Metal-insulator semi-conductor LEDs have fared little better, an exception being the dark green emission zinc telluride devices.

Lacking an efficient material capable of producing blue, a full colour LED display is at best a possibility for the distant future. Gallium phoshide which has served as the basis for efficient green, yellow, orange and red single colour LEDs has been the subject of very modest multi-colour LED display research [6.5]. Two terminal multi-colour hybrid and monolithic LEDs utilising current density to control the mixing of the red and green primary emission colours and using pulse duration to control luminance have been demonstrated. Attempts to achieve spatially uniform mixed colour for application to large area multi-colour displays proved to be beyond the means of the modest development efforts conducted.

More recent multi-colour display research has concentrated on monolithic LED arrays utilising independent red and green primary colour diode junctions (to provide yellow and orange mixed colours) in a superimposed geometric structure. This structure is suitable for low resolution displays because the colour mixing occurs within the LED. The technique has also provided LED efficiencies approximately the same as those of single colour red and green LEDs. Colour dot-resolutions of 1.6, 2,5 and 5 colour dots mm^{-1} have been demonstrated, the latter in up to 126 mm square monolithic chips. Improvements in colour uniformity and demonstration in the four-edge abuttable 26 mm square modular building block format needed for large-area head-down displays, still must be accomplished to prove the feasibility of this multi-colour display approach.

6.6.7. Viewing angle

The viewing angle associated with LED displays depends on the optics used to form the final display product. The use of restricted viewing angle lensed LED indicators and displays is common in consumer product applications where it is assumed that legibility will be achieved by orienting the display. In head-down aircraft display applications the displays are attached to instrument panels at locations which can result in viewing angles of up to about 45°. The latter application is satisfied using flat LED arrays. These arrays may be viewed at angles approaching 90° from a normal to the display surface. This is possible because the luminance of the characters depicted using these LED arrays remains approximately constant as a function of viewing angle, although the apparent foreshortening of display characters experienced at large viewing angles (i.e. $> 60^{o}$) should be avoided.

6.7. State of development

6.7.1. Special applications of LEDs

At the present state of development light emitting diode displays are likely to fulfil a niche market. The absence of efficient blue emitters prevents their use in displays demanding a full range of hues. Their expense tends to make them uncompetitive for large-area direct view displays. However there are a number of applications in which they are likely to find continued use.

In touch panels with light emitting legends a limited range of symbology is needed, and there is advantage in using devices of small size. Counter/pointer instruments using a combination of analogue strips or arcs with odometer-type numeric readouts have secured acceptance for a number of functions. Direct replacement by dedicated LED arrays provides an economic and reliable replacement for such presicion indicators.

Other applications where other technologies have failed to penetrate are small size, high resolution, image projection systems. Here high intensities are commonly required. Illuminated passive displays tend to be poor contenders as there is no suitable substitute as illuminant for the filament bulb, not withstanding its poor reliability and life. LED arrays have therefore found use in helmet sighting systems, night vision goggle annotations, displays for scanning infra-red imagers, simple head-up displays and sights, and in the non-airborne sectors, projected displays for automobiles and printer heads.

6.7.2. Head-down programmable pushbutton switches

A typical operating programmable pushbutton array is illustrated in figure 6.6. Each module of 40x30 mm contains a LED array of 48 horizontal by 24 vertical elements which may be addressed in a number of fonts, from 6x9 to 18x24. Using

Fig. 6.6 Multi-function key pad (programmable text on keys) 1986 (Photograph courtesy Litton Systems Canada).

green LEDs peaking at 565 nm a contrast ratio $(L_f-L_b)/L_b$ of 2:1 is obtained at $10.^5$ lux luminance by use of filters and black areas between the emitters. Night vision goggle compatibility is achieved by controlled spectralcontent and wide-range dimming controls.

A display refresh rate of over 400 Hz, avoids vibration-induced break-up; an operating temperature of 55 C is possible with conduction cooling alone.

6.7.3. Control/display unit

A control/display unit for direct interfacing with a MIL STD 1553 data bus is illustrated in figure 6.7. On a ceramic substrate 8,400 discrete LEDs are laid down in 240 character positions of 7x5 elements each. Typical display content leads to a display surface dissipation of 15 W to achieve a 2:1 contrast ratio in high ambient light conditions. This heat is conductively removed via an aluminium backplate on which the display drive circuits are also mounted.

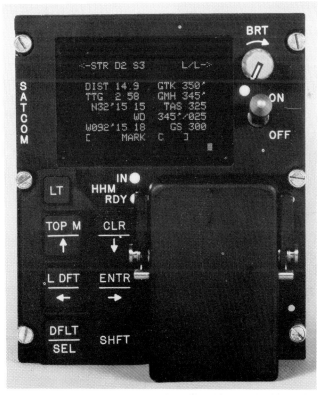

Fig. 6.7 Control/display unit for MIL STD 1553 bus (1986) (Photograph courtesy Litton Systems Canada Ltd.).

A contrast enhancement filter including circular polariser is used, and the refresh rate approaches 300 Hz.

6.7.4. Counter/pointer instruments

The problem of the flickering least significant digit of a numeric readout has been solved by making the least significant digit to appear to roll-up or down the display, by suitably driving a LED matrix, providing rapidly-appreciated trend information (figure 6.8), [6.6]. Hybrid construction isused, with discrete LEDs laid down to form a small matrix for the digit counters and concentric areas for pointer simulation. Auto-brilliance control is used operating from photodiodes sensing ambient light levels; fixed markings are illuminated by integral tungsten filament lamps which are adjusted to match LED brilliance. Contrast ratio in a 10^5 lux environment (emitted plus reflected over reflected) is of the order 3! Refresh rates over 200 Hz ave proved sufficient to overcome flicker.

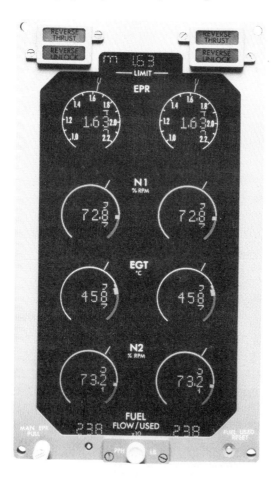

Fig. 6.8 Primary engine panel; note scrolling of least significant digit (Photograph courtesy Smith Industries).

6.7.5. Imaging display Systems

An early application to be considered was the provision of a standby sight for a head-up display. It is advantageous, for the purpose of optical combination, for the standby reticle to be of a different colour to the green cathode ray tube and they have conventionally been implemented in red. Since red LEDs are particularly efficient and crisp definition is required, a fixed-format light emitting diode device is very suitable and has been successfully evaluated in flight. A device for a typical stand by sight is shown in figure 6.9, before the processed slice has been cut into individual devices, and consists of a central 'piper' diode 0.17 mm in diameter surrounded by a separately addressable segmented circle 1.7 mm in diameter. It can be seen that the aluminium metalisation has been defined as grids over the diode areas to spread the current and give uniform luminescence.

The luminescence of the two types of diode as a function of current density at

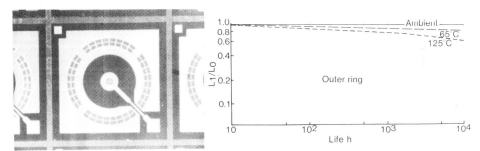

Fig. 6.9 A typical head-up display stand-by sight (Photograph courtesy G.E.C. Aviation).

Fig. 6.10 Stand-by sight lifetest results.

20 C is shown in figure 6.3. The required maximum luminance for stand-by sights is approximately 30×10^3 cd m^{-2}, which can be achieved with current densities below 1 A mm^{-2}. In fact brightnesses much in excess of this figure are achievable with the 'piper' diode, since its small area involves currents of less than 50 mA. The same is not true for the outer-circle diodes. With an area of just below 1 mm^2 currents of the order of 1 A are required to achieve the specified luminance. Even with massive heat sinking, the thermal impedance of the GaAs results in a significant junction temperature rise which causes the luminance to saturate.

Lifetest results of the outer diode areas, driven at current density of 0.80 A mm^{-2}, corresponding to an initial 20 C luminance in excess of 35×10^3 cd m^{-2}, are summarised in figure 6.10. Even at 125 C times to half-luminance in excess of 10^4 hours are obtained.

6.8. Thermal imaging systems

Linear arrays of LEDs are used to reconstruct area displays of the outside
world detected by linear arrays of cadmium-mercury-telluride photo sensors.
As the scene is scanned mechanically in the direction normal to the linear
photo sensor array a linear LED array fed from the photosensor array is
similarly scanned by a mirror thus placing the emitting array in the
appropriate line in the reconstructed image plane.
This application also calls for fine resolution; 40 μ areas are used at 100 μ
pitch. The wide range brilliance control available allows reconstruction of a
IR-sensed image with a very wide dynamic range without resolution loss.

6.9. Multi-colour and monolithic matrix arrays in development

A series of papers [6.7,6.8] describe the construction of monolithic matrices,
in which isolation has been successfully carried out by etching, and displays
in which multi-colour emitters are fabricated in GaP material at a fundamental
stage in the diffusion process. A 320x320 del multi-colour flat panel display
device has been demonstrated. Automatic die-bonding and wire bonding enable 1
mm pitch to be reliably realised.
The monolithic arrays realise dels of 300 μ x 250 μ. However luminance levels
demonstrated do not exceed 170 cd m^{-2}.
A printer head application demanded a 265 mm long array, achieved by bonding 64
dot chips to a substrate, the dot size being 50 μ and pitch 105 μ.
To enable symbol registration chip dimensional tolerances had to be maintained
to 10 μ.
High resolution high luminance devices for printer heads are reported [6.9]
using GaAlAs; 40x70 μ active areas at 55 μ pitch are achieved.

6.10. Acknowledgements

The authors wish to acknowledge the contributions of dr. D. Wickenden of the
General Electric Company p.l.c. (UK), mr. D. Kennedy of Optotek Canada. Part of
the UK aspect of this work has been carried out with the support of the United
Kingdom Minsitry of Defence and part of the United States aspect with the
support of the Flight Dynamics Laboratory, US Air Force.

REFERENCES

[6.1] Varon, J. et al, High brightness GaAlAs heterojunction red LEDs, IEEE Transactions Electron Devices, vol. ED-28, no. 4, (1981), pp. 416-420.

[6.2] Burnette, K.T., Moffat, A.J. and Wareberg, P.G. Multi-mode matrix (MMM) modular flight display development, Proceedings of the Society for Information Display, vol. 21, N2, (1980), pp. 143-156.

[6.3] Burnette, K.T. and Melnick, W., Multi-mode matrix (MMM) flat panel LED vector graphic concept demonstrator display, Proceedings of the Society for Information Display, vol. 21, N2, (1980), pp. 113-126.

[6.4] Riley, T.M., Multiple images as a function of LEDs viewed during vibration, Human Factors, vol. 19, no. 1, (1977), pp. 79-82.

[6.5] Kennedy, D.I., Fabrication and properties of gallium phosphide variable colour displays, Micro-electronics, vol. 5, no. 3, (1974), pp. 21-29.

[6.6] Smiths Industries, MD-80 series engine and systems display, Solid State. Display system description report TSP, 3584, (1984).

[6.7] Nüna, T. et al, High brightness GaP green LEDs, IEEE Transactions Electron Devices, ED-30, no. 4, (1983), pp. 264-267.

[6.8] Yamaguchi, I., Nüna, T. et al, A high brightness GaP multi-colour LED, IEEE Transactions Electron Devices, ED-28, no. 5, (1981), pp. 588-592.

[6.9] Himiro Takasu et al, Optical printing head with one-line LED arrays and driver IC, Proceedings of the 3rd International Display Conference, paper 12.2, (1983), pp. 448-450.

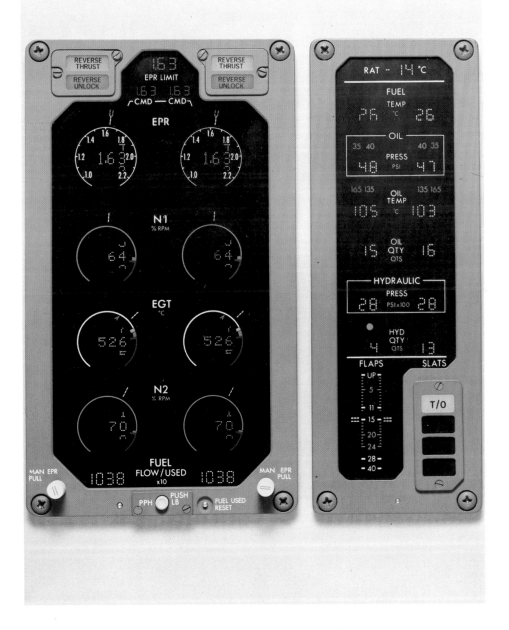

LED Primary Engine Display. Photograph: Smith Industries.

DISPLAY ENGINEERING: D. Bosman (Editor)
© Elsevier Science Publishers B.V. (North-Holland), 1989

II.7: MISCELLANEOUS TECHNOLOGIES

D. BOSMAN
University of Twente, The Netherlands

7.1. Historical survey

To understand the drive towards separate and cheap display systems the course in the developments of measurement channels is briefly examined. Displays used to be an integral part of measurement devices and were in early instruments individually calibrated. Progress in fabrication uniformity of the sensing/transducing parts of the devices made it possible to use preprinted scales. Next, the development of electrical transmission methods for transducer signals enabled system designers to group the display instruments in one convenient location together with system control devices, the transducers remaining at the sensing locations. These display instruments were dedicated to the type of transducer and ranged from simple ammeters to sophisticated servo recorders providing continuous traces. Gradually the complexity of the system increased to include requirements for graphic display and pictorial information.

At about the same time, the cost of specialist workmanship associated with the dedicated display instruments became prohibitively high; and the emerging digital technology provided methods for powerful standardisation of data signals. This opened the way to search for electrically addressable displays which could be fabricated at low cost, using large scale manufacturing techniques with a minimum number of process steps.

Obvious candidates which conform to both the increased display requirements and the desired manufacturing processes are the technologies described in chapters II.4-II.6; all have in common that they are single step electro-optical transducers. Passive electro-optical transduction (chapter II.3, this chapter) operates in at least a two step mode:

- a local change in electrical state brings about a change (at the same location) in an intermediate variable such as particle orientation, polarisation, charge transport;
- the change in this intermediate variable produces a change in luminous state; either directly (scatter) or indirectly (optical density results from integration of charge transport);
- sometimes followed by a third step (polarisation angle to transmission coefficient, using an analyser).

Another example of the three step mode is:

- local piezo-electric deformation,
- magnification of the deformation by e.g. cantilever action,
- movement of a local reflecting surface by the cantilever.

It is natural that the number of fabrication steps is at least proportional to the electro-optical transduction mode, with the consequence that many multi-step mode designs are too costly for general application. For this reason many of the cascaded effects investigated (almost every possible combination!) have been abandoned except for some very specialised applications. This may also be the fate for some of the designs described in this chapter.

Also, early investigations concentrated on simple alpha-numeric displays, e.g. Nixie tubes, 7 and 16 segment displays and status indicators, without much attention being paid to matrix addressability. That is still evident in some of the technologies discussed below.

Passive displays can be categorised as follows:
- electro-optical   [7.3-7.6]
- electrochemical   [7.7-7.15]
- suspended electric particles [7.16-7.21]
- electromechanical   [7.22-7.24]
- suspended magnetic particles   [7.25]
- magneto-optical   [7.26-7.27]

None of these developments are expected to become a pervasive technology. In this chapter the operation of a selection of newer developments is briefly explained. For discussion of devices not covered here the reader is referred to [7.1] and [7.2].

7.2. Principles of operation

Many early designs capable of displaying complex images were quite elaborate systems using projection techniques to obtain the required image size. For instance:
- controlled local surface deformation (of an oil film as in the Eidophor) producing diffraction patterns which are made visible by Schlieren optics [7.3]
- photochromic response of certain materials, induced by UV-CRT images and read out by the visible light projection system [7.4]
- electronically induced rotation of the plane of polarised light (like in LCD) in a solid under the influence of an electric field (Pockels effect) [7.5]
- digital deflection of a light beam by means of a Kerr cell in combination with a birefringent crystal [7.6]: solid state addressing system.

The newer technologies comply with the large scale fabrication technology, i.e. can be manufactured in a number of fabrication steps in which all dels are simultaneously processed.

Unlike the previously discussed effects, many convert the intensive electrical input signal (voltage, current) into an extensive output variable with optical properties (e.g. the amount of particles rotated/transported/deposited, change of energy level). Such conversion with its inherent storage has the advantage of complete absence of flicker and, when the amounts involved are controllable, may provide several levels of luminance.

The storage feature also introduces a particular problem. Erasure not being spontaneous, it becomes a separate operation when the display is to be refreshed: either complete display reset, or selectively where the changes are required. Especially for moving images (e.g. scrolling) this can be a complicating factor.

Moreover, when optical density relates to particle density, the high contrast sensitivity of the eye makes small erasure defects disturbingly noticeable.

7.2.1. Electrochemical displays

The features of electrochemical displays are: good contrast without angular dependence, no need of polarisers, inherent storage effect due to the charge transport process, low input voltage, wide operating temperature range. Resolutions of up to 20 lines per mm are achievable. The display panel is essentially an electrolytic cell consisting of a transparent front electrode on a glass substrate, a back electrode on a glass substrate and a suitable electrolyte in between, much like the liquid crystal cell.

In the case of electro deposition displays (EDD) or electrolytic/electroplating displays [7.7], a thin light absorbing metallic film is deposited by the electric current on the transparent front electrode; a thickness of 5 nm to 15 nm of silver film is sufficient for an optical density of 0.3 with grey to black appearance. The transparent front electrode could be Indium Tin Oxide (ITO), the counter electrode Ag. The liquid electrolyte contains silver ions (e.g. silver iodide AgI in methanol). A charge of 50 μC mm^{-2} is necessary to "write"; "erase"; is achieved by reversing the polarity. The actual state is preserved after switching off the voltage (blocking the current). If a galvanic path remains so that current can flow,

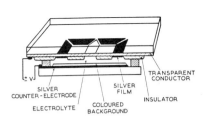

SILVER COUNTER-ELECTRODE
ELECTROLYTE
SILVER FILM
COLOURED BACKGROUND
INSULATOR
TRANSPARENT CONDUCTOR

Fig. 7.1 Silver Iodide electro-deposition display [7.29].

the cell in the written state acts as a battery causing discharge current to flow which automatically erases written information after some minutes. Lifetime is about 10^7 cycles. A prototype EDD has been described [7.8], requiring $\sim 10^{-3}$ A mm^{-2} at 1 V, achieving reflective contrast ratios of 4 to 1 with switching times from 50 to 200 ms. Contrast can be enhanced by adding an appropriate colour to the electrolyte or by using a coloured back electrode.

7.2.2. Electrochromic displays (ECD)

The electrochromic effect is based on a reversible reaction with different luminous absorption spectral distributions being produced in the oxidised and the reduced states of the material. Depending on the material used, the appearance can be switched from colourless (transparent) to coloured or from one colour to another; with very crisp contrasts because polarisers are not used. Appropriate materials are inorganic materials such as tungsten oxide (WO_3) and iridium oxide (IrO_2); and organic compounds like viologens and rare earth diphthalocyanines [7.9–7.15]. Figure 7.2 shows the construction [7.10] of a tungsten oxide display; it is essentially a galvanic cell with the front electrode of the electrochromic material used (in this case 0.5 μm layer of WO_3) at the locations where colouring must occur, and transparent elsewhere (e.g. ITO). In most cases $LiClO_4$ in an organic solvent is used as electrolyte. In reflective cells the contrast is improved by adding a white filler or film of titanium oxide (TiO_2). The cell thickness is not critical.

The colour change from transparent to deep blue of the tungsten oxide (WO_3) is obtained by accepting simultaneously electrons (e^-) and Li^+ ions, preserving the electrically neutral state:

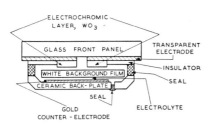

Fig. 7.2 Construction of WO_3 watch display [7.28].

$$ne^- + WO_3 + nLi^+ \rightarrow Li_nWO_3$$

The charge transport involved is 50 μC mm^{-2} as for the EDD; the response for a 2 to 1 contrast ratio slow, about 1 s [7.10]. A similar system using iridium oxide as the electrochromic layer produces black/white contrast ratio of 2:1 in 0.25 s. Lifetimes of $> 10^7$ cycles are reported, which is not good enough for clocks (10^7 s $\sim$ 3000 h) but acceptable for instruments with 3000 h time between overhaul (TBO).

Similar charge densities are involved for viologen [7.11] based ECDs, although the ratio of optical density over time constant can be larger, depending on the luminous absorption characteristics of the chosen electrochromic material. The electrochromic reaction is

$$V^{++} + X^- + e^- \rightarrow VX \downarrow$$

where V^{++} is the viologen material, X^- an anion from the electrolyte and e^- the one-electron reduction term; the precipitated salt can be re-oxidised electrochemically to erase the display. The switching times are $t_{on} \sim 4$ ms, $t_{off} \sim 20$ ms. An improved viologen display system with active matrix addressing

Fig. 7.3 Photograph of an image displayed on a viologen (320x300 dels) display using 8 grey levels [7.12].

Fig. 7.4 Packaged 320x300 dels
viologen display with active matrix
addressing on back electrode
integrated circuit [7.12].

is described in [7.12]. The poor
threshold characteristic inherent
in charge transfer devices has
been overcome by using a large
area Si integrated circuit as the
back plate of a reflective cell.
The dels of the display of figure
7.3 are selected one line at a
time, while constant current
pulses with magnitude corresponding
to the desired grey levels (out of
8 available) are applied to the
columns. The line pulse duration is
2 ms; the 320x300 del array thus is
written in 600 ms. In figure 7.4
the (prototype) device is depicted,
with the density, determined by
the integrated circuit, at 20
dels mm^{-1}.

The third group of electrochemi-chromics uses rare earth diphthalocyanine as
the active material; this is made-up of two diphthalocyanine rings (dye)
coupled by chemical bonds to the rare earth (often lutetium) ion. Its luminous
absorption characteristics are strongly affected by the addition or removal of
one electron in the rings. The display front plate is fabricated as usual with
an electrochromic material layer of 0.1 μm [7.13]. Again the cell thickness is
not critical. The colour of the layer is violet when electrons are added to the
structure, at a cell potential of − 1 V, changing to blue and green as the
potential is raised to zero; at increasing positive potentials, the colours
become yellow, orange and, at + 1 V, red [7.14]. All the coloured states are
insoluble in the electrolyte and exhibit very long term open-circuit memory.
The switching charge density required is about 10 times lower than that for
WO$_3$, due to the effective light absorbtion of the layer; switching time is
about 10 ms. Active matrix addressing is necessary because of the potential
difference between dots of different colour which would cause equalisation
currents through the common connectors of the direct multiplexing scheme.
Cycling studies [7.15] show that cell lifetime can be better than 10^7 cycles.
No data are available on colour uniformity (standard deviation of chromaticity
coordinates).

7.2.3. Suspended particles displays

The features of the suspended particles displays are comparable to those of the electrochemical displays; good contrast (1:3 to 1:30) without strong angular dependence, switching time ~ 30 ms, resolution 4 lines per mm for a 50 μm thick cell, life 10^8 cycles. Drive voltages are higher (30 V – 100 V) [7.16]. The side effects are markedly different because the particles are macroscopic as compared to the size of ions and problems encountered were: flocculation; sedimentation; transverse migration between electrodes due to fringing fields. Additionally the absence of a natural threshold in the voltage response curve can cause problems as discussed in section I.1.5. Although solutions to these problems have been found, maintaining strict requirements during fabrication is costly and the rapid developments in LCD have superseded these developments.

The electrophoretic display (EPD) works on the principle that charged particles colloidally suspended in a non conducting fluid can move under the influence of an electric field. The structure of the cell is given in figure 7.5. The front electrode is transparent. The particles ranging is size between 0.2 μm and 2 μm, are a pigment which when packed onto the front electrode, behaves as a diffuse reflector $(0.2 < \rho < 0.6)$. In the suspending fluid a dye is dissolved; the parts of the front electrodes not covered by particles appear in the (absorptive) colour of the dye solution. The charge associated with one switching action ranges from 0.1 nC mm^2 to 1.0 nC mm^{-2}. The

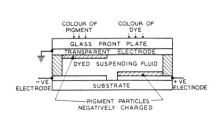

Fig. 7.5 Electrophoretic display cell structure, showing "on" and "off" states [7.17].

addition of stabilisers, in the form of surfactants adsorbed to the particle surfaces and dissolved in the fluid, imparts electrophoretic activity to the particles while also maintaining dispersion. Sedimentation is a problem with the heavy TiO$_2$ pigment: the particles and the fluid should be equally dense. It seems that dispersion is also helped by turbulent fluid flow caused by switching.

Achieving reliable threshold response is a problem [7.17]. With the addition of an extra layer on top of the column electrode layer but insulated from it, an artificial sharp threshold has been introduced [7.18]. This so-called control layer is patterned into lines orthogonal to the columns; at the so formed display elements the line electrodes are perforated into a dense array of holes in which the charged particles can be locked with a potential difference between line and column electrode of 30 V. At the other side of the cell is a continuous electrode (anode) at a high potential. With the potential difference between line and column electrode lowered and reversed, the charged particles in the del leave the holes and are transported to the anode.

This matrix display, at 16 rows of 32 5x7 characters and a resolution of 2 dels mm^{-1} was demonstrated at an addressing speed of 5 ms per line. Selective erasure is a problem.

Since the memory mechanism is related to interparticle and particle/electrode forces, the storage time is influenced by the type and amount of stabilisers. Images can be stored for months without refresh.

Very little quantitative data is available on the visual appearance of EPD devices. The steps required to achieve usable stability and life in these devices have a detrimental effect on both luminance and contrast.

The "suspended particles display" (SPD) [7.19] sketched in figure 7.6, consists of a colloidal suspension of long, thin, light absorbing particles in a transparent liquid. With no (AC) voltage applied all particles are randomly orientated and the system is strongly absorbing. When an adequate AC field is applied the particles align parallel to the field and the cell becomes transparent; the contrast ratio in transmission is 10 to 30, the "on" witching time 0.1 s, the "off" switching time 0.3 s, with 10 V RMS drive. As with EPD, most developments have concentrated on

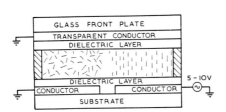

Fig. 7.6 "On" and "off" states in a suspended particles display [7.28].

density matching to maintain sufficient dispersion, and on drive methods and particle surface treatments to avoid (local) coagulation. No reference was available on low temperature performance or multiplex drive.

The electrical twisting ball display (ETBD) is like the SPD but with the difference that the long thin particles are replaced by glass spheres with a dipole moment [7.20,7.21], each in their own fluid filled cavity. The fabrication of the spheres, dye coated over 2π steradians, can be surprisingly simple; the dye, with proper choice of the fluid, provides the electrical charge. The spheres of 50 μm diameter, are in double layers locked in a

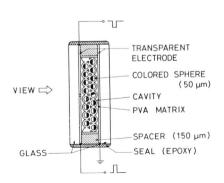

Fig. 7.7 Simplified structure of the electrical twisting ball display [7.21].

polyvinyl alcohol (PVA) sheet of 150 μm thickness, the cavities between PVA and spheres filled with toluene with additions. Obtainable contrast is 4:1 for drive voltages of 100 V during 30 ms; the switching time being about 25 ms and lifetime about 10^8 operations.

7.2.4. Electromechanical displays

The eye-brain system is most sensitive to changes in pattern and before the digital era the choice to bring about such changes was very limited indeed. The range of magnitudes of a variable was translated into an angular displacement or translation of a pointer along a dial. When more reading accuracy was required a second pointer was added (hands on a clock) or even a third (second generation of altimeters). When electrical measurement systems were developed the same skills were applied, and the data in the electrical signal were again converted to simple displacements by means of d'Arsonval motors (ammeters) and, for accuracies needing more than one pointer (long scale indicators), servo instruments. The latter were also used to drive a counter consisting of several wheels, one for each digit, or a moving scale travelling past a fixed pointer. It was found that counter indicators required less reaction time for the required reading accuracy than pointer instruments, but are less suitable for qualitative reading and tracking.

Digital indication with a bank of ammeters or a servo driven counter is clumsy and costly and soon digital indicators were made using arrays of neon lamps or incandescent lamps, or special designs thereof (Nixie tubes, 7 segment incandescent indicators). This stage seemes to mark the end of the era of electromechanical indicators, although a few attempts have been made since to use electromechanical transduction to produce genuine digital indicators, many designs based on the use of electrostatic forces. Advantageous features could be low power consumption, size not being a strong constraint (large area displays), good viewing characteristics, no need to contain toxic material, and they can be designed with memory and switching threshold for matrix addressing. Difficulties are in the manufacturing process: fabrication tolerances strongly affect the reliability.

In one version, the "electroscope" [7.22-7.23], small metal foils are moved up or down by an electric field between transparent conductors in a cell filled with dye solution. This development is not brought to the production stage, even though manufacturing technology for low cost and high yield fabrication used ingeneous etching techniques which define foil mirrors, springs, electrical connections and so on. The driving signal can be AC, since the electrostatic force is a quadratic function of the applied voltage, thus minimising electrochemical reaction in the fluid. On-state reflectance of 80 % with at least 10:1 contrast ratio and wide viewing angle is possible; resolution of a few dels per mm was achieved [7.23].

Another interesting development of the deformable surface variety is the deformable mirror device (DMD): a hybrid integrated circuit consisting of an array of metalised polymer mirrors bonded to a silicon address circuit [7.24].

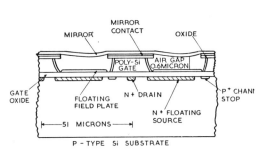

Fig. 7.8 Cross section of deformable mirror device [7.29].

The mirrors are one electrode of an air gap capacitor, the other electrode being a fixed, electrically floating plate bonded to the Si surface and capacitively coupled to the floating source of the MOS drive transistor which forms part of the integrated addressing circuitry. An experimental design of 128x128 dels has a pitch of 51 μm, air gap area of 23 μm x 36 μm and a depth of 0.6 μm. Driven from + 30 V and – 10 V, the average contrast ratio (averaged over the entire del area) is better than 5:1, at switching speed of 25 μs. The floating source storage time is about 200 ms so that image refresh rate should be 20 ms or higher.

The gates are connected to the address lines of the matrix; between the (short) line pulses the drains are brought to the required potential. During the line pulse all the sources are charged to the potentials of the corresponding drains. Data rates of better than 10 MHz are possible.

The prototype (1982) contained 5.8 % defects, of which 2.3 % were attributable to one drain line and two gate line faults. The remaining defects were largely due to fabrication teething troubles, which can be reduced to produce a defect count of less than 0.5 %.

7.2.5. Magneto-optical displays

The application of magnetic effects to display technology has been a very
popular subject for many decades, from the crude call systems of the turn of
the century to the sophisticated use of Faraday rotation of polarised light
in transparent magneto-optic
materials. The most successful
might also be categorised
"electromechanical", being arrays
of small lightweight objects
(e.g. discs) that flip over on
application of a magnetic pulse
of correct polarity and then are
held in position by a weak
magnetic field. The energy
consumption per display element
is rather high; power
requirements are kept reasonable
by e.g. addressing one display
element at a time which is fine
for large area public
announcement boards but
unsuitable for interactive use.
The subject of magneto-optical
displays is still pursued and two
examples are given below.

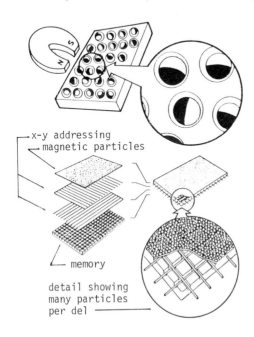

x-y addressing
magnetic particles

memory

detail showing
many particles
per del

Fig. 7.9 Magnetic particles display [7.25].

Magnetic particles displays (MPD) consist of a thin layer of permanently
magnetised, 200 μm diameter, particles, as sketched in figure 7.9. One side of
each particle is strongly reflecting and the other side absorbing. Several
realisations of this device have been published, but in the preferred version
[7.25] each particle was encapsulated in a transparent micro-capsule with a
layer of oil and so was free to rotate according to an externally applied
magnetic field. This external field was supplied by a layer of moderate
coercivity ferrite powder immediately behind the capsules. The direction of
magnetisation of the ferrite, and hence the orientation of the display
particles, was controlled by an x-y matrix of current carrying conductors. Thus
the conductors could address each element of the display individually, the
ferrite particles locally stored the data, and the encapsulated particles
rotated to present either their bright or dark sides on the observer.
Display matrices with up to 120x120 dels and resolutions down to 300 μm have
been made. Data on the reflectance, contrast and angle of view were not

disclosed, nor was information on temperature range. It was claimed that to reverse the magnetisation of each ferrite memory required 20 μJ (3 A for 0.5 μs), but no data on the reorientation speed of the display particles was given. Earlier versions of this device, using less sophisticated construction methods, demonstrated wide viewing angles and contrast ratios up to 15:1. Unfortunately these earlier versions had poorer sensitivity or suffered from particle migration in the plane of the display with consequent non-uniformity of appearance.

Another type of magnetic particle display is based on the Faraday effect: when polarised light passes through a suitable material its plane of polarisation is rotated. The ideal situation produces a 45° clockwise rotation with one direction of magnetisation, and a corresponding 45° anti-clockwise rotation with reversed magnetisation. When viewed through a correctly oriented analyser the combination can be switched between maximum and minimum transmission.

A demonstrated version of this type of display [7.26] used an array of small islands (~ 100x100 μm square, x 5 μm thick) of magneto-optic iron garnet grown epitaxially on a gadolinium gallium garnet substrate as shown in figure 7.10. A combination of heat and external magnetic field was required to reverse the magnetisation of the magnetic islands. Temperatures of between 50 C and 60 C combined with fields of ~ 2.5 A m^{-1}, were found to be adequate. For data storage devices the heating is supplied by a focussed laser pulse. In the display case, heating is generated by a minute resistive heater evaporated on the surface of each display element. The magnetic field is also locally generated by current pulses through a grid of conductors on the substrate. A well defined threshold exists for both the magnetic field and temperature, so matrix addressing without crosstalk was readily achieved. To switch a single element in 10-20 μs requires 10 μs pulses of heat and field and dissipates 280 μJ. To switch 10,000 elements per second in a 1000x1000 display consumes 2.6 W, but thanks to the permanent memory of the device the mean power can be much lower.

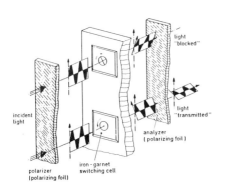

Fig. 7.10 Operating principle of magneto-optic display elements [7.28].

The largest display prototype described [7.27] contained 256x64 elements on a 3 cm x 0.75 cm substrate. Because of the small size it was viewed in projection with considerable magnification.

Contrast ratios up to 20:1 were claimed, but no data was given on the transmission efficiency of the device, Because of the thermal aspects of the device operation it must be assumed that its temperature range is very limited. At high temperatures (i.e. above 50 C) the operating effect is lost, whereas at low temperatures the power requirements probably become excessive.

7.3. Conclusion

Many new and ingenious physical/chemical/optical effects have been seriously considered for display purposes. Only a few have found application, specifically the electro-optical effects, for dedicated systems. In tables 7.1.a and 7.1.b, borrowed from [7.28], the relative merits of the remaining techniques are summarised in an attempt to facilitate comparison. With line-at-a-time addressing, only two are applicable to television type signals but the number of dels per display is still modest. For video phone applications (64 kHz bandwidth) with low resolution and low frame rate, the choice may also include electrochromic displays. However, it is doubtful whether the additional property of intrinsic memory in the display can be used to advantage in view of the increasingly lower costs of digital storage and of high performance LCD displays.

The data on contrast in the table are also open to questioning. Often these data refer to normal incidence, the dependence on viewing angle not quoted. Also the line on grey scale capability is omitted, although some effects have scope for grey scale. Response speeds quoted generally are measured at 20 C with the exception of the silver electro deposition display where operation to - 40 C at constant writing speed was achieved.

7.4. Acknowledgement

Many of the data obtained for this section were acquired in the framework of an AGARD Avionics Panel Working Group on "Modern Display Technologies and Applications" [7.29] where several in experts in the field reported to the group at specially organised meetings. In particular the contributions to this compilation of dr. G. Meier (FhG, Freiburg) and dr. A.J. Hughes (RSRE, Malvern) are acknowledged.

Table 7.1.a

Type	Electroplating		Electrochromic		
Material	Silver	Viologen	WO_3	Ir Oxide	Lu Diphth.
Reference	[7.8]	[7.30]	[7.31]	[7.32]	[7.33]
Size	–	25x25 mm	5 mm	75 mm	–
Number of					
Elements	–	64x64	7-bar	7-bar	7-bar
Contrast	4:1	–	2.5:1	2:1	Multi-colour
Direct View or					
Projection	Direct	Direct	Direct	Direct	Directy
Voltages	1	–	<1	±1	±1.2
Current/charge	<1mA/mm^2	<.5mA/mm^2	4mC/cm^2	30mC/cm^2	–
Memory?	10 min	Memory	48 hrs	Memory	Memory
Matrix Address	No	Yes	No	No	No
Active Substrate	No	Si FETs	No	No	No
Operating					
T-range	–40 to +80	–	–	? to +70	–50 to ?
t_{ON}	100 ms to	4 ms	0.9 s	20 ms	<20 ms
t_{OFF}	300 ms	20 ms	0.9 s	20 ms	<20 ms
Liquid/Solid	Liquid	Liquid	Solid	Liquid	Liquid
Complexity	Simple	Complex	Fairly Simply	Fairly Simple	Simple
Lifetime	>1 year	–	>3 years	–	–
	10^7 cycles	–	>10^7 cycles	>10^7 cycles	5x10^6 cycles

Table 7.1.b

Type	Electrophoretic		Suspended Particles	Magneto-Optic	Magnetic Particles	Deformable Mirrors
	Direct Drive	Varistor				
Reference	[7.34]	[7.35]	[7.19]	[7.27]	[7.25]	[7.24]
Size	25x25 mm	25x25 mm	-	30x7.5 mm	40x40 mm	65x65 mm
Number of Elements	32x32	32x32	-	256x64	120x120	128x128
Contrast	-	-	10-30 in projection	20:1	15:1	5:1
Direct View or Projection	Direct	Direct	Either	Projection	Direct	Projection
Voltage	<80V	±70V	<10V	15V	3V	30V
Current/ Charge	-	100mC/cm^2	-	1.5A for 10 µs	3A for .5 µs	-
Memory?	Memory	Memory	Refresh	Memory	Memory	Refresh
Matrix Address	Yes	Yes	No	Yes	Yes	Yes
Active Substrate	No	Varistor	No	No	No	Si FETs
Operating T-range	-	-	? to +85°C	? to +50°C	-	-
t_{ON}	>20 ms	20 ms	100 ms	10 µs	-	25 µs
t_{OF}	>40 ms	20 ms	300 ms	10 µs	-	25 µs
Liquid/Solid	Liquid	Complex	Liquid	Solid	Lubricated Solid	Solid
Complexity	Simple	Complex	Simple	Complex	Comples	Complex
Lifetime	-	-	- >10^5 cycles	-	-	-

REFERENCES

[7.1] Sherr, S., Electronic displays, Wiley Interscience Publications, New York (1979).

[7.2] Tannas, L., Flat panel displays and CRTs, van Nostrand Rheinhold Company, (1985).

[7.3] Sponable, E.I., The Eidophor system of theatre television, Journal SMPTE, 60, no. 4, (1953), pp. 343–377.

[7.4] Stettin, K.J., Real-time CRT photochromic projection system, Proceedings 4th National Symposium SID, (1964), pp. 59–72.

[7.5a] Marie, G., Light valves using DKDP operated near its curie point: Titus and Phototitus, Ferro electrics, 10, (1976), pp. 9–14.

[7.5b] Marie, G. and Donjon, J., Single crystal ferro electrics and their application in light valve display devices, Proceedings IEEE, 61, no. 7, (1973), pp. 942–958.

[7.6] Kulcke, W. et al, Digital light deflectors, Proceedings IEEE, 54, no. 10, (1966), p. 1424.

[7.7] Duchene, J. et al, Electrolytic display, Conference Record SID Biennial Display Research Conference, Cherry Hill (1978).

[7.8] Meyer, R. et al, Operational temperature range of the electrolytic display, Proceedings Eurodisplay, Munich (1981).

[7.9] Faughan, B.W. and Crandall, R.S., Electrochromic displays based on WO$_3$ Topics in Applied Physics, 40, (1980), pp. 181–210.

[7.10] Miyoshi, T. and Iwasa, K., Electrochemical displays for watches, Digest of Technical Papers, SID International Display Symposium, (1980), pp. 126–127.

[7.11] Barltrop, J.A. and Brid, C.L., Organic electrochromic systems – a basic approach, Proceedings Eurodisplay, (1981), pp. 97–100.

[7.12] Barclay, D.J. et al, Electrochromic displays, Internal report, IBM Hursley Park.

[7.13] Nicholson, M.M., Electrochromic flat panel multi-colour displays, Information Display, Feb. (1984), pp. 4–14.

[7.14] Moskalev, P.N. and Kirin, I.S., Effect of electrode potential on the absorption of a rare earth diphthalocyanine layer, Opt., Spectrosc. 29, (1970), p. 220.

[7.15] Bessonat et al, Investigation of lutetium diphthalocyanine display lifetime, SID Digest, (1982), p. 102.

[7.16] ..... Special issue on particle type displays, Proceedings SID, 18, no.'s 3/4, (1977), pp. 233–282.

[7.17] Chiang, A., Electrophoretic displays: the state of the art Conference Record of the SID Biennial Display Research Conference, Cherry Hill (1980), pp. 10–12.

[7.18] Lieberts, R. et al, A 512 character electrophoretic display, Conference Record of the SID, BDRC, Cherry Hill (1980), pp. 26–30.

[7.19] Saxe, R.L. et al, Suspended particle display with improved performance, IEEE International Display Research Conference, Cherry Hill (1982), pp. 175–179.

[7.20] Sheridon, N.K. and Berkovits, M.A., See [7.16].

[7.21] Saitoh, M. et al, A newly developed electrical twisting ball display, Proceedings of the SID, 23. no. 4, (1982), pp. 249–253.

[7.22] Goodrich, G.W and O'Connor, J.M., Dye-foil digital display, Digest of SID International Display Symposium, (1980), pp. 130–131.

[7.23] Velde te, T.S., A family of electroscopic displays, Digest of SID International Display Symposium, (1980), pp. 116–117.

[7.24] Hornbeck, L.J., 128x128 Deformable mirror device, IEEE International Display Research Conference, Cherry Hill (1982), pp. 76–79.

[7.25] Lee, L., Magnetic particles display, IEEE Transactions Electron Devices, vol. ED-23, no. 9, (1975), pp. 758–765.

[7.26] Hill, B., X-Y Addressing methods for iron-garnet display components, IEEE Transactions Electron Devices, vol. ED-27, no. 9, (1980), pp. 1825–1834.

[7.27] Hill, B. and Schmidt, K.P., X-Y Addressed iron-garnet display components with integrated magnetic control, Proceedings Eurodisplay, Munich (1981), pp. 213–215.

[7.28] Hughes, A.J., Other types of display. In: AGARD Lecture Series LS-126, Modern Display Technologies and Applications, Paris (1983).

[7.29] Bosman, D., Modern display technologies and applications, AGARD Advisory Report AR-169, Paris (1982).

[7.30] Barclay, D.J. et al, An integrated electrochromic data display, SID Digest, (1980), pp. 124–125.

[7.31] Giglia, R.D. and Haacke, G., Performance improvements in WO_3 based electrochromic displays, Proceedings SID, 23, no. 1, (1982), pp. 41–45.

[7.32] Dautremont-Smith, W.C. et al, Iridium oxide electrochromic seven segment display, SID Digest, (1980), pp. 122–123.

[7.33] Bessonat, Y, et al, Seven bar numeric display using lutetium diphthalocyanine, Proceedings Eurodisplay, Munich (1981), pp. 104–106.

[7.34] Chiang, A., A matrix addressable EPD, Proceedings Eurodisplay, Munich (1981), pp. 107–110.

[7.35] Chiang, A. and Fairburn, D.G., A high speed electrophoretic matrix display, SID Digest, (1980), pp. 114–115.

PART III

APPLICATIONS

DISPLAY ENGINEERING: D. Bosman (Editor)
© Elsevier Science Publishers B.V. (North-Holland), 1989

III.1: DISPLAYS IN THE OFFICE

A.E. VAN DER MEULEN AND I. PLANCENCIA PORRERO
Océ Nederland, The Netherlands

1.1. Introduction

During the last decade the character and nature of work in the office has changed dramatically. Rapid developments in areas such as, computing-power, memory-capacity, communication systems, printing, and display technology have given rise to a new era in office automation. The need to keep ahead of one's competitors has been the driving force behind the rapidly increasing demand for even faster and more efficient methods of transfering information.

The development of office automation has led to the introduction of the word-processor and, more recently, to the automation of less formal office tasks. The combination of all these advances has meant that Visual Display Units (VDUs) are now employed for prolonged periods of time in almost every office.

The initial use of VDUs by engineers, who were usually highly skilled and self-motivated, led to few complaints about the viewing characteristics of the VDUs. However, the present-day, widespread, continual use of the VDU by non-scientific operators has highlighted several physiological problems. Obviously the visual effort required for working with paper is different to that needed for VDUs. In particular the varying viewing distances, the difference in background and luminance levels of visual displays compared to paper might be cause of health problems for office workers. Even today, these problems have still not been investigated fully. Furthermore, in considering published research results in this area, one has to take care to distinguish between poor VDU design, poor job-design and poor job-motivation.

The performance of an office display is strictly related to its application. For example, a high resolution A_4 VDU will not be used for the entry of rough text; on the other hand, the preparation of a document will require a state-of-the-art A_4 or A_3 high resolution screen. In general, VDUs can be categorised by referring to their softcopy quality just as printers can be classified by their hardcopy quality. Images are composed of picture elements (pixels); the inverse of the pitch from pixel centre to pixel centre is the resolution. In displays the display elements (dels) do not always coincide with the image pixels, but the resolution (dels mm^{-1}) gives a first impression of their image quality, as is customary with printers (pixels mm^{-1}). Just as printers can be categorised in terms of their resolution, so can displays, for example low resolution half page VDUs for terminal and personal computer applications, high resolution full page A_4 displays for word-processors and state-of-the-art A_4/A_3 displays for document preparation and retrieval. The manufacturers of office equipment often refer to the "WYSIWYG" approach, which means "What You See Is What You Get". However, todays laser printers outperform any state-of-the-art visual display.

Environmental conditions are important. Displays could place tremendous constraints on the arrangement and furnishing of offices. Unfortunately, in practice it is more often found that the equipment is placed in a pre-existing office structure which gives rise to discomfort and health complaints, making the tasks that have to be performed more difficult. Although great improvements have been made in matching the visual characteristics of softcopy with hardcopy, the main problems are caused by the fact that the one usually is a light emitting and the other a light reflecting medium. As long as visual displays do not have the same visual characteristics as hardcopy, they impose their own constraints. In addition to environmental conditions, ergonomics can have a significant influence on the use and performance of a VDU. Ergonomic design is very dependent on the display technology used in a VDU or workstation. For example if one considers the volume of a display there is a remarkable difference between a high resolution A_4 CRT and a modern flat panel display with chip-on-glass technology.

The font (chapter I.2.2, this chapter section 5) is the design of a letter including the required spacing. As most displays build up their images from basic dels, display fonts can be characterised as dot-matrix letters. These are not directly comparable to well known analogue fonts used in publishing processes, typewriters or daisywheel printers. The fonts are a very important aspect of the complex interface between form, perception, abstraction, language and comprehension. They influence the legibility, reading errors, efficiency of information transfer and the aesthetics of a display. The limited resolution of displays places special constraints on font design and this is a matter of deep concern.

About ten years ago people predicted that the CRT would be replaced as the workhorse of the display technologies for office applications, however this change has not (yet) occurred. We could repeat this prediction but we have learned from history that CRT technology is improving as fast as flat panel technology, so we better play safe. At this moment one can envisage a very cautious introduction of flat panel displays into the office, usually for applications requiring portable, or removable workstations [1.36].

In the following sections an overview of the state-of-the-art in the different applications and technologies will be given, together with an idea of the probable situation a few years hence. The rapid improvement in the technologies involved necessitates a continuous evaluation of this situation [1.39].

An interesting spin-off of modern display technologies is the bar imaging systems mainly used in electrophotographic printers. These will be discussed in the last section. When compared with laser printing this technology has the important advantage of producing a solid state device.

1.2. Applications for the displays

Displays are used in the office as the output devices of electronic information handling systems. The various tasks undertaken in an office result in different methods of handling this information, and display devices with different characteristics being needed [1.28].

When the various types of work in the office are considered, the information handling techniques can be divided into three main categories:

- Data-processing. Here emphasis is placed on the transformation of the information and not on its method of input or output. The majority is numerical information processed by a computer system. The system concentrates on automatic processes based on straight-forward rules.

- Word-processing. Alpha-numeric text is introduced into the work station and is ultimately visualised on hardcopy. Displays act as the intermediate media to help achieve the desired result.

- Document preparing. This is a logical extension of the previous category. As well as the alpha-numeric information, the option of graphics facilities is required to produce documents with typesetting quality.

Each of these three categories uses a CRT as the most functional display method. An additional category includes the office tasks in which displays have not yet been widely introduced probably because they are better supported by Flat Panel Displays. These include for instance copiers, printers, video phones and portable computers.

1.2.1. Data-processing

Due to decreasing hardware prices, micro, mini and personal computers become available for a continuously growing number of office applications. This avoids the user being dependent on large computer centres which sometimes require the information to be sent in advance and then collected after processing. In early systems, each task was processed as part of a batch without any direct interaction with the user. Often the input to the system was by punched cards and the output was by printed listings. User interaction with the system has been provided by the addition of a display unit which allows partial visualisation of the information contained in the system. Clearly, the display units are not essential to the computer system but they are a valuable aid to the user [1.31].

Displays never appear alone or independently of a system, they form part of a VDU. Displays are the output device of the terminal and a keyboard is often used as the input device. Information introduced to the computer via the keyboard is rapidly visualised on the display screen to allow the accuracy of the data to be checked.

Displays are usually referred to as softcopy devices in contrast to paper printers which are known as hardcopy devices. Although the size of these displays varies, with some manufacturers including only a single row of characters displayed in their computers, most screens are in the range of 12 inch (300 mm). This refers to the length of the diagonal of the screen.

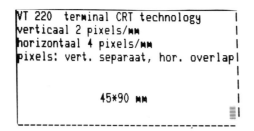

Fig. 1.1 Picture of data-processing CRT display 65 % of real size.

In the display unit the usable size is represented by the number of lines (rows) and the number of characters per row (columns) that can be displayed on the screen. This is known as the display format. The commonly used format for display terminals is 24 lines by 80 characters width. The 80 columns width was established to maintain compatibility with punched cards. Today lines of 40, 64 and 132 characters can also be found, although they are not so common.

Alpha-numeric characters are usually represented by matrices of dots. In early devices, 5x7 matrices were used but nowadays they are being replaced by 6x7 and 7x7 matrices which give clearer and sharper definition of the characters.

With these sizes it is easier to define lower and upper case characters, as well as special symbols needed for communication with the system.

Some displays used as computer terminals also offer graphics facilities. The resolution of the graphics is measured by the number of independently addressable pixels and is defined by the number of lines and the number of dots per lines. Graphics packages allow bargraphs, charts and diagrams to be created and viewed on the screen.

The colour of the screen depends on the phosphor used in the CRT, though sometimes the orginal colour can be changed by using filters. The most widely used screens are those which have green coloured letters on a dark background. This combination is preferred by the majority of operators and it is claimed that it reduces eye strain and fatigue. However some countries, such as Sweden, recommend the use of amber characters on a brown background. Sometimes the display monitor offers the option of inverting the image giving dark letters on a lighted background.

To improve the image and to reduce reflections, filters or coatings can be used. The option of altering the intensity of the CRT to suit the user is often present. To adapt the monitor for each user it has to be possible to move or tilt the display screen so the operator can choose the best viewing angle.

Some of these monitors have a Random Access Memory (RAM) that permits the storage of data for further viewing using scrolling facilities.

1.2.2. Word-processing

A word-processor is a computer which is dedicated to edit texts, giving the option of storing it or transmitting it further. Basically, it is composed of one input device the keyboard, with which the text is entered to the system, and two output devices, one being a display unit and the other being a printing facility. Some storage capability is also added to the system, as well as intelligence to provide functions to perform the texts editing tasks. These editing capabilities and the presence of the display unit distinguish word-processors from electronic typewriters. When typewriters do have a screen, it is usually limited to a single line. This displays the text line just entered by the keyboard allowing it to be checked and corrected, after which the line is immediately printed on paper.

The first displays used for word-processing applications were limited to a few lines of text and the editing process was performed line by line, the origin of this being that first generation text editors were meant as a tool for software programming. They soon evolved to word-processors in which a larger part of the text could be viewed and handled simultaneously. This development implied a direct modification of the display units. The primary modification was in the size, they became larger. The actual size usually varies from 25 to 30 lines of 80 characters on a screen size of about 12 inch diagonal. The main technology used is the CRT.

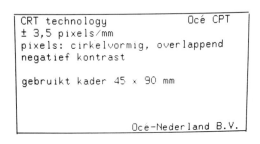

```
CRT technology            Océ CPT
± 3,5 pixels/mm
pixels: cirkelvormig, overlappend
negatief kontrast

gebruikt kader 45 × 90 mm


                    Océ-Nederland B.V.
```

Fig. 1.2 Picture of word-processing CRT display 65 % of real size.

The size of the character matrices is usually 7x9 and each system usually has several fonts available. This allows the user to, for example, write some words of the text in italics in order to draw the attention of the reader or to write the titles of paragraphs in a bold typeface to distinguish them from the rest of the text. This characteristic is one of the primary distinctions between word-processor and data-processor monitors. The latter type is limited to only one set of standard simple characters.

The colour of the letters on the screen depends as in data-processing terminals on the phosphor type used. To avoid flicker, high refresh rates are preferred to long persistence phosphors.

In comparison to data-processing, word-processors are distinguished by the wide variety of functions that must be performed to handle the texts. This implies certain prerequisites of not only the screen, but the controlling software and hardware which make-up the unit. The number of functions performed by the software determines the intelligence of the VDU. It varies from very simple systems where the display is only used to present the information to the computer, to some VDUs which have sophisticated word-processing capabilities able to operate in a stand-alone mode. It is useful to describe some of these functions.

At every moment the user must be able to control the text he can see on the screen by changing it, adding new text, storing it or printing it. Scrolling is one of the first basic functions required of the display unit. Due to the small screen size, only a portion of the total text can be viewed at a time. To see the entire version the user needs to be able to scroll in both horizontal and vertical directions. Up to 3 or 4 pages of the text that cannot be seen are stored in a RAM in the unit. Larger amounts of text may be stored if diskettes, tape units, etc. are connected to the system. In order to view different parts, a window facility may be used. This allows small text units to be displayed at the same time on the screen. These units can be edited and scrolled independently.

The display unit must also provide the means whereby some text on the screen may be highlighted. This calls the attention of the user to important information and is achieved by underlining, blinking, reverse video or bold typeface.

Some displays allow the user to choose the size of the characters, the space between the lines and the number of columns in which the text is to be displayed. One selection never offered however, is the variation of the distance between the letters as this feature is determined by the character font and is fixed for the different sets.

To facilitate the use of a word-processor system, the screen is often also used to display control messages, a status area or a menu with the editing functions available at that time: insert, delete (character, word, line), underline, replace, the type of text, etc.

1.2.3. Document preparing

The aim of this category of systems is to produce documents of almost typesetting quality. The display monitor is intended to produce an image which perfectly reproduces the form of a document printed on paper. This concept is known as the "What You See Is What You Get" aproach. Due to technological limitations it is not yet possible to obtain images on a CRT with the same quality as that offered by printers. In terms of resolution alone, non impact printers giving 20 dots per mm are available on the market, while with CRTs it is difficult to obtain more than 8 dels per mm [1.33]. Other display technologies are currently limited to 4 or 5 dels per mm due to addressing problems. However it must also be noted that the technology has improved considerably in the past few years and this difference between hard and softcopy is becoming smaller [1.8].

Other important problems related to the prolongued use of the display terminals such as fatigue and eye strain have decreased almost to printed paper levels by designing better fonts. This means that now the problems are more associated with the quality of the image than with the display medium itself.

The first direct implication of the need to have the displayed image resemble printed paper influences the colour of the screen [1.3]. Screens must be produced having black characters on a white background. From the ergonomic point of view this is very important because the eye of the operator does not then need to readapt its perception capacities when alternating between a paper document in the background and a displayed one. To achieve this it is vital to maintain a uniform intensity profile.

The second implication is for the size of the screen. This needs to be a minimum of A_4 size in order to allow a complete page of a document to be viewed. Some systems use a larger screen to allow a control area to be defined and others are large enough to allow two A_4 pages to be viewed simultaneously. The preparation of documents implies that the information to be displayed is no longer only of the alpha-numeric type. Text of different fonts, graphics and real images (coded as binary information) might be gathered together on a single page of the document [1.30].

To be able to handle the text, all the facilities and functions described for the word-processors should be present in these systems. It must be also possible to address all the dels of these displays separately. The minimum acceptable number of dels on the screen is about 1000x1000 and there are systems which use up to 2280x1728 [1.11].

State of the art CRT technology   MYFRA S.A.
A4 scherm 1728 * 2288 pixels, 8 pixels/mm.
Dit kader is 45 * 90 mm groot.
Charactermatrix 7*9, 8*16 incl. spacing.
N.B. Op het scherm worden steeds 2*2 pixels
     gebruikt voor een element van de matrix!

Er is een kadertje van 45 * 95 mm afgebeeld.
Pixelvorm: cirkelvormig, overlappend.

Oce Ned. B.V.

Fig. 1.3 Picture of document preparing CRT display 65 % of real size.

The need to deal with real images on the screen requires an easy method of handling them, and therefore the integration of simple image processing-functions to the system. This refers to functions like compressing, image enhancement, filtering, scaling, rotating and repositioning of images on the screen [1.42].

To produce documents which approach printed quality, a subset of the facilities that are available in typesetting have to be available in the system. These include, for instance, the choice of font size, the type of font itself (classic, italic, bold) and also mathematical and Greek characters for scientific applications. The different types of font and the users perception of them are improved by the use of grey levels. These can be obtained by adding one or more bits to the display memory for every del [1.10].

With these requirements the hardware of the display unit needed for document preparation becomes more complicated [1.37]. The fact that the pages of the document have at least bi-level, sometimes even grey level information associated with every del, increases the amount of memory needed for storage and thus the size of the RAM unit provided with the system [1.19].

With these afore mentioned levels of resolution and with a white background on the screen, a flicker free image is produced by storing the image in a memory and refreshing the screen at a rate of the order of 70 or 80 Hz (chapter I.2.4). The refresh rate is also influenced by the fact that the images on the screen have to be visible under office work conditions. This can mean luminance levels of the order of 100 cd m^{-2}, implying a contrast of about 6:1 between the screen and the surroundings.

1.3. Working environment

The working environment of a VDU operator is considered to be all the characteristics which describe the physical surroundings in the office. These characteristics are very important because they influence the performance of the user the VDU [1.16].
The three most important factors are lighting, temperature and noise. Of these, lighting is the first one to be considered because it directly affects the user while reading or looking at the display screen. The other two characteristics, although not so vital, also have a significant influence on the comfort of the operator [1.7].

1.3.1. Lighting

The lighting environment includes the light emitted and reflected from the VDU. Large differences in luminance levels produce disturbances to the eye of the operator. When the light comes from an illumination source, either sunlight or artificial sources, the disturbance is called direct glare. When it comes from reflected light it is called indirect glare. Both types should be avoided. When this is not possible, they should be reduced to a minimum since they disturb the effectiveness of the eye in receiving information.

Reflections on the screen can be reduced by a screen surface treatment, screen filters (micromesh or dark etched) or an anti-reflection spray which makes the reflected image diffuse, but at the expense of a decrease in the luminance of the characters on the display (chapter I.1.4). As it is important to reduce reflections it is desirable that the reflectance of the desk is kept around 0.4 to 0.6 to maintain the luminance level as similar as possible to the luminance of the screen.

The light received is described by the illuminance (luminous flux per unit of surface area). The illumination required at a work station depends on the task to be performed. For printed material an illuminance of about 500 lux is

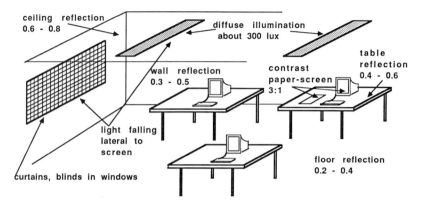

Fig. 1.4 Office environment distribution.

needed and for a luminescent screen a luminance of about 100 lux. A compromise
has to be achieved because usually both tasks are performed simultaneously. An
illuminance of about 300 lux is recommended, as a lower one has been shown to
create social problems due to the low level of visibility. The user receives
light from reflections from the surfaces of the objects in the office,
producing a non-uniform luminance distribution. Luminance ratios between the
screen and surroundings make the eye adapt constantly from one to the other
which involves a visual effort. At an illuminance of about 300 lux, it has
been found that the optimum character background contrast was around 10:1,
while the luminance ratio between screen and paper should be around 3:1.

Some recommendations are given for the position of the VDU screens in
workplaces. The best place for a VDU is far away from the windows and other
light sources. If this is impossible, it should be in such a position that the
line of vision of the user is kept parallel to the windows and to luminous
sources so that light falls laterally onto the screen and reflections are
minimised. The lighting installations for VDUs in an office environment should
meet standard regulations, such as those of the Work Environment Fund (ASF) in
Sweden or the German Standards for display workstations [1.17].

1.3.2. Temperature

The room temperature required for a sedentary job such as working with a VDU
is considered to be about 21 C. A person produces about 100 W in this
situation and a workstation can emit the same while it is operating. Hence,
when studying the heating conditions in a work environment, apart from factors
like population in the room and radiation from light sources, radiation from
the equipment should be included. The VDU and other office equipment usually

have fans to ventilate the systems and these may redirect hot air towards the operator resulting in discomfort. The presence of any hot or cold spots in the office should be avoided. When possible, air conditioning equipment should be installed to regulate the temperature and humidity of the office. Here also special care should be taken to ensure that the air is not directed towards any operator. The speed of air movement should be less than 0.1 m.s^{-1}. At higher speeds draughts are apparent and cause a nuisance. The relative humidity factor should be mentioned here because it is closely related to temperature, although it is not influenced by the VDU itself. It needs to be maintained between 50 % and 70 % and held at as constant a value as possible.

1.3.3. Noise

Most of the noise experienced in a VDU workplace is not due to the VDU itself. The rest of the equipment in the office, such as the fans, telephone bells, typewriters and printers and of course colleagues produce this noise. Printers should be located in a different room if possible, or at least placed in a noise reducing cabinet. Ordinary human conversation produces a noise level of about 60 dB, while the rest of the environment can produce more than 70 or 80 dB which is very disturbing. If the noise level attains 120 dB, there may even be irreversible damage to the ear. This situation is seldom reached, but nevertheless excessive noise produces distraction and perhaps irritation, lowering the productivity of the operator.

It is not only the level of noise but also the type which determines the influence in the environment. High frequency noises affect the young in particular, while low frequencies disturb older people.

It is also important to remember to insulate rooms. External noises from other rooms, or even from traffic, sirens, etc. in the street, can result in a higher noise level than that inside the office.

1.4. Ergonomics

Displays in the VDU are the medium by which the operator can check and control the information going into or through the system. This flow of information in a VDU system can be characterised by the different office tasks which are related to it, such as data input, check reading, data output, or a kind of dialogued type work mostly connected with document preparing. This gives an idea of the importance of the VDU and it is clear that the information displayed on it should be both legible and readible (chapter I.2.2.). There is a consequent need to adapt the display units to human physical and mental characteristics. Adapting or improving VDU characteristics suggests that technical designs and features should be considered, but it also implies consideration of the layout of the information on the screen [1.6].

The former can be called the ergonomics of the hardware, while the latter is related to the software and is called the user interface. Hardware involves concepts like luminance, contrast, flicker and the type of phosphors. Software involves the screen layout, the format, coding and shape of the characters, the size of the screen and the cursor shape and performance. Other concepts like colour and the use of positive and negative images are difficult to classify, but colour will be considered as an aspect of the user interface while using reverse images will be treated as a hardware factor. Apart from these ergonomic factors, other effects on the people working with VDUs should also be considered [1.1]. The relation of physical problems such as eye fatigue or posture pain to the use of the VDU will be discussed, but it is also important to consider the mental factors involved (i.e. job satisfaction and work organisation) and their influences on these complaints.

1.4.1. Technical aspects

Most of the VDUs found in the office environment are the display media for personal computers, computer terminals, text-processors and document preparing systems. Although there have been great improvements in a variety of display technologies, the CRT is still the main device used for these purposes. Flat panel technologies are now beginning to be introduced into the office, usually as screens for portable computers, however only the CRT characteristics will be discussed here [1.29].

Two of the most important factors involved in the perception of a CRT screen are the luminance level and the contrast. Characters on the screen of a CRT are made visible by the light they emit in the case of negative images, or by the difference between the lighted background and the dark characters in the case of positive images. Positive images present more or less the same visual characteristics as printed paper. This is an advantage when working alternately with the VDU and with printed paper because the eye perceives almost the same luminance levels with both. In display engineering terms, positive and negative images are usually described as negative and positive contrast respectively.

The measure of diffuse light is known as luminance and is expressed in candelas per square metre (cd m^{-2}). Associated with luminance is brightness, which describes the visual sensation produced by an amount of light emitted by a given area (chapter I.2.4.). In a CRT, the brightness of the characters or background is defined by the technical and physical characteristics of the unit. It is very difficult to obtain an exact measure of the luminance of the screen because it varies from one point to another on the image, however commercial equipment is available to measure it. In normal lighting office conditions, it is desirable to have a character luminance of between 80 and 160 cd m^{-2} for positive contrast, with less than 45 cd m^{-2} being unacceptable. The character should remain sharp at the maximum level. For negative contrast a background luminance of up to 170 cd m^{-2} should be available. This higher luminance value can result in flickering, so particular care should be taken to maintain a flicker free image. However, there is also the advantage that the reflection problem is practically solved in this case. All these luminance levels should be adjustable to allow an optimum value to be chosen to suit the background luminance on the screen and the lighting level of the environment. This suggests that the background luminance should also be adjustable. For positive contrast it should remain between 10 and 15 cd m^{-2}.

Contrast is the ratio between the character luminance and the background luminance. Hence, once these levels are known, the contrasts can be calculated. The optimum contrast in the case of a positive contrast display varies from 8:1 to 10:1, with a minimum admissable value of 3:1. Of course,

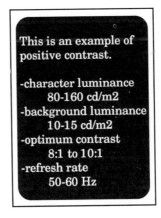

Fig. 1.5 Differences between positive and negative contrast.

these contrast levels decrease when the room illumination is increased. For a negative contrast display, the contrast between the characters and the background should be between 1:8 and 1:12, with a maximum of 1:20. It is essential to provide the operator of the VDU with the option of adjusting the contrast as appropriate, independently of the brightness. The human eye is more sensitive to changes in contrast than to changes in luminance, hence it is important to consider the contrast carefully.

Other factors to be considered in this section are the regeneration rate and the type of the phosphor. When the phosphor on the screen is excited, the characters become visible and remain there for a time period dependent on the persistence of the phosphor. It is therefore necessary to refresh or rewrite this image to prevent it disappearing from the screen and producing flicker. The frequency with which it is necessary to sweep the screen with the electron beam depends on the type of phosphor and whether there is positive or negative contrast. For positive contrast it is necessary to have 50 Hz to 60 Hz to obtain good performance and to avoid flicker on the screen. For negative contrast a higher refresh rate of between 70 and 80 Hz is needed to ensure a flicker free image on the display. The refresh rate may be reduced by using a long persistence phosphor, but this produces shadowing on the screen when the image changes quickly. It should be noted that long persistence phosphors easily burn in because they need more energy than the short persistence phosphors to produce the same average brightness of the characters. This is translated into a shorter tube lifetime. It is therefore recommended that a short to medium persistence phosphor is used. If these lifetime characteristics are not available from existing phosphors, mixtures may be made to achieve the desired result. Choosing a type of phosphor means choosing the colour of the image on the screen, or at least a range of colours. The actual colour is determined by the spectral distribution of the radiation emitted by the phosphor. The colours recommended for a positive contrast display are green, white, yellow and red in order of preference. For a negative contrast display black on white is preferred, for the remaining colours the choice is not so obvious.

The final parameter to be considered in this section is the resolution. The resolution of a CRT is limited by the density of the scanned lines and by factors such as spot size and addressability. Normal CRTs used as computer terminal displays exhibit a few hundred lines, independently of the size of the screen. CRTs used for producing high resolution documents can have over 2000 lines, each with just under 2000 dels. From a practical point of view, a minimum of 10 raster lines per character is required and a significant improvement in appearance can be achieved by increasing this. This number of lines per character is required when different fonts or graphics are to be displayed with accuracy and reliability. Most of the font designs used in CRTs are based on a dot-matrix and their visual impact is very dependent on the dot size. A dot size of 1.2 scan lines is recommended for non-interlaced monochrome displays and 1.7 for interlaced ones. In this way it is possible to produce a continuous image by merging the dots (example 1.1, chapter I.1), resulting in a sharper and more clearly defined character. Increasing the size of the CRT does not improve the resolution if the same number of scan lines and dots per character are used. The number of dots on the screen is limited by the ability to address them. This in turn depends on the speed (data rate) with which information can be sent to the CRT.

1.4.2. User interface

The technical specification of a CRT should be considered in conjunction with the software specifications. In judging the VDU performance, factors such as the shape of the characters, coding and the grouping of the information need to be included. One of the first characteristics to be determined is the screen size. This is always larger than the actual viewing area because the area at the edges of the CRT produces a distorted image and is not used. The choice of screen size depends on the amount of information to be displayed and the format, size and spacing of the symbols which, in turn, depend on the viewing distance and luminance conditions under which the screen is to be used. Once they have been chosen, the optimum performance of the VDU is fixed and increasing the density of information by reducing the size of spacing of the characters only results in a poorer performance.

The shape of a set of characters and their spacing is known as a font and will be discussed in more detail in section 1.6. Here we describe their effect on the users perception of the screen. One of their most important parameters is height. At 500 mm viewing distance, a height of 3 mm is desirable, and less than 2.5 mm is unacceptable, resulting in a visual angle of 16 to 20 minutes being subtended. Smaller heights make it difficult to discriminate between the characters. The width of the character should be about 80 % of the height. A stroke width of about 12 % of the height is recommended, slightly thinner being acceptable for negative contrast displays. The legibility of the text is largely determined by the spacing between the letters. On average, the characters should be separated by between 20 % to 50 % of the character height, while the vertical line spacing should be 100 % of the height. With high resolution displays, these spacing values may be reduced to 10 % of the character width and height dimensions respectively. This is equivalent to one del of the dot-matrix from which the character is formed. The distance separating words of the text should be approximately the width of one character. If documents are being prepared, all these figures may vary according to the layout required.

Once the character has been designed, the problem arises as to the best way to present the information on the screen such that it is comfortable and easy to read for the user and results in good display performance [1.43].
It must be appreciated that all the VDU users will not be system experts.

There are many different tasks to be performed such as text editing, information storage, document preparing, etc. and these may use different styles of dialogue, menus, filing schemes and graphics. All these variations result in a need for a simple and clear format for organising the information to allow all users to handle it. All related information should be grouped together on the screen and there should be enough space between the items to allow each one to be easily distinguished. A good display design increases the efficiency with which a task can be performed. The sequence in which information is asked for and presented on the screen should be marked by the task development and not by the internal characteristics of the system.

When text is being displayed for editing, it is usually displayed in narrative form with the lines covering the entire screen. Sometimes it may be helpful to have the screen divided into two columns containing complementary information. For text editing and document preparation it is often very useful to have an area of the screen displaying the typographic commands which are used to format the text. When working with both paper documents and with the VDU, it is desirable that the format of the screen and the format of the documents are as similar as possible. A tabular format may be used to organise items into rows and columns on the screen. A header explains the context and relationship between the fields. Various formats are available depending on the task to be performed.

When there is more information being stored than can be presented on the screen, as usually occurs, facilities such as scrolling and the use of windows are necessary. To increase the amount of information being displayed on the screen, the information can be coded to reduce the space needed by each item. If the method of coding follows logical rules, the user may derive the code himself.

The various types of code include an alpha-numeric system whereby data is represented by numbers or letters which may be present in the information. The graphical code allows a piece of information to be represented by a shape.

Coding is also used to enhance some of the information, to allow it to be easily recognised on the screen. By varying the size of an item, its importance is indicated by its dimensions relative to those of the other items. This purpose may also be achieved by using different luminance levels, though it is recommended that no more than two or three are used or the eye will not be able to discriminate easily between them. Reverse video may be used to emphasise some of the information. Blinking with a flash rate of between 2 and 4 Hz attracts the eye to particularly important information. The efficiency of blinking may be enhanced if it is combined with, for example, a size coding method. Finally, colour may be used as a code to group items on the display or to call the users attention to a certain area or item on the screen. The use of colour complicates the technical characteristics of a VDU. Due to its importance, colour will be discussed separately later.

When discussing displays for use in an office system, the use of the cursor must be emphasised. It is present in all alpha-numeric interactive display systems and indicates the position on the screen where the next information to be typed will appear. It must be easily found anywhere on the screen and also moved and controlled without much effort. The cursor is usually moved by an set of keys on the keyboard, but more sophisticated and direct methods like a mouse, joystick or graphics tablet are also available. The cursor should be designed so that it cannot be confused with any other displayed item and so that it does not disturb the reading or writing process. Cursor recognition can be enhanced using some of the above coding techniques: blinking, reverse video or even colour. In general, the use of colour on the screen needs careful consideration from the ergonomic point of view.

Colour is achieved in CRT displays by grouping the three primary colours, red, blue and green. Colour displays are being increasingly used for a variety of applications. We are mainly interested in text manipulation here as this is the dominant application in an office environment. Colour can be used to indicate the errors in a piece of text, to indicate changes made to update the content, to enhance different paragraphs, or to delineate windows on the screen [1.38]. Different colours should be used for different applications. Clearly, red should be used to attract the users attention or when reporting a dangerous situation. Blue characters should be avoided as blue is difficult to discriminate, but it is a very good background colour. It should be noted that the resolution of the text is influenced by the colours of the characters and of the background. The optimum contrast ratios are between 3:1 (red and purple) and 7:1 (yellow and white).

It is recommended that no more than seven colours are used together as the eye has difficulty to distinguish between them [1.41]. A better performance is obtained when the colours are not very saturated. Various experiments have shown that colour does not affect the time needed to perform a task and that dark characters on a light background are still preferred for document preparation [1.23].

Concepts like legibility and readibility are the result of the parameters discussed. Legibility can be defined as the characteristic of alpha-numeric characters which allows each one to be distinguished from the others (low confusion probability). Readibility refers to the recognition and interpretation of information defined by groups of characters. A measure of legibility is obtained from the number of errors produced while discriminating between characters. This depends on VDU characteristics such as resolution, luminance and contrast ratio, and also on environmental characteristics such as viewing distance and luminance levels. The individual operator must also be considered. It is hence very difficult to give an absolute value of legibility and it is preferable to describe the performance of the VDU using more objective and measured properties of the display.

1.4.3 Human factors

The number of VDUs in the office is still growing. However, the performance of the displays is still not as good as it should be. The VDUs workplaces are often not designed according to ergonomic principles, are badly positioned and offer inadequate illumination. Also as the amount of information increases even faster than the quality of the presentation methods, the user is not able anymore to process all that information. All of these factors cause physiological and psychological problems for the user of the VDU.

Physiological problems such as eye fatigue and visual discomfort are obviously caused by a poorly designed VDU, but complaints about the effect of CRT radiation are not so clearly answered. Mental problems associated with limited job satisfaction, fatigue and monotony are related to the type of work undertaken using the VDU and are induced by fear, lack of understanding and disinterest in the work [1.44].

1.4.3.1. Physiological problems.

Four main problems are to be discussed in this section: eye problems, postural problems, radiation and dermatitis.

VDU users complain about a variety of eye problems. It has been shown that working with a VDU can strain the eye but that many of the problems are due to the aggravation of existing uncorrected eye defects, or are related to poor office conditions, such as lighting. The type and frequency with which the work is performed may also cause problems. These eye problems are transitory and disappear after a short time.

Visual discomfort and eye fatigue are the two fundamental eye problems. They are caused by the effort that the eye has to exert in changing from one level of luminance (on the display) to another (on the document) and by the effort involved in reading from the screen. This latter effect may be influenced by poor lighting conditions or caused by working for too long with the eyes focussed continuously at a fixed distance. Although this feeling of fatigue will be relieved in time, it should be avoided as it may cause other psychological problems. More important symptoms are a burning sensation in the eyes, tiredness and headaches. These obviously reduce the effectiveness of the operator [1.13].

These type of eye problems may be relieved by matching the condition of the screen, the luminance levels, the format and contrast ratio and the work environment with the ergonomic recommendations. It might also be necessary to consider a limitation of the time spent working with the VDU and planning the work more effectively.

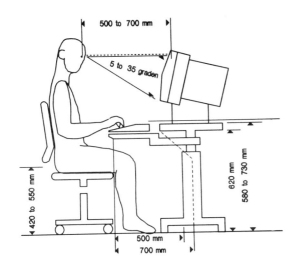

Fig. 1.6 Position of VDU user.

Whilst working with the VDU, the user is obliged to retain a fairly rigid position in order to remain at the same visual distance. The CRT based VDU cannot readily be moved because of its physical dimensions and this is a disadvantage for the user as compared to paper or flat panel displays. The enforced static position results in constrained postures which can produce pain and fatigue in the muscles, tendons and joints of the neck, shoulders, arms and hands. The discomfort can also result from repeating the same movements over an extended time period. These problems are not restricted to the users of VDUs and are caused by most machines used for office automation. Postural problems can also result from a badly designed workstation and can be relieved by using a more flexible design. The main recommendations are that the keyboard and VDU should be independently movable and that it should be possible to adjust the heights and positions of the chairs, desk and documents.

CRTs produce ionising (X-ray) and non-ionising (ultraviolet, infrared and radiowave) radiation. The radiation levels have been measured by different experimental means and all the results fall in the permitted safety levels; moreover, that the radiation is confined to the immediate vicinity of the VDU. It can clearly be concluded that the radiation is not harmful to the user either in the short or the long term [1.9].

CRTs also produce a static electric field on the front of the display screen. This influences the relative humidity and has been related to the complaints

of some VDU users of itching and the appearance of a rash even after working
for half an hour with the machine. These symptoms disappear a few hours after
leaving the vicinity of the CRT. It is also possible that the symptoms are a
consequence of psychological problems, fatigue or postural discomfort.

1.4.3.2. Psychological problems.

VDU work introduces a social isolation among the users. The need for social
contact in the work is reduced, while the human need for social interaction
remains, this is one source of psychological problems. To this we have to add
the danger of requiring the user to perform a monotonous task whilst at the
same time increasing his efficiency. This causes mental strain and is evident
as anxiety and irritation. The design of a more flexible work plan which gives
the user independence can help to change an individuals attitude and introduce
greater job satisfaction. The introduction of rest periods to avoid fatigue
and to promote social contact should not be forgotten.

1.5. Fonts

1.5.1. Introduction

Immediately after the introduction of the first generation of VDUs and
keyboard entry equipment, research began into the legibility problems and
reading errors introduced by these devices. Clearly, in the early days many
problems were caused by the poor design of the VDU such as low resolution,
luminance and refresh rates. The early VDUs were based on raster scan TV
technology.

While these machines were being used by engineers, who were highly skilled and
motivated, few complaints were voiced. This situation changed completely when
the VDUs were introduced to a broader range of users. Today VDUs are used
almost everywhere and to understand the impact of a typical screen it is
necessary to distinguish between the applications.

In the order from low performance to high sophistication VDUs are computer
terminals, graphics workstations and word-processing or document preparing
workstations. With the document preparing workstations we are referring to
those which offer WYSIWYG (What You See Is What You Get – although fortunately
we usually get more than we see). The lower performance VDUs place most
emphasis on legibility, while at the higher end of the range the emphasis is

on showing realistic document layouts, using various fonts in such a way   that
they approach the final paper result.

1.5.2. **Fonts**

In the information technology era, each aspect of the process of providing
information to the user needs to be optimised. There is a continuously growing
number of office tasks which are undertaken with the aid of a computer. The
visual display is the interface between this equipment and the user. Today
most office equipment uses a raster scan CRT, with flat panel displays being
slowly introduced alongside. As long as the fonts are built-up from pixels
[1.5] arranged in a matrix, the size, shape and nature of the del will be a
constraint on the font design. With CRTs for example, the dels can be either
overlapping or separated. This is in contrast to flat panel displays on which
the dels are always separated due to the inherent electrode arrangement. Only
the amount of separation differs from implementation to implementation.

Figures 1.7 to 1.9 show font designs for displays having various   resolutions.
A font design for a high resolution laser printer is shown in figure 1.8b.
Each letter has a capital height of about 3 mm. and has been   reproduced with
an enlargement factor of about 14. The elements of the matrix have been drawn
as non separated squares, rather than the separated square dels of the   actual
flat panel display implementations. The letters shown in figure 1.7a are
simple crude characters and the term font is more justifiably attributed to
the designs shown in figures 1.8 to 1.9.

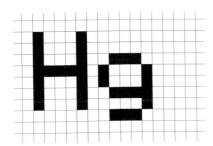

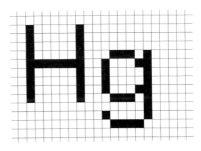

Fig. 1.7a A 5x7 font as used
on an early PC.

Fig. 1.7b A 7x9 font on a full
page word-processor.

A letter designed for use with a personal computer display is shown in figure
1.7a. This is the most basic design available and is often referred to as a
5x7 character matrix. The letter design could be used with a 6x8 matrix

arrangement, in which case the descenders of letters such as 'g' would touch the capital letters on the line below.

Today computer terminal and personal computer displays usually use 7x9 or even 9x11 character matrices. Figure 1.7b shows the design of characters used on standard full screen A_4 word-processors. They are more detailed and aesthetically more pleasing. They are more legible and correspondingly induce less fatigue. A character matrix of up to 7x12 pixels is used and in practice 9x14 pixels are allocated to the matrix in order to allow two pixels for the separation between the characters.

A font which could be used on a state-of-the-art display, for example for

Fig. 1.8a Similar to a Times Roman corps 12; font designed for a 8 dels mm^{-1} display.

Fig. 1.8b Similar to a Times Roman corps 12; font designed for the 20 dels mm^{-1} Océ 6750 laserprinter.

document preparation, is shown in figure 1.8a. Since such fonts are used on equipment with typesetting facilities, size and position of the character matrix is no longer fixed and is in fact a function of the font design. The spacing depends on the letter and in the example shown, the "H" uses 27x24 dels. Therefore bitmapped hardware is a necessity. The screen for which it is designed has a resolution of 8 dels mm^{-1} and with this degree of resolution a wide variety of fonts may be used. The example is similar to a Times Roman corps 12 which has a capital height of about 3 mm.

Figure 1.8b shows a slightly smaller font which was designed for the Océ 6750 laserprinter which has a resolution of 20 pixels mm^{-1}. The differences between these designs and those of figure 1.8a are obvious and when the laserprinter fonts are reduced to their original size (which is 14 times smaller) the staircasing is no longer visible.

The height of the capitals produced by most typewriters is slightly smaller than this. After the font has been designed it must be adapted to a specific resolution and a particular display. The element size, point-spread-function and overlap of individual elements are all taken into account.

The main difference between the dels of different types of display technology has already been emphasised: the dels of CRTs almost always overlap, whereas for a flat panel display they are separated due to the spacing between the horizontal and vertical electrodes. Today users increasingly demand negative contrast displays and this form of display is also supported by research into ergonomics which has emphasised the importance of the display having as much as possible the same visual characteristics as paper. In figure 1.9a the same font design as in figure 1.8a is shown, but the raster is now deleted in order to make a fair comparison with figure 1.9b. This shows the appearance of that font on a backlighted Liquid Crystal Display. The dels on the display

Fig. 1.9a Similar to a Times
Roman 27x24 font on a CRT.

Fig. 1.9b Similar to a Times
Roman 27x24 font on a backlighted
LCD.

are assumed to be 80x80 μm^2 with a pitch of 125 μm. It is clearly expressed that the letter contrast is highly influenced by the amount of separation of the individual elements in this example.

In the case of light emitting flat panel technologies used in negative contrast mode, such as plasma or electroluminescent displays, the appearance is rather different to that shown in figure 1.9b. With these displays, a part of a letter which is parallel to one of the electrodes, is seen to be continuous. However, parts which are at an angle are seen as discontinuous and jagged due to the non-overlapping nature of the dels. The light emitting background is also not continuous because of the dark lines, which form the separation, between both line and column electrodes.

Over the last decade there has been an evolution from word-processing applications towards document preparing. As this is now one of the most important office functions, it must be supported by an effective workstation and peripherals. The quality of the output from printers has improved enormously and this has acted as one of the driving forces for demands for improvements in display quality. The printers now offer the user considerable flexibility in output, but this is not supported by the capabilities of today's workstations. Non-impact printers can already print at 20 pixels per mm – which is approximately 4000x6000 pixels on an A_4 page. In contrast, the displays are only capable of up to 8 dels per mm (1728x2288 dels for an A_4 display). Even when the dimensions of the CRT are increased to allow an A_3 format, the total amount of information displayed does not change. The limiting factor is the rate of information transfer to the CRTs electron gun – a video rate of 300 Mbit s^{-1} is already being used today.

In our opinion, it is an overstatement to refer to WYSIWYG with respect to current displays and workstations. These cannot match the output of non-impact printers. Even if state-of-the-art CRTs are used with suitable high speed driving electronics, which are just commercially available today, the displayed image quality is still inferior to the printed one. In such a system with an A_4 tube, a 6 point font (with a capital height of 1.5 mm) has only 12 dots in the height available to define the character form. Clearly the resolution of the CRT is limiting the application of a good quality font in workstations.

Today fonts are all designed for applications using a bi-level display concept, even though a CRT is easily capable of handling grey levels, of course at the expense of the need for additional screen memory capacity, as is the case for a screen with higher resolution. For instance increasing the memory by a factor of four offers the option of either
- increasing the resolution of the CRT by a factor of two in both directions, or
- improving the CRT from a bi-level system to one with 16 levels of grey scale.

Since the latter option does not need the CRTs electron gun optics to be improved, or faster access to the screen memory, the rapidly falling cost of memory facilities means that the implementation of grey scale is a viable possibility. Grey scale fonts are being developed in several research laboratories, with the first results looking promising. Grey scale technologies could almost allow the WYSIWYG approach to be implemented, although the display units would be correspondingly more expensive.

Considering the commercially available equipment, most word-processing systems have an A_4 screen, a resolution of about 3 dels mm^{-1} and occasionally they are fully bitmapped. Bitmapped means that all dels are individually addressable, rather than the characters being addressed in a fixed grid. Bitmapped screens allow the manipulation of different fonts, character spacing and line spacing and often offer graphics capabilities, for example for business graphics applications. These screens just allow the distinction between font designs if the capital height is more than about 2.5 mm. They still differ considerably from the original masterfont designs, but a typographer would be able to recognise a design as, for example: "probably a Times Roman".

The German display standard DIN 66234, recommends a minimum viewing angle of 18 minutes of arc, which corresponds to a 2.6 mm capital height at the normal viewing distance of 500 mm. Obviously this depends on factors such as the luminance contrast between the character and its background, the background luminance itself and the illumination of the environment. It must also be appreciated that these figures are based largely on research into legibility and reading errors. For terminals and rough text entry equipment research has shown that fonts with 9x11 matrices are already satisfactory for producing an A_4 page of 60 lines by 70 characters. In our opinion the problems to be overcome in font design for sophisticated word-processing and document preparing equipment, using state-of-the-art CRTs, are no longer legibility problems but are related to the aesthetics involved in achieving WYSIWYG.

1.6. Flat panel displays in the office

Today the vast majority of office display systems are still based on the mature cathode ray tube technology. CRT technology has proven its ability to accept the challenge of a continuously increasing demand for improved image quality in office workstations over the past decade.
The first examples of flat panel displays in office workstations were not entirely successful for two main reasons: poor image quality and high cost compared with the competitive CRTs. This situation will probably change during the next decade for several reasons:
- maturity of CRT technology,
- cooperative flat panel technology display research programmes,
- the demand for higher information content displays.

Maturity of CRT technology means that in practice improvements beyond the current state-of-the-art (1800x2200 dels, 300 Mbit s^{-1}, 100 Hz frame interlaced) are quite difficult to realise. With a monobeam CRT this means video rates of over 300 Mbit s^{-1}, which requires technological breakthroughs. In multibeam technology where the video rate decreases with the number of applied beams, further improvements can be expected. Although, these will result in a need for very complex control circuitry to keep, for example, all the beams at an equidistant spacing over the entire screen.

When introducing the flat panel displays into workstations, it will be a necessity to avoid the acceptance problems which were encountered during the introduction of the first generation of visual display units.

The concept of softcopy quality is not yet fully understood and there is no clear translation into quantitative measures. It is obvious that the different visual characteristics of flat panels compared with CRTs will greatly influence the softcopy quality of VDUs. Contributing factors to those differences are, for instance, the different point-spread-functions, being approximately Gaussian for a CRT, almost square for Electroluminescent and Liquid Crystal displays and more or less crater like for plasma displays. The rise and decay times and different spectral response ranges are also important factors.

There is a continuing demand for higher information content display units for a broad range of applications, varying from high definition television and computer aided design, to document preparing and desktop publishing systems with a WYSIWYG approach. Everyone who is familiar with desktop publishing using one of todays PCs, such as a Macintosh with a standard screen size of 512x342 dels, will recognise this need. Unfortunately, with the realisation of higher information content CRTs, this already bulky device tends to grow even bulkier. A state-of-the-art "19 inch" monochrome CRT is easily 60 centimeters deep, which requires a considerable amount of desk space.

1.6.1. Simulation of flat panel technologies

National and international cooperative technology development programmes are being carried out. In Europe the RACE and ESPRIT programmes are sponsored by the European Commission and there are several similar programmes in the USA, for example ERADCOM.

As the importance of softcopy quality has already been stressed, it is interesting to mention one of the ESPRIT projects which is addressing this particular area [1.22]. The project is called "Modelling and simulation of the visual characteristics of modern display technologies under office work conditions" and is headed by the authors of this section. The consortium consists of five European companies and one University. The objectives of the project are:

- to design and build a real-time flat panel simulator, this will be a tool for the flat panel display designer in which design parameters can be optimised in an interactive way by experimenting,
- to identify from "ergonomic" experiments designed to simulate prolonged office work conditions, user requirements of flat panel displays,
- to give a lead to European display manufacturers in designing user acceptable flat panel displays, both by using the simulator tool and by increasing the cooperation between manufacturers, ergonomic experts and office system experts,
- to describe an engineering model of the visual system-technology interface which takes into account the visual interactions between spatial and temporal properties.

The complete hardware simulator will consist of a general purpose VME bus/UNIX V.2/68010 computer, interfaced to a modified image-processor. The image-processor will drive high speed dedicated hardware which will allow real-time simulation of display switching characteristics and point-spread-functions, with calculation speeds up to 20 Giga operations per second. Convolutions of up to 9x9 kernels will be calculated in real time, which is 9 nanoseconds per pixel. The system will drive two specially developed high resolution monitors, one being monochrome and the other being colour.

The databank of the system will contain models of display technologies (LCD and EL) and functional models of visual perception, as well as transmission and reflection data of typical display covers.

By simulating flat panel technologies, rather than experimenting by trial and error with parameters in the particular technologies and necessarily implementing each variation, one can optimise the different displays before they are even built and study their performance, user acceptability and quality.

1.6.2. Removable workstations

So far, in office applications, flat panel displays have only been applied in the family of portable, or removable, workstations. Since in recent years only multiplexed LCDs were priced competetively with standard monitors, this was the only technology to be found in the portable workstations. Due to the poor image quality of these early LCDs, it was not a pleasure to work with them; they did not invite you to write reports at airports, on aeroplanes, or in hotelrooms!

Today the situation has changed significantly. In LCD devices the contrast improved dramatically with the development of the Super (and Highly) twisted Birefringence Effect displays: SBE and HBE (now HBE versions are available [1.15] from all major Japanese suppliers). Examples of reasonably (640x200) sized Electroluminescent devices also became available (Sharp, Planar etc.), at decreasing prices. Devices using Plasma technology with up to 640x200 dels became available (AT&T, Fujitsu, Panasonic) at reasonable prices, some of them even having grey scaling options. All of these new devices have already been applied in present day portable workstations. These are now so sophisticated that most of them are more or less PC compatible – even a removable Unix machine is offered.

1.6.3. Liquid crystal displays

Liquid Crystal Displays were the first flat panel displays to be used in portable workstations. There were two main reasons for the dominance of this technology.

First of all they were cheap and rather easy to produce. This refers, of course, to the multiplexed versions. Actively addressed versions are more complex to produce and therefore more expensive. Furthermore LCDs require modest power what make them ideally suited to truly portable (not just removable) workstations, as long as a backlighting facility is not needed.

The main reason that they did not result in a tremendous breakthrough initially was their poor image quality, mainly expressed as a very limited contrast. For 200 line displays this was approximately 2.5:1. Another important drawback is their switching behaviour. Rise and decay times of the order of a few hundred milliseconds precludes the use of the devices in applications in which interaction is a prerequisite, such as usage of mouse devices, scrolling etc.

Although the supertwisted birefringence effect gave a significant improvement in contrast, (in 200 lines displays, contrasts of the order of 10:1 can be achieved), the rise and decay times did not improve. The effect, first reported and demonstrated by Philips [1.35] was rapidly modified to highly twisted birefringe effect, which unfortunately gave a lower contrast than the original SBE device, but were easier to commercialise due to the improved reproducibility of this device.

An example of a portable workstation in this category is Epson's Equity LT, delivered with an interchangeable 640x400 del LCD. Two options are available: a backlighted and a reflective LCD, the interchangeability offers the option that if technology advances are made, updates are easily installable.

Devices which will also allow interactive applications will either be active matrix addressed or ferro electric types. The field of active matrix addressing in connection with LCDs is currently the subject of extensive research. It promises very high contrasts and fast switching characteristics, with the possibility of moving video.

1.6.4. Electroluminescent displays

Over the past few years electroluminescence as a display technology was considered to be inappropriate for applications in portable workstations for the following reasons: first of all it was much more expensive than LCDs, there were lifetime problems, the devices had a higher power consumption and production of large sized displays has only been possible since a few years.

The advantages of the devices are very clear: high contrast (20:1), a wide viewing angle, fast switching characteristics, good luminance etc. They give a general impression of better image quality than the LCD devices. Since the prices are now falling due to mass production by more than one company, initial applications will be discussed.

Data General, who firstly introduced the Data General One portable computer with a 640x200 dels LCD, recently (in the latter half of 1986) changed the display unit to an electroluminescent display from Planar. The LCD version had not been very successful, one of the reasons being the barely readable screen which, by the way, was not one of the super or highly twisted birefringence devices.

Hewlett-Packard introduced a removable workstation with a 256x512 EL display. Not only the choice of display was interesting with this machine, but also the fact that it was one of the first examples of a portable Unix machine.

1.6.5. **Plasma displays**

A first attempt to force a major market breakthrough was made by IBM with an
OEM display product. This is an AC plasma display with 960x768 dels and a
resolution of 2.8 dels mm^{-1}. The unit was available in early 1983. Apart from
OEM deliveries, it is used in the IBM 3290 information panel display terminal.
Later an interfacing kit for the PC became available.

With the availibility of rather cheap new devices, most of them in the range
of 640x400 dels at 3 dels mm^{-1}, a range of new portable PCs reached the market
place. Although portable is not equivalent to removable, most of the devices
still need a mains supply due to the plasma display. Some of the available
plasma displays are capable of handling grey levels, as for instance Toshiba's
T5100 which is supplied with a 640x400 dels, four grey level, display.

1.7. State-of-the-art; CRTs in office workstations

It will be clear from the previous section that flat panel displays can only
be used today in a very limited class of workstations. Portable or removable
workstations are still niche applications with the users being confronted by
the constraints of todays flat panel technologies. The application of CRTs in
offices has evolved from the obsolete TV raster scan visual display units to
the more specialised and adapted monitor systems of today. The cost factor is
obviously the most encouraging one in CRT applications. In applications where
state of the art monitors are required, (which now means about 2000 lines of
about 2000 addressable dels), it might appear that the depth of the devices
causes too many constraints on the furnishing of offices. In these high
resolution CRTs, the deflection is limited to 70° and in order to reach the
high resolution, the electron gun and optics require a considerable depth; for
a 19 inch tube this will easily result in an overall depth of more than 55 cm.
Together with the space required for a keyboard, this might be too much for
some office environments. The display units currently in use in offices can be
roughly classified into three major groups, the boundaries are of course not
sharply marked.

The first class to be considered is that of terminals and personal computer
displays. In general they use a 12 to 13 inch CRT, often referred to as a half
page monitor. The resolution, or rather, the number of addressable dels, lies
in the range 200x560 to 400x640, and the video rate is of the order of about
25 MHz. Although the majority are characterised by a positive contrast the
trend is towards negative contrast. Often the display memory of these devices

is limited to a simple character memory, which means that the screen is divided in submatrices of 8x8 dels on which characters and/or limited graphic sets (mosaic graphics) can be defined. The more sophisticated devices are equipped with bitmapped memories. This means that every addressable del has memory allocated to it and is obviously a much more flexible approach. With bitmapped screens it is possible to perform graphics tasks to define different lettertypes and to implement a more friendly user interface, making use of, for instance, desktop metaphors as is done in the Apple Macintosh. This trend is initiated by, on the one hand, applications like graphics packages and desktop publishing, and on the other hand driven by increasing ergonomic demands. Falling memory prices are also a contributing factor.

The second class to be considered is that of full page word-processing workstations and engineering workstations. The main differences between these and the first class are the size and the information content of the displays. The screen sizes in this class lie somewhere between 15 and 20 inches, with the number of addressable dels being in the order of 700x1000 dels. The video rate is usually in the order of 70 MHz, with a non interlaced frame refresh rate of about 70 Hz. Almost all of the devices in this class use negative contrast, as most of the work done with the workstations is closely related to paper; word-processing, document preparing and publishing are terms that characterise the applications. Figure 1.10 gives a typical example. In this class most of the devices use bitmapped memories. It must be appreciated that we are dealing with dedicated workstations used by professional employees who normally spend a significant part of their working day performing a specialised job. This justifies the extra expense of a dedicated workstation. The WYSIWYG approach is mentioned by almost all suppliers and obviously there is the possibility of giving an impression on the screens of how the ultimate paper result will appear. This will not be a one-to-one copy due to the differences in resolution between non-impact printers and the screens. Sometimes the smaller sized fonts are unreadable, but nevertheless they give a better effect than the personal computer class of displays.

The third class to be considered is a new generation of displays which are, from a technical point of view, equipped with state-of-the-art CRTs. The application of this type of display lies in the area of publishing and document retrieval and in general it is used for the more sophisticated applications for which its high price can be justified. Almost all of the commercially available design implementations [1.40] centre around an advanced CRT (Philips M38-200) design. This is probably because it is supplied with complete application schemes and because it is compatible with Fax group III. CCITT standards. All devices have the following characteristics. The number of addressable dels is 1728x2288 bi-level (i.e. black or white), resulting in a resolution of 5.6 or 8 dels mm^{-1}, depending on whether a 15 or 20 inch tube version is being used. The usable luminance is about 70 cd m^{-2}. Video rate is high, being 288 Mbit s^{-1}, the line frequency is about 125 kHz, the frame

Fig. 1.10 Photograph of a full page A_4 word-processor (Courtesy of Océ-Nederland B.V.).

refresh rate is 100 Hz and there is an interlaced scanning system. An example of such a device is shown in figure 1.11. Suppliers of these monitors are e.g. Cimsa-Sintra, Digivision, Fimi, Myfra S.A. Considering the technical specifications it may appear to be very difficult to achieve a stable and proper operating design. This is also clearly shown by some of the device implementations. The stability of the two interlaced frames with respect to each other is particularly difficult to maintain.

What else can we expect from CRT technology?

1.7.1. Multibeam scanning CRT technology

This technology benefits from a renewed interest [1.2], probably initiated by the initial technical problems in developing higher resolution monobeam scanning. IBM reported at the SID 1985 Research Conference about a CRT using a multibeam scanning system with 16 beams, but the device is not yet available commercially. Tektronix, Inc. together with Azuray Inc. developed a multibeam scanning monitor with 8 parallel beams [1.26]. The monitor, which is produced commercially by Azuray has 2000x2000 addressable dels and 256 grey levels.

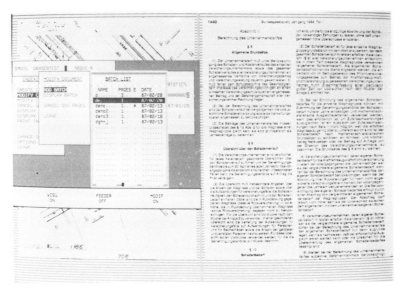

Fig. 1.11 Photograph of the screen of A₃ monitor with 1728x2288 dels (Courtesy of Myfra S.A.).

1.7.2. Ultra high resolution monobeam CRT technology

Recently a major breakthrough in monochrome CRT monitors was demonstrated (Siggraph 1986). Mega Scan Technology Inc., a two years old Pensylvania based young start up company, succeeded in the development of a 12 dels per mm, 19 inch diagonal monitor, shown in figure 1.12. The resulting addressable resolution is 4096x3300 dels. The applied line frequecy is 244 kHz using a non interlaced 72 Hz scanning system. The video rate reaches 1.5 Giga bit per second and despite of the advanced specification the overall power demand is as low as 180 watt. The monitor produces a remarkable stable and high resolution image.

Fig. 1.12   Photograph of an incredibly high resolution monitor (Courtesy of
Mega Scan Technology Inc.).

1.8. Bar imaging systems

Bar Imaging Systems (BIS)   are   one of todays most interesting spin-offs of
modern display technologies.   These   devices   are   increasingly   used   as   the
imaging system in non-impact printers based on electrophotographic principles.
Principally   a   BIS   might   be   regarded as a one dimensional array of display
elements, either in the realisation of shutters or as light emitting elements.
Non-impact printers basically use the technology of photocopiers. The standard
optical light path is replaced by a digitally controlled imaging device   which
is   capable   of addressing the photoconductor with a resolution of at least 12
pixels mm^{-1}. Almost all of   the   non-impact   printers   produced   today   use   a
scanning   laser   beam as the imaging device. Although the reliability of these
non-impact printers has improved considerably in recent years, they are   still
fragile   high   precision   systems   and   even   more   is   demanded   from them as
resolution and speed are improved. Laser printers evolved from a system   which
used   a   gas   laser   and   high speed rotating polygon mirror, to a holographic
scanning system using   a   laser   diode.   The   next   generation   of   non-impact
printers will use advances in solid state technology.

The Bar Imaging Systems such as thin-film electroluminescent edge emitter arrays, light emitting diode arrays and liquid crystal light shutters are all spin-offs from flat panel display technologies. The general device requirements (depending on resolution and speed) lie in the following ranges:

dot dimension: 20x20 μm^2 to 80x80 μm^2

switching speed: 50 μs to 500 μs

light output: 2 to 50 mW for a full bar of 200 to 300 mm length

contrast: greater than 5.

Even though these are fairly demanding specifications, some devices are already available commercially (e.g. the Agfa P400 LED printer, the Casio LCS-2400 LC-shutter printer, etc.) with some restrictions in meeting all of the requirements.

1.8.1. Liquid crystal light shutters

Today the main constraints on Liquid Crystal Light Shutter (LCLS) technology are related to the switching speed and contrast required for devices. In practice, the limiting factor on the printing speed is the switching speed of the shutters, rather than the available amount of light which depends on the transmittance of the shutters and the available light sources. The contrast required by the device depends on the slope of the gamma curve of the electrophotographic process. Practically useful response times are often achieved by using a dual frequency addressing mode [1.32]. This can result in response times of the order of half a millisecond. This is adequate for shutters in non-impact printers which produce an average of 12 dots mm^{-1} and about 6 A_4 pages each minute. Figure 1.13 shows the concept of a non impact printer using a liquid crystal light shutter.

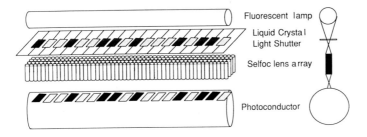

Fluorescent lamp

Liquid Crystal Light Shutter

Selfoc lens array

Photoconductor

Fig. 1.13 The general concept of a printer using a LCLS.

The demand for increased resolution and printing speeds means that a new approach is needed to increase the maximum switching speeds. Significant improvements have already been achieved using ferro electric liquid crystals [1.25]. These use thinner cells and allow much greater switching speeds to be attained. This has been implemented in the printer manufactured by Hitachi which produces 30 A_4 pages per minute. The NEC corporation has also made significant advances in this field by proving the feasibility of 1/4 multiplexed devices which have less complex driving electronics. Two realisations of LCLS devices are shown in figure 1.14 and figure 1.15.

In general, the advantages of the liquid crystal light shutter technology may be summarised as follows:

Fig. 1.14 Photograph of a LCLS printhead of NEC Corporation (Courtesy Society for Information Displays).

- the imaging device is produced from a single cell,
- the photoconductor sensitivity may be matched simply by choosing the appropriate light source,
- there is negligible heat production,
- there are benefits from the advances in flat panel display technology.

1.8.2. Light emitting diode arrays

Fig. 1.15 Photograph of a LCLS printhead of Casio Computer Corporation (Courtesy Society for Information Display).

As in the case of the liquid crystal light shutters, the light output of the LED array is coupled to the photoconductor using a Selfoc^{TM} lens array. The system configuration is similar to that used in the liquid crystal light shutter case, although now no additional light source is required [1.34].

Research in this field concentrates mainly on:
- light output,
- non-uniformity problems associated with light output from different LED pixels,
- the production of arrays without missing pixels,
- a higher resolution, i.e. more than 16 pixels mm^{-1}.

The amount of radiant power received by the photoconductor depends on several factors: the angular characteristics, the emitted power and the numerical aperture of the optical system. Light emitting diodes have an almost Lambertian angular characteristic, and combined with the standard Selfoc^{TM} lens arrays this results in the following figures values for the fraction of light which can be coupled to the photoconductor:

SLA6 :   1 %
SLA9 :   2.4 %
SLA20:   11.7 %

It must be noted that the resolving power of an SLA20 (which has a modulation transfer function (MTF) of 50 % at 4 line pairs per millimeter) is significantly less than that of the SLA6 (which has an MTF of 50 % at 8 line pairs per millimeter). This means that despite the higher light transfer efficiency, it is less suitable for high resolution printing.

The efficiency of LEDs depends on the spectral range in which they are operating [1.21]. It is low in the green and increases in the red and infrared. This suggests that photoconductors should be sensitised to the red and near-infrared region.

Uniformity problems can be overcome by laser trimming the thick film transistors or by compensation, which involves scanning the output of a bar and modulating the driving pulse width to compensate for the characteristic of each individual dot. This requires more complex grey scale drivers.

A more common problem encountered with bar imaging systems which use different modules is that of achieving precise alignment and connection. This needs considerable effort and will be particularly difficult when the resolution becomes greater than the present figure of 16 dots mm^{-1}.

Some printers which use this technology are already commercially available. Agfa produced a printer with a resolution of 16 pixels mm^{-1} and a speed of 18

A_4 pages per minute based on LED bars and Nippon Telegraph developed a 10 pixels mm^{-1} printer.

1.8.3. Electroluminescent arrays

Electroluminescent arrays have not been used for non-impact printers to date, but future applications in the field of Bar Imaging Systems seem probable so they will be discussed here. Thin-film technology and photolithography form a sound base for cheap fabrication. The light output from electroluminescent arrays has so far been considered to be too low for practical devices. The light output may be increased by increasing the thickness of the phosphor layer, however this is constrained by the driving voltage needed and the limitations imposed by IC technology.

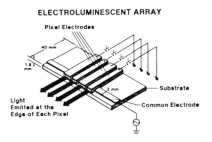 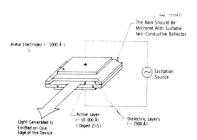

Fig. 1.16a Schematic arrangement of an edge emitter array of Westinghouse R&D centre, (Courtesy Society for Information Display).

Fig. 1.16b The TFEL edge emitter concept of Westinghouse R&D centre (Courtesy Society for Information Display).

Westinghouse R&D center used a novel approach whereby the light is emitted from the edges of the phosphor layer, rather than from the face [1.18]. This increases the depth of the light producing area by a factor of more than a thousand. The figures 1.16a and 1.16b show the general concept of the electroluminescent edge emitter. The light guiding phenomenon in this Thin-Film Electroluminescent (TFEL) structure causes the light output per unit area to be much greater for edge emission compared with face emission. Light output of the order of 5 $mW.mm^{-2}$ can now be achieved, which is a factor of three more than the light output from LED bars. It should however be noted that although the active dots in the case of a TFEL edge emitter have the same width as an LED bar, their height is only 1 micrometer. This is a factor of 50 to 100 lower and implies that the active area and correspondingly, the total light output in a Bar Imaging Device is decreased by the same factor. Apart from this, the angular characteristic of the TFEL edge emitter means that emitted light is more concentrated in the forward direction and improves the coupling to the photoconductor via an optical system. Furthermore, due to this phenomenon the necessity of an optical coupling system is not certain. One of the first realised samples is shown in figure 1.17.

Fig. 1.17 Photograph of an edge emitter array of Westinghouse R&D centre (Courtesy Society for Information Display).

Another advantage worth mentioning arises from the limited pixel height. This causes the MTF of the printer to be the same in both directions. Usually the MTF of a printer is slightly higher in the directon perpendicular to that of the photoconductor motion due to the print of the pixels being smeared in the direction parallel to the motion.

In conclusion, although the application of this technology for printing purposes is still in its infancy, it does look promising [1.20]. A light emitting bar made from a single component, without connection and alignment problems, and with low complexity, must be a very marketable commodity.

REFERENCES

[1.1] Baldauf, D., Invited technology challenges for improved human factors, Record of the 6th International Display Research Conference Japan Display 86, SID, Tokyo (1986), pp. 526-529.

[1.2] Beck, V.D., et al, Multiple beam cathode ray tube. Design overview, Conference Record International Display Research Conference, SID, San Diego (1985), pp. 186-191.

[1.3] Blado, S. von, Printed versus displayed information, Nachr.f. Dokum, 36, (1985), p. 172.

[1.4] Bruce, M. and Foster, J., The visibility of coloured characters on coloured backgrounds in viewdata displays, Visible Language XV14, (1982), p. 382.

[1.5] Bunell, E.H., Understanding digital type, J. M. Post Graphics Corporation, New York (1978).

[1.6] Cakir, A., Tagesgesprach: Ergonomie bei der Informationsverarbeitung, Management-Zeitschrift io, 51, nr. 9, (1982), pp. 333-336.

[1.7] Cakir, A. et al, The VDT manual, IFRA, Darmstad (1979).

[1.8] Chamberlin, D.D. et al, Janus: An interactive system for document composition, Sigplan. notices (1981), pp. 82–91.

[1.9] Cox, E.A., Electromagnetic radiation emissions from visual display units: a review, Displays, $\underline{4}$, 1, (1983), pp. 7–10.

[1.10] Gould, J.D. and Grischkowsky, N., Doing the same work with paper and CRT displays, International symposium digest of technical papers, SID, San Francisco (1984), pp. 280–283.

[1.11] Gutnecht, J. and Winiger, W., Andra: The document preparation system of the personal workstation Lilith, Software–practice and experience, 14, (1984), p. 73.

[1.12] Helander, M.G. et al. An evaluation of human factors research on visual display terminals in the workplace, Human factors review, The human factors society, Santa Monica (1984), pp. 55–129.

[1.13] Kalsbeek, J.W.H. et al, How specific is VDT–induced visual fatigue? Proceedings of the SID, $\underline{24}$, no. 1, (1983), pp. 63–65.

[1.14] Karov, P., Digital Formats for Typefaces, URW Verlag, Hamburg (1987).

[1.15] Kiyoshige Kinugawa, et al, 640x400 pixel LCD using highly twisted birefringence effect with low pretilt angle, International symposium digest of technical papers, SID, San Diego (1986), pp. 122–125.

[1.16] Knave, B.G. et al, Ergonomic principles in office automation, Ericsson Information Systems AB, Stockholm (1983).

[1.17] Koch, H., German standards for display work stations, Displays, $\underline{5}$, no. 2, (1984), p. 131.

[1.18] Kun, Z.K. et al, TFEL edge emitter array for optical image bar applications, International symposium digest of technical papers, SID, San Diego (1986), pp. 270–272.

[1.19] Lannamico, M., The integrator. Publishing system streamlines document production, Mini–micro systems, 11, (1984), p. 137.

[1.20] Leksell, D. et al, The construction and characterisation of a 400–lines per inch TFEL edge emitter, Eurodisplay proceedings, SEE, Paris (1987), p. 86.

[1.21] Librecht, F.M., Text and image printing with LED arrays, Journal of imaging technology, $\underline{11}$, no. 6, (1985), pp. 306–310.

[1.22] Meulen, A.E. van der, Architecture requirements and specifications display simulator, ESPRIT'87 achievements and impact, North Holland, Amsterdam (1987), pp. 1313–1324.

[1.23] Murch, G., The effective use of colour: perceptual principles, Tekniques, $\underline{8}$, no. 1, (1984), pp. 4–9.

[1.24] Murch, G., Human factors of displays, Seminar, SID, San Diego (1984), pp. 2.1-2 – 2.1-34.

[1.25] Naemura, S. et al, Multiplexed ferroelectric liquid-crystal printing head, Proceedings of the 6th international display research conference Japan Display 86, SID, Tokyo (1986), pp. 442-445.

[1.26] Odenthal, C.J. et al, A gatling-gun multibeam CRT, International symposium digest of technical papers, SID, San Diego (1986), pp. 53-55.

[1.27] Parris, J., Laser variations, Computer Systems, no. 10, (1986), p. 51.

[1.28] Peltu, M., A guide to the electronic office, Associated business press, London (1981).

[1.29] Radl, G.W., Ergonomische und arbeitspsychologische Erkenntnisse bei der Textverarbeitung mit Bildschirmterminals, DOZ, Koln (1983), pp. 3-20.

[1.30] Oron, J.P., Informatique et lecture de documents, Correlative systems France (1950),

[1.31] Saffady, W., The automated office. An introduction to the technology, National micrographics association, Silver Springs Maryland (1981).

[1.32] Saitoh, K. et al, Development of a high-speed and high-resolution Liquid Crystal Shutter (LCS) for printer based on dual-frequency addressed G-H mode, International symposium digest of technical papers, SID, San Diego (1986), pp. 262-265.

[1.33] Sasaki, T., Information displays for the age of new media technologies, Proceedings of the 6th international display research conference Japan Display 86 Sid, Tokyo (1986), pp. 2-5.

[1.34] Schairer. W., Properties of LED-array for electrophotographic applications, Journal of Imaging Technology, no. 12, (1986), pp. 76-79.

[1.35] Scheffer, T.J. et al, 24x80 Character LCD panel using the supertwisted birefringence effect, International symposium digest of technical papers, SID, Orlando (1985), pp. 120-123.

[1.36] Sherr. S., Electronic displays, Wiley Interscience publication, New York (1979).

[1.37] Shuto, M. et al, Integrated document editing and organising system, Proceedings of SPIE, The International society for optical engineering, Washington, 435, no. 8, (1983), pp. 214-220.

[1.38] Smith, W.J. and Farrell, J.E., The ergonomics of enhancing user performance with colour displays, Seminar lecture notes, SID, Orlando (1985), pp. 5.1/1.16.

[1.39] Tannas, L., Flat panel displays and CRTs, Van Nostrand Reinhold company, New York (1985).

[1.40] Thoone, M.L.G., A very high resolution document display, Philips Technical Review, (1985), pp. 6-10.

[1.41] Walraven, J., Kleuren op het beeldscherm, Instituut voor zintuigfysiologie TNO, Soesterberg (1985).

[1.42] Wong, K.V. and Sanz, J.L.C., A review of digital processing techniques for document handling and other applications, Seminar lecture notes, SID, Orlando (1985), pp. 6.2/1-26.

[1.43] Yamamoto, S. and Noro, K., An evaluation of a displayed layout on a VDT screen, Proceedings of the 6th international display research conference Japan Display 86, SID, San Diego (1986), pp. 534-537.

[1.44] Yamamoto, S. et al, The adaptation of VDT operators to VDT work, Proceedings SID, 25, no. 3, SID, (1984), pp. 231-237.

DISPLAY ENGINEERING: D. Bosman (Editor)
© Elsevier Science Publishers B.V. (North-Holland), 1989

III.2: DISPLAYS IN MONITORING AND CONTROL TASKS

D. BOSMAN

University of Twente, The Netherlands

2.1. Applications

Monitoring and control of its environment has always been important to life: to adjust the immediate and future living conditions and to increase the state of knowing. By creating artificial environments and processes (e.g. aviation, chemical plants) man has introduced many new foreign variables to be observed. Also their number and their rates of change often are beyond the capabilities of natural comprehension. Therefore instrumentation aids are applied to acquire appropriate measurands through transducing and signal processing; and to transform these into suitable observables by (model based) calculations and data format conversions.

In a well designed man-machine interface (MMI) three kinds of mapping take place:

a) first, all the variables describing the process and its environment are converted into one engineering format, suitable for further processing; possibly simple preprocessing computations are performed like decoupling, noise removal, bandwidth limitation and low level fusion of several variables into one or more observables:

b) second, the so obtained observables of common format are transformed into analogue signals and/or digital data, representing information to the observer after:

c) third, the signals/data in b) are transformed into natural stimuli (visible, audible, tactile), taking into account physical and psychophysical requirements and constraints imposed by the human receiver. The domains involved are electrical, optical, ergonomic and, of course, the physics of the electro-optic transformation itself.

In the early days of display engineering, this mapping was constrained to the grouping of variables of the same kind (e.g. a pressure distribution) or of dissimilar but closely related variables (modelling e.g. temperature and consumed power, pressure and flow rate). The flexibility of electronic data processing (EDP) has opened a whole new area of ergonomic matching of the process state and its behaviour as displayed, to the inner representation (or mental model) [2.1,2.2,2.3] of the process as perceived by the observer of the display.

The first two mappings depend very much on the type of process and its environment (in other words they are application oriented); thus are outside

the   scope   of   this   text which is devoted entirely to the third. However, the
display engineer concerned with   this   third   mapping   cannot   succeed   without
careful consideration of all aspects pertaining to a) and b).

Considering   the   range   of   man's   sensors   and   their   information   handling
capabilities, it is logical to channel the bulk of   the   data   to   be   observed
through   a   visual   display   unit   (VDU)   or monitor, because the eye is better
supported by the brain than all other natural receptors.
The simplest form of visual display consists of one display element (del)   with
binary   (on-off)   exitance   (indicator   lamp   with fixed legend, flag, blinking
cursor).   In an instrument panel or on the screen of a process   control   monitor
many   such   indicators/warnings can be integrated, the pattern of their (fixed)
locations being chosen to maintain control/display association and to appeal to
the operators' mental model of the process structure: an example   is   given   in
figure 2.1.   The   coordinates (X,Y) of these locations then are associated with

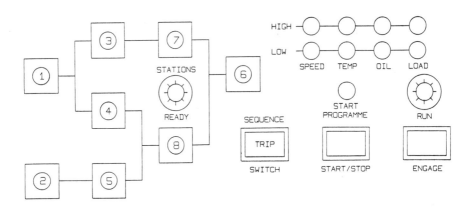

Fig. 2.1   Functional lay-out of a manual power-up panel.

the   function   of   the   indicators in relation to the process control; i.e. the
structural information (section I.1.2). Quick look of correct operation is also
provided by the panel layout. Similar designs are possible with a   2-D   display
with   touch screen. When the indicator lamp is dimmable, or the flag/cursor can
be presented in several intensities to change its attention value, the metrical
information increases.   The   next   step   in   sophistication   is   the   (electro)
mechanical   instrument   with   binary reflectance, which transforms the metrical
range of the variable in question into a range of   relative   positions   of   the
pointer/dial   combination.   More   complex displays require both positional data
and a broad variety of shades of grey. The dependency of the structural and the
metrical   information   on   the   type   of   application   leads   to   the   display
classification below, adopted   by   several   authors and users. Although not a

perfect guide, the engineering problems of drive and interface design, fabrication, connection, and packaging all increase sharply with the number of picture elements.

- Video devices are those characterised by today's raster television where high- data-rate (at least 25 times a second) information is presented in a pictorial form with multiple shades of grey. This would be utilised for display of sensor information in natural and/or artificially generated format, utility television and scan-converted imagery (laser, ultrasound scan, radar). Presently used raster formats of 525, 625, 875 and 1024 lines are considered typical;
- vector-graphic displays with very large numbers of dels but small number of simultaneously selectable grey/colour shades (computer aided design, graphic information from measurements, and process control with added alpha-numeric data);
- message displays for low data rate text, caution or warning legends (alpha-numeric symbols in dot-matrix form, ordinarily not more than two levels of exitance in addition to the off-state, but at most eight), used on push buttons (figure II.6.6) and small screens up to 15 rows of 25 characters;
- separate data devices in instrument panels or special purpose screens such as the trace display in oscilloscopes (figure II.3.19).

These four categories are depicted in figure 2.2. where video and vector-graphic devices are taken as one class: images.

Another type of classification is according to the environment in which the display is to operate. Environmental factors are: light intensities, temperature, vibration, air pressure etcetera. The first factor affects the legibility of caution lights and messages, and the quality of video images (contrast, colour). The other factors determine mostly the operating conditions and useful life of the equipment. The categories are:

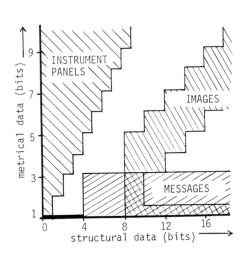

Fig. 2.2 Information capacity classification.

- office work conditions (offices, typist rooms, command and control rooms, home);
- severe workplace conditions (land vehicles; marine and aircraft instrumentation; agricultural machines, cranes, outside notice boards).

A third classification [2.4] is in the type of display (see section 2.4):
- head-down display (HDD), i.e. the operator concentrates on the data displayed on an instrument panel with no interference from real world data;
- head-up display (HUD), i.e. the displayed data are superimposed over the real world scenery through the aid of e.g. a collimating optical system;
- process/mission management displays (PMD/MMD), i.e. data pertaining to the type of operation and its performance quality; the data rate is lower and the display is likely to be positioned to one side, rather than centrally. They include mimic or map data, status data, subsystems operation data;
- helmet-mounted displays (HMD), for dynamic primary data: the display projects directly into the eye's pupil through a suitable optical system, much like that of the HUD but potentially with wider field of view. Situated at or in the helmet, the display rotates with the head and therefore is not spatially confined as are other types of display;
- keyboard displays, i.e. the display is used in conjunction with a touch switch device as a programmable keyboard. This category usually is shared with others like HDD, PMD/MMD;
- alpha-numeric displays, multi-legend switches. Modular read-out devices, specifically alpha-numerics and bargraphs: programmable, variable format control display components.

2.2. Historical survey

The first electronic displays came in the thirties, cathode ray tubes which served the dual purpose of transforming the signal (raw radar pulses) into the desired image structure and of displaying of this image. Some were of conventional design, in other tubes the integral transformation function in the instrument was even more evident in dedicated e.g. (circular deflection) construction. In the fifties the extruded beam CRT was introduced wherein the beam was first shaped to form a character by forcing it to pass a selected location of a mask containing the choice of characters, before being deflected to the appropriate location on the screen [2.5]. Several other designs for the same purpose were developed [2.6].
At about the same time electronic character generators were developed [2.6], directing the beam of conventional CRTs to stroke-write the symbols, which found application in displaying the memory contents of computers and in the HUD

[2.7]; the latter became operational in combat aircraft from the sixties onward. Stroke writing still is extensively used, also in combination with raster addressing, in modern CRT displays requiring very high spatial resolution at high refresh rates (graphics workstations, avionics displays).

But it was not until digital technology matured that electronic displays became a general commodity, being able to provide windows on the state of processes as represented in digital storage devices. Computers proliferate in almost every kind of system and plant, and so does the visual display unit (VDU).

At present the display capacity in dels sr^{-1} seems to converge to a common format in the different applications from control room to cockpit, largely because both the visual and the cognitive capabilities of the human side of the interface have given characteristics.

2.3. Operating conditions

The environmental conditions (operating) may differ considerably for the various fields of application. In offices the illumination, temperature, humidity and pressure vary little. At the other end of the operating conditions range the cockpit is the more severe environment where only limited protection can be afforded due to, among others, requirements of operational nature combined with size, weight and power constraints. The display must be able to operate within its specification under large pressure variations, severe shock and vibration, large (and sudden) temperature variations from arctic cold to tropical heat, be sealed for water and hydraulic fluids, protected from sand, dust and fungus, accept voltage transients and power fluctuations, be immune to electromagnetic interference (EMI) and magnetic fields, etcetera.

At these conditions the display must remain legible. Relative vibrations between the display and the eye, at good contrast and temporal frequencies up to about 12 Hz, give complicated response in terms of reading error rate; at still higher frequencies and peak-to-peak amplitude of more than 25 minutes severe blur occurs [2.8] of displayed characters of recommended size which are defined in a too small matrix block (5x7), even when displayed with intrinsic memory; for displays with both horizontal and vertical scanning in short persistence technology (LEDs, plasma) the image may break-up to become completely incomprehensible. Such considerations can restrict the choice of the technology and of the image formation techniques. Likewise the vast illuminance range with eye adaptations from low scotopic (dark night) to very high photopic states (noon, high altitude). For instance, for a display to remain legible [2.9] with sunlight (10^5 lux) directly on the instrument panel, with a required contrast ratio of 1.42 (see section I.1.4.3), at a display and display surround reflectance of 0.1 %, the foreground/background luminance difference of the display should be at least 45 cd m^{-2} leaving little room for additional levels of luminance, especially restrictive in colour displays.

The display requirements for aircraft displays are, of necessity, spelled out in great detail to ascertain their usefulness under various combinations of operating conditions; an excerpt of an excerpt for transport type aircraft is given in table 2.1.

Table 2.1

Display visibility	: clearly readable from night time conditions up to 10^5 lux at 60° incidence
Display refresh rate	: symbology and graphics > 60 Hz filled regions > 80 Hz both non-interlaced
Luminance uniformity	: within 20 %
Manual luminance control (set point)	: 5 % to 100 %
Automatic luminance control	: maintain contrast around manual set point within 10 %
Minimum contrast ratios	: red stroke: 1.5; red raster: 1.2; green stroke: 3.3; green raster: 1.5; blue stroke: 1.5; blue raster: 1.2
Line width (symbology, graphics)	: 0.2 to 0.5 mm over full luminance range
Symbol positioning error	: < 1 % of diagonal
Relative symbol accuracy, local	: better than 0.5 mm
Symbol/line jitter	: < 0.2 mm
Display dimensional stability, including spatial effects of EMI	: better than 1 %
Display face plate reflectance	: diffuse < 1 % specular < 0.1 %
Luminance ratio unaddressed/addressed dels	: < 3 %
Luminance ratio moving/stationary dels	: within 30 %.
Colour variation	: radius on 1960 UCS chart < 0.013

Operating conditions are not limited to effects in the physical and psychophysical domains; the diversity of information sources and of decision rates also make demands on the cognitive abilities of the operator [2.2,2.10]. The displayed information must be absorbed and mentally processed at meaningful pace and at acceptable error bounds. The latter tend to grow sharply when a certain display/control density (which depends on available visual search time) is exceeded. In modern aircraft and some control rooms this is already the case as depicted in figure 2.3. It has become necessary to replace the cluttered

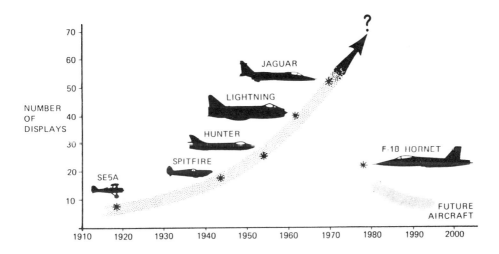

Fig. 2.3 Growth in number of aircraft displays [2.11].

instrument panel by a design with multi-function displays and programmable switches, discussed in the next section.

The displayed information should depend on the task at hand. For aircraft the average task distribution curve is given in figure 2.4, with exceptions occuring due to momentary shifts of workload in critical phases of the task and at change-over in operations. The display function selection may be effected manually or under control of an expert system.

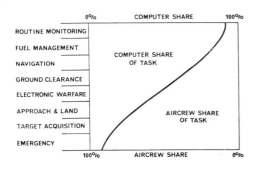

Fig. 2.4 Typical distribution of task sharing between aircrew and controlled systems [2.22].

Depending on the process to be controlled by the operator, the tasks can be very different in nature. In e.g. chemical plants and aircraft under autopilot control, the task is most of the time characterised as "operation by exception". The kind of data needed for such operation is twofold: local deviations from desired setpoints and trend data (traces, time histories) for variables having strong correlation with future states of the process. Such formats are calculated to enable the operator to perceive at a glance the total picture with many variables involved (figure 2.5), e.g. by mental template matching of the presented pattern with memorised

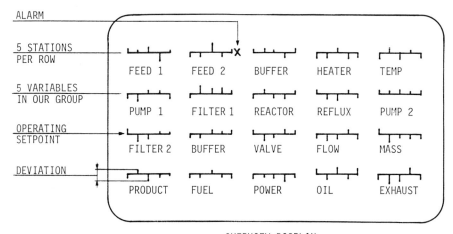

OVERVIEW DISPLAY
DEVIATION MAGNITUDE SCALED TO IMPORTANCE

Fig. 2.5 System state overview display [2.23].

acceptable patterns gained from experience. The data entry method is not so important in non-time critical regimes.

In active control (high decision rates) more stringent display requirements exist, e.g. director displays which guide the operator in the activation of the appropriate controls, necessitating more intelligent conditioning of the available arrays of data, combined with automatic selection of display functions. The preferred data entry method is by dedicated (sometimes programmed) push buttons, touch panels and analogue controls rather than coded entry through a general purpose keyboard.

Delays incurred by the reaction time of the operator can be partly compensated by predictive displays, where a system model extrapolates response based on the inputs and present states/conditions, indicating future system behaviour.

All these considerations have their impact on the choice of display technologies, formats, locations relative to the position of the operator, display data conditioning and the operator environment. In figure 2.6 a typical control room lay-out in depicted. The secondary data shown on displays further away from the operator must be sampled periodically, either by the operator doing rounds, or by shifting the data from the "home display" to a scanner display in front of the operator. While examing the secondary data, it may be necessary to attract the operator's attention to a change in the primary data. This can be accomplished by a window in (all) the secondary data display(s), or by providing the operator with a helmet-mounted display (HMD) which super imposes the window on the scene, providing much more freedom to move about. For similar applications the HMD has already found acceptance in military aircraft where they also reduce pilot workload in timecritical phases of flight.

2.4. <u>Display type</u> [2.4]

In the context of modern, programmable (i.e. multi-function) display technologies the displays discussed below are electronic displays, computer driven; display image format and information content is effected under software control, either manually or by a programme or expert system. In control rooms of plants and, more recently, control systems of flexible production cells and

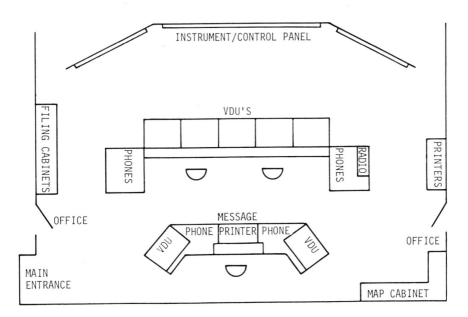

Fig. 2.6 Typical control room layout [2.24].

of robots in plants, considerable progress has been made. Historically, the
application of programmable displays in the aircaft cockpit is the more
advanced development and also more consistent with general (aircraft)
operation. Therefore we will focus attention to aircraft displays with the view
that spin-off will eventually benefit other applications as already experienced
in automotive displays (chapter III.3). The situation of the electronic
displays (CRTs) in a typical transport aircraft instrument panel, for captain
and co-pilot, is shown in figure 2.7. Many of the traditional electro-

Fig. 2.7 Typical transport aircraft layout (Fokker F100).

mechanical displays are still in use alongside the CRTs for various valid
reasons, among others as a back-up system.

2.4.1. Head-Down Displays (HDD)

Head-down displays are direct view, flexible information format displays having
the capability to display both graphic and video information. The Horizontal
Situation Display (HSD) and the Vertical Situation Display (VSD) are two
classes of HDD that are associated with two specific types of flight
information. Reference [2.12] discusses the use of map displays in high
performance aircraft, [2.13] discusses HDD in military transport aircraft.

The HSD is a multi-function display that, at a minimum, displays the heading information: Horizontal Situation Indicator (HSI). In essence, it has a designated information function, that being the display of information giving aircraft orientation and any related situation information with respect to a position in a plane parallel to the earth's surface. In particular the heading indication, bearing pointers, distance, and course deviation indications, as

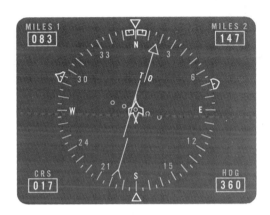

shown in figure 2.8. The display provides in the top left and right windows the distances to two ground based reference points (Digital Measuring Equipment, DME); in the lower left window the selected course and in the lower right window the selected heading. The situation display within the compass rose, shows the heading, the course and the course deviation with respect to the fixed aircraft symbol. The HSD should also be able to provide combination of:

Fig. 2.8 Horizontal Situation Indicator (HSI) (Thomson–CSF).

• Aeronautical charts and/or electronically generated maps.
• Navigation, target/drop zone identification.
• Electro-optical and radar sensor video.
• Flight control cues.
• Electronic warfare information.

The VSD also is a multi-function display which, as a minimum, displays the attitude information portrayed on an electromechanical Attitude Director Indicator (ADI). The VSD can provide information similar to the HSD but with respect to a plane that is perpendicular to the earth's surface. In figure 2.9 an example is given of the vertical situation indicator configured for the landing phase. With relation to the fixed wing symbol in the middle it shows: roll angle 0°, pitch angle 4° up, indicated airspeed 130 knots, and the runway aligned for normal touchdown with the star predicting the touchdown point. At the left in the window is the indicated air speed (130 kts), correctly aligned between the outer speed cues; the vertical scale on the right is the vertical speed (100 FPM down); in the boxes barometric altitude (948') and radar altitude (98').

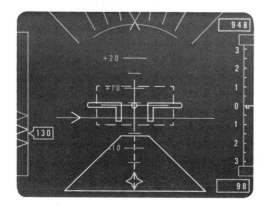

Fig. 2.9 Attitude Director Indicator
(ADI) (Thomson-CSF).

In the several modes available in the HDD there is a requirement for some grey levels and several colours (red, yellow, brown, green, blue). Furthermore, in imagery from video sensors and stored maps the desired number of grey levels can vary from e.g. five to thirty for daylight conditions; judicious placing of that type of display is essential.

Primary displays must be located in front of the pilot(s), minimising the effort of moving head and eyes. It is desirable to maximise the ratio of display active area to the total panel area occupied, subject to the constraints imposed by clutter and the need for other displays, light sensors, controls and switches. Conversely, minimum separations between functionally different types of display information and the need for instrument panel structural integrity both suggest limiting the proximity of adjacent displays.

Nominally, 10 % of the active area linear dimensions of the larger of two viewed displays is needed to separate different but related task information. Larger distances are needed to separate information for unrelated tasks. Thus, the displays for aircraft systems management are placed aside or, with the 2 seat side by side cockpit as in transport aircraft, in the middle.

2.4.2. Helmet-Mounted Systems

Helmet-mounted systems have been used for over ten years but only until recently they are generally accepted. Helmet-mounted systems are discussed in [2.14-2.17]. The displayed images are collimated, i.e. focused at infinity. An example is shown in fig. 2.10.

These systems fall into two related categories: sights and displays. A Helmet-Mounted Sight (HMS) is designed to measure the pilot's line of sight to a target in relation to the airframe and to process that information for use in direct control of remote sensors and weapon delivery systems. The Helmet-Mounted Display (HMD) provides the crew member with a head-up display monitor which is lightweight, low powered, gives high resolution and can also be used as a sight.

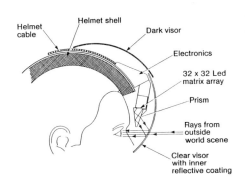

Fig. 2.10 Cross section of a helmet-mounted display [2.25].

In sighting applications the attitude of the helmet must be measured relative to a selected reference frame. There are, in general use, three systems which are used to measure helmet angles and position. One is optical, one uses infra-red and the other is magnetic in operation. For the helmet to operate as a sight it is necessary to have an aiming mark. The simplest possible sight is an illuminated cross or circle. The image source can be a graticule illuminated by a miniature lamp or a matrix array of any suitable technology (e.g. LEDs, chapter II.6). This image is then projected and collimated by the optics and presented to the eye after reflection from the visor. Latest advances include the use of a diffractive optical element in the visor which both reflects and collimates the image.

An advantage in using a matrix display is the flexibility to change the image. In addition to the aiming mark, discrete data such as speed, range, altitude and pressure can be digitally displayed. Also, flashing direction indicators can direct the crew member where to look for a target.

The helmet-mounted sight allows the pilot to acquire the line-of-sight to a target outside the normal field-of-view of the head-up display. This data is then immediately available for use by the weapon aiming system or to update the navigation system by spotting way-points. In addition, both slewable weapons and sensors can be slaved to the helmet. The pilot's head, in effect, becomes a very sophisticated and ergonomically attractive direction controller, thereby integrating the pilot's visual/monitor skills with the specialised accuracies of weapon and navigation systems.

The Helmet-Mounted Display (HMD) as opposed to just a sight, combines a helmet angle and position sensing system with a helmet-mounted high resolution display and optics to provide a highly flexible system incorporating many of the facilities to be found in both head-up and head-down displays. Because it is a quality imaging system, it is possible to present both pictorial and symbolic information together offering a combination of synthetic imaging and sighting information.

The recent development concerned with miniature display devices has ensured that in spite of their small size there is no reduction in performance. Also by mounting the display on the helmet close to the eye this effectively provides the equivalent of a much larger panel-mounted display with the advantage in saving both weight and power in the process. Under good viewing conditions the eye can resolve 0.25 milliradian and it is quite obvious that the performance from the display will fall short of this. It is, however, very important to get the maximum benefit from the HMD by having high resolution for the picture. Another very important factor associated with helmet-mounted displays is that the picture scale and viewing angle should not be compromised by limitations in the optics. The image to be displayed will be collimated to appear at infinity since it should register with the outside world. The instantaneous viewing angle should preferably be 40 degrees in azimuth and 30 degrees in elevation with an exit pupil of 2.5 cm to accomodate a range of users. An attractive solution appears to lie in the use of a diffractive optical element preferably embodied in the visor which will both focus the image at infinity and operate as a high efficiency (90 %) reflector at the narrow band wavelength of the CRT phosphor but offering see-through ability at all other visible wavelengths without producing noticeable colouration of the outside scene. An adjustment to ensure that the center-line of the optics is aligned with the user's eye-ball may reduce the exit pupil size with corresponding reduction in the size of the diffractive optics.

There could be considerable advantages in providing images to both eyes. This would combat the inevitable binocular rivalry which will otherwise occur and would ensure harmonised viewing by both eyes. However high vibration can cause convergence problems and flicker detection may be increased with binocular viewing [2.18].

2.4.3. **Mission Management Displays (MMD)**

Mission management displays include map displays, aircraft situation data, and aircraft subsystems data such as engine, fuel, hydraulic, oxygen, ordnance, navigation, communications, countermeasures, etc. One example is a flexible information format to display computer generated vector graphic aircraft systems data as depicted in figure 2.11. Another example is an optically combined display of tactical data overlaid on the microfilm image of an aeronautical chart.

The first example is a graphic display that has no requirements for video presentations. A two- to three-shade of grey display can take advantage of sizes from 100 mm x 100 mm for the display of the more conventional data-up

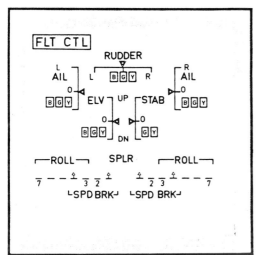

Fig. 2.11 Typical aircraft systems display. Faulty units shown in different colour.

to 300 mm x 300 mm for the display of terrain contours, extracted from video images, combined with aircraft state data; symbolic map data such as threats, weapon engagement perimeters, computer synthesised targets, computed optimal flight paths, and navigation displays. For a scanned matrix display a minimum resolution of 2.5 dels mm^{-1} and up to 250 Hz refresh rate for the display of vector graphic display imagery is necessary. The requirement is based on the minimum requirements to provide both the appearance of smooth continuous image motion and simultaneously the precise scales, pointer positioning, vector rotation, and vector translation positioning accuracies needed for the rapid, accurate presentation of vertical and horizontal scale, bargraph, round dial, graphical, symbolic map, and perspective information formats. This del density also allows the presentation of upright numerics and alpha-numerics in highly legible and identifiable 10x14 dual stroke width for larger fonts. Small, visually complex characters like alpha-numerics, to be rotated, must have stroke widths that contain a minimum of three dels and simultaneously satisfy the 15 % of character height requirement of good legibility (e.g., 15x21 array font size or greater independent of display element density). Presently being used are monochrome CRTs utilising two discrete luminance levels in addition to off. Current sizes are approximately 120 mm x 180 mm, oriented as a normal page of a book. This size is a compromise between that requested by human factors studies and limitations based on airframe installation requirements. Present studies of systems utilising CRTs indicate that an alpha-numeric character must be at least 5.5 mm high, with an ideal height being 6.5 mm (assuming a 700 mm viewing distance). The graphics capability should include simple graphs, geometric shapes and lines generated by a display with uniformly spaced horizontal and vertical display elements. Colour is desirable to accentuate information.

Certain applications may favour the display of actual aeronautical charts which have resulted from a long interplay between cartographers and users [2.12]. Although they are not always ideal for display purposes, charts from this mainstream of development are likely to form the basis of future displays. Their optimum display demands high resolution, full colour, for the projected map although the overwritten data can be monochrome.

An optically combined display provides high resolution colour as well as being compatible with a full electronic display suite. It provides a form of dual redundancy and installs in a single unit. An alternative using the same data base is the remotely scanned map, which depends on the availability of a full-colour high resolution display embodying a proper solution to the problems of high ambient light.

The all digital solution is a possibility which awaits a viable solution of the editorial and logistic problems of a digital data base together with a proper full-colour display head. There are digital hardware and software problems and the storage required could be massive, even in terms of devices such as optical discs and bubble memories. Bearing in mind the complexity of cartography as an activity and the long history which has led to present mapping material, it would be a considerable task to demonstrate that such a system is necessarily superior to a combined display. It is however possible to filter information with an all digital data base, e.g. show SAM sites or all radar on demand. In the longer term the rapid change of data base content, the ability to declutter cartographic features, and display unit compactness may well cause to displace film-type systems.

2.4.4. Head-Up Displays (HUD)

The HUD is a collimated (focused at infinity), optically projected display (figure 2.12) designed specifically for airborne use, although proposals for other vehicles already have been made [III.3.1]. The HUD is placed above the primary data HDDs. The HUD combining glass, located between the windscreen and glare shield, transmits the real-world scene directly while reflecting the display image so that the two superimposed images are viewed simultaneously by the pilot/driver. The optical magnification of the HUD is scaled one to one for symbols which need to match the real world scene. Indications such as pitch angle may be greater than one to one. Because the reflected image is collimated, the angular size and position of symbol cues projected within the display field-of-view remain constant with pilot head motion. The HUD is used to project vector graphic and video information. The primary functions of the HUD are flight guidance and weapon aiming, therefore a high quality optical system is required that can maintain accurate symbol positioning. The HUD is also used to display symbol cues for other mission modes such as take-off, landing, navigation, terrain following and ground collision avoidance. Electro-optical sensor displays such as Forward Looking Infra-red (FLIR) can also be

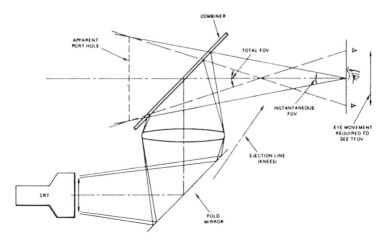

Fig. 2.12 Functional diagram of refractive HUD optical system [2.27].

presented on the HUD. Figure 2.12 shows a functional diagram of HUD optics.
The combiner is located between the windscreen and glare shield with sufficient
clearance to minimise secondary reflections from the windscreen. In aircraft,
the HUD must clear the pilot's ejection line. The HUD can extend beyond the
surface of the instrument panel, but it must not obscure the pilot's
line-of-sight (LOS) to other surrounding displays and controls or the pilot's
"over-the-nose" LOS. These geometric constraints determine the specific optical
performance that can be obtained in any particular aircraft installation. A
total field of view from 20 to 40 degrees is desirable for fixed wing aircraft.
A contrast ratio of 1.2 against a 10^5 lux (10^4 fc) background is required for
graphics. Symbol position accuracy of one milliradian over the central 5
degrees of the field-of-view is required. A three times larger resolution per
picture element is required throughout the field of view.

Recent   advances   in holographically formed optical elements enable the
construction of combiner elements which have very high (90 %) reflectivity over
a very narrow band of optical wavelengths. This enables a bright image to be
presented while imposing very little attenuation of the outside scene. Narrow
band CRT phosphors are available to which the spectral response of the
holographic plates are matched. In addition to the above advantages the angular
selectivity of the holographic elements enable compact optical units to be
constructed with wide fields of view. Most existing military HUDs are
monochromatic. While several hues of colour may be desirable colour is unlikely
if holographics become fashionable. Reference [2.19] provides an in-depth
literature review of 293 publications relating to the HUD for aircraft cockpit
application.

2.4.5. **Keyboard displays**

Keyboard displays (integrated in the push buttons, figure II.6.6) afford the possibility of providing at the same location several commands, thereby integrating the control of several sets of functions presently realised by separate dedicated control panels and the management of several primary multi-function displays. They can reduce operator real-time workload both by allowing preselection of flight tasks on a functional or mission segment basis and by providing flexible but orderly procedures in the conduct of the total crew station management. Through the incorporation of programmable switches and information processing, the plethora of dedicated control switches can be sharply reduced.

It is typical to have a legend area for each switch which consists of two rows of four to eight characters per row. The individual legend should use as a minimum a five by seven matrix, although a seven by nine matrix is preferred. Experiments have established [2.6,2.9,2.20] that a 5x7 matrix is the minimum required while matrices larger than 7x9 do not lead to meaningful improvements. Fifteen to thirty switches are desirable. The legends should be viewable under bright sunlight at a minimum contrast ratio of 1.2 and for some applications be usable with night vision goggles.

2.4.6. **Alpha-numeric modules**

Modular readout devices, specifically alpha-numerics and bargraphs, are needed to provide legible, reliable, variable format control display components. Readouts will be needed both in discrete and mosaic formats to allow variable fonts and positioning in alpha-numeric modes and/or line positioning for bargraph applications. Discrete readouts take the form of individual or grouped alpha-numerics. They are used in applications which range from one or two digit displays and legend lamps, to message displays of several rows of up to 20 or more characters per row. Reference [2.21] discusses these applications in more detail.

These devices can be developed as modules employing three basic forms:
• Segmented numerics.
• Dot-matrix alpha-numerics.
• Small area mosaics.

It is desirable that these modules should be constructed so they can be abutted on all four sides and incorporate all necessary drive and address circuitry. This concept will allow a wide variety of form factors and display areas which are appropriate for application of these devices. Applications include: control panel numeric and alpha-numeric readouts; flight, mode, and caution legend lamps; and multi-legend display switches. The devices need to be readable in bright sunlight as well as with night-vision goggles. The resolution should be from one to six dels mm^{-1}

2.5. Conclusion

The desired resolution of process management displays are likely to increase to about 100 dels mm^{-2}, and about 400 dels mm^{-2} in demanding applications (del size 0.5 minute corresponding to eye resolving power). Local geometric accuracies should be even better. For professional A_4 type displays the del count will become about 2000x4000 (4000x6000). Presently these figures seem to be the limit of technology, except for some projection type displays. Possible contenders are the CRT, the laser display, liquid crystals and the combinations thereof. Other display technologies are less likely to meet the requirement of 10 dels mm^{-1}.

Moreover, the displays must provide a number (probably $\leq$ 32) of recogniseable contrast levels, a mixture of brightness and tint, with a minimum of 5 hues maintained under unfavourable conditions. The update rate (not: frame rate) must allow fusion of successive images of moving objects/images; recommended is an update rate > 50 s^{-1}. Light output of active devices shall be dimmable over a substantial range while maintaining dominant characteristics.
In the following table 2.2. the major display parameters are compared for 11 technologies.

Table 2.2

	del density	contrast	update rate	light output	dimmability	grey levels	colour	size small	large
CRT	++	+	++	++	++	+	++	++	++
VFD	+	+	++	+	++	+	+	++	-
Laser	++	++	++	++	++	++	-	++	++
LED	+	++	++	++	++	++	+	++	-
PD	+	++	-	++	-	-	-	++	++
LC	+	+	-	na	na	+	+	++	+
LV	+	++	+	++	++	++	+	+	++
EC	-	++	--	na	na	--	--	++	-
EP	-	++	--	na	na	--	--	++	+
EE	-	++	-	na	na	--	-	++	++
EL	+	+	-	+	-	-	-	++	++

2.6. Acknowledgement

The author is indebted to prof. W. Hollister who provided much of the material on aircraft displays for this chapter, and to AGARD for permission to reprint many illustrations from their publications. As many of the references are from AGARD, and readers many wish to obtain back copies, the address is: Publications Officer, 7 Rue Ancelle, 92200 Neuilly-sur-Seine, France.

REFERENCES

[2.1]   Rasmussen, J., The human data-processor as a system component: bits and pieces of a model, Danish Atomic Energy Commission, Report R-8-74, Risö, (1974).

[2.2]   Kalsbeek, J.W.H., The human side of the operator, Proceedings Symposium "The operator-instrument interface", Institute of Measurement and Control, Middlesbrough (1978), 14 p.

[2.3]   Bosman, D., Systematic design of socio-technical systems. In: Stress, Work Design and productivity. Eds E.N. Corlett, J. Richardson, John Wiley and Sons, Chichester (1981), pp. 165-195.

[2.4]   Hollister, W.M., Applications. In: Modern display technologies and applications, ed. D. Bosman, Advisory Group for Aerospace Research and Development (AGARD), Advisory report 169, ISBN 92-835-1438-6, Paris (1982), pp. 181-192.

[2.5]   Peterson, R.M. and Ritchart, R.C., Recent developments in shaped beam display and recording techniques, IRE Internat'l Convention Record, 6, Pt 3, (1958).

[2.6]   Sherr, S., Electronic displays, Wiley Interscience, ISBN 0-471-02941-6 New York (1979).

[2.7]   Freeman, M.H., Head-up displays, a review, Optics Technology, 1, (1969), pp. 63-70 and 175-182.

[2.8]   Lewis, C.H. and Griffin, M.J., Predicting the effects of vibration frequency and axis, and seating conditions on the reading of numeric displays, Ergonomics, 23, no. 5, (1980), pp. 485-501.

[2.9]   Wharf, J.H. et al, A comparative study of active and passive displays for aircraft cockpit use, Displays, 1, no. 2, (1980), pp. 115-121.

[2.10] Hopkin, V. David, Human factors in the ground control of aircraft, AGARDopgraph 142, Aerospace Medical Panel, Paris (1970).

[2.11] Lyons, J.W. and Roe, G., The influence of visual requirements on the design of military cockpits, AGARD Advisory Report 255, Paris (1980).

[2.12] McKinlay. W.H., Evolution of tactical and map displays for high performance aircraft, AGARD Advisory Report 255, (1980).

[2.13] Chorley, R.A., Electronic flight deck displays for military aircraft AGARD Conference Proceedings 32nd Symposium, Stuttgart (1981).

[2.14] Task, H.L. et al, Helmet-mounted displays, Design considerations, AGARD-AG-255, Oct. (1980).

[2.15] Laycock, J., A review of the literature appertaining to binocular rivalry and helmet-mounted displays, Royal Aircraft Establishment, TR 76101, (1976).

[2.16] Stanley, R.D., Limited flight evaluation of a helmet-mounted tactical manoeuvring display system in the NT-38A aircraft, SY-115R-79, (1979).

[2.17] Wesley, A.C. and Blackie, I.T.B., Integration of sensors with displays, AGARD-AG-255, Oct. (1980).

[2.18] Laycock, J. and Chorley, R.A., The electro-optical display/visual system interface, human factors considerations, AGARD AG 255, Oct. (1980).

[2.19] Shrager, J.J., Head-up displays, a literature review and analysis with an annotated bibliography, FAA-RD-78-31, April (1978).

[2.20] Shurtleff, D.A., Legibility research, Proceedings of the SID, 15, no. 2, (1974), pp. 41-51.

[2.21] Gurman, B.S., The impact of a multi-function programmable display unit in affecting a reduction of pilot workload, AGARD Guidance and Control Panel 27th Symposium, The Hague (1978).

[2.22] Hunt, G.H., Introduction to "Modern display technologies and applications", AGARD Lecture Series 126, (1983).

[2.23] Reynolds, B., Operator interfaces for centralised control rooms, Proceedings Symposium "The operator-instrument interface", Institute of Measurement and Control, Middlesbrough (1978).

[2.24] Francis, R.F. et al, Developments in the use of a complex control system. Ibid.

[2.25] Price, D. and Burnette, K.T., Light emitting diodes. In: Modern display technologies and applications, ed. D. Bosman, AGARD Advisory report 169, Paris (1982).

[2.26] Hussey, D.W., Wide angle raster head-up display design and application to future single seat fighters. In: Impact of advanced avionics technology on ground attack weapon systems, AGARD Conference Proceedings 306, Agheos Andreas (1981).

DISPLAY ENGINEERING: D. Bosman (Editor)
© Elsevier Science Publishers B.V. (North-Holland), 1989

III.3: AUTOMOTIVE AND MARINE APPLICATIONS

P. BEATTY

Ferranti Computer Systems Ltd., United Kingdom

3.1. Introduction

3.1.1. Rationale

The need for application of electronic displays as vehicle instruments deserves some explanation. For many decades, electromechanical pointers, or sometimes ribbons or digits, have performed adequately and have blended well with overall styling. Light guide techniques evolved for uniform lighting at night. High contrast white on black paint, together with coloured graphics and legends visible at all times, gave the designer and stylist plenty of scope. Why then should this be changed? Reasons can be listed as follows.

- Lower cost: inflation escalated the cost of materials and labour in assembly of several hundred small components of a modern instrument cluster package. The smaller number of parts and process steps for some flat panel display products offered prospects of a less expensive alternative, based on extrapolation of trends from the late '70s. Lower inflation in the '80s and viability of only the more expensive technologies, actually prevented realisation of this aim.

- System compatibility: other parts of the car are undergoing similar changes as well. Electronic ignition and other functions are emerging, also newer transducers using electronics often in digital form. The advent of electronic instrumentation would allow easier interfacing with the electronic monitoring and control devices.

- Reliability: solid state electronic displays are less likely to develop faults, compared to conventional instruments using moving parts. Past experience of low voltage semi-conductor transistors and light emitting diodes helps to encourage this view, although some newer technologies did have serious life limitations.

- Marketing: extra features are thought to appeal to the end user and so help to sell the car. This includes for instance the computational facilities of trip computers, navigation and diagnostic aids, as well as multi-functional gauges with scale changes. More uniform appearance at night and accuracy of digital information are additional factors. Against this must be set the preference of traditionalists, the loss of angular indication provided by a pointer and the black hole effect of non working displays when the car is switched off.

In the event, marketing and system compatibility have led to the introduction of these systems and reliability is being evaluated; first indications suggest that vacuum fluorescent (VF) and liquid crystal (LC) display panels are satisfactory. The original main aim of lowered costs is some way off still, with complete primary instrument dashboards costing up to three or four times that of the conventional instrument it replaces, albeit usually with the extra features of a trip computer and sometimes with speech synthesis warnings. Thus, the market penetration so far has been limited to higher line models where sophisticated instruments and accessories are expected. Improved and larger scale manufacturing methods with considerable automation and high yield, as well as advances in electronic addressing, are lowering system costs. Size is not generally recognised as a constraint, although flat panels are favoured for least depth even though edge connections may limit the useable area. In the future, smaller cars might force the move to smaller multi-function displays.

3.1.2. Environmental requirements

In table 3.1, ambient specifications are indicated which vehicle display systems must withstand. The electronic components must comply with regimes almost as severe as for aerospace and military markets, yet with the minimum cost.

Table 3.1

Car environmental requirements

Temperature	$- 30$ to $+ 85^{o}$ C operational $- 40$ to $+ 85^{o}$ C storage
Humidity	90 to 95 % RH (38 $-$ 42° C)
Vibration	0.5 g, 100 $-$ 1000 Hz, 2 hours
Shock	50 g, ½ sine, 9 $-$ 13 ms
Illumination	50,000 lux sunlight

Combinations of high temperature and humidity are especially difficult for panel seals and any polarisers or filters. Presence of moisture from use after iceing up is a serious hazard; temperature cycling can put particular stress on connectors and laminated components. More will be said on illumination levels later but it is evident that strong sunlight can both raise the temperature and cause wash-out illegibility. Operating life, for private cars projected at

5,000 hours, can be met by many display technologies. The requirement of around 20,000 hours for some commercial vehicles (CVs) can be troublesome depending on levels of high brightness operation required. Storage life is less of a problem. Vibration is generally an easy parameter to meet, but care is needed for some constructions, such as displays with rigid pin connections.

3.1.3. Car instruments

Automotive displays are predicted to increase at 64 % a year in Europe; they are a major growth area for display technology with about 15 % of the 2.2 billion dollar USA automotive electronics market in the second half of the '80s and approximately some 20 million vehicle displays worldwide, which accounts for nearly half of all the flat panel displays. One of the major manufacturers of VFDs claims the bulk of this market with predictions of rising sales rather than a gradual saturation. They consider complete automation of production likely, with improved performance at lower cost, to hold off the upsurging by LCDs. For 1983, some 21 models of 5 Japanese car manufacturers were using VFDs for the speedometer and most other functions. The proportion of very large dashboards, as opposed to car clocks and radio tuning indicators, is relatively small in Europe; probably well under 15 % of the total vehicle instrument market. Future growth will depend on user acceptance of appearance and extra features in high line cars, and on the availability of standard low cost display systems for the mass produced smaller models.

In spite of the high growth rate predictions for automotive displays, very few European models were available in 1987 with full electronic instrumentation, even as an option. In the USA, 13 % of Delco instruments were electronic displays and this could rise to about 55 % by 1996. Most manufacturers were reporting on concept vehicle demonstrators of prototype panels to decide how to use both fixed format and dot-matrix or CRT flexible display presentations. Hopefully, the display hardware will be able to match the requirements of new transducers, navigation and computer diagnostic systems in terms of legibility and reasonable cost. Most likely, the final fight in high line models will be between ruggedised CRTs and dot-matrix LCDs (either multiplexed using new materials such as ferro-electric or some types of supertwisted liquid crystals, or driven directly by active semiconductor switches at each del). In the meantime, VFD technology continues to have the most widespread use for fixed format displays or very simple small dot-matrix panels. One of the major factors affecting market penetration is the man-machine interface. Given the generally higher cost of electronic displays compared to conventional instrumentation, drivers will only be interested in obtaining more value such as more information or control as in the trip computer and navigation aids.

This in turn means greater user interaction to select such information and some ergonomists argue that the driver is becoming overloaded. Work by Zwahlen and De Bald at Ohio University suggests that there is an adverse effect on steering precision whilst a driver uses a CRT touch panel radio control. Overreliance on visual rather than tactile or auditory feedback is probably a contributory reason. In the UK, there is also a law forbidding placing of any TV screen where it can be seen by the driver (orginally not intended for CRT instrumentation of course).

It should be remembered also that displays can be auditory as well as visual. The first examples of voice synthesis in the early '80s were not well received mainly because sensible human factors design principles were trailing the technological ability. Mere repetition of the legend on function keys when depressed added little of value and yet, used properly, audio feedback information can assist in reducing driver workload, such as in route guidance systems, and in cases where longer messages are required which otherwise cause too much time diverting gaze away from the road. Extra cost and the problem of multiple languages apart, the auditory channel should be considered in combination with visual displays. Conversely, the ability to control functions, or select display information, by speech recognition might alleviate the driver overload situation whilst avoiding difficulties with overlay touch switches which can hinder display visibility.

Another aspect of ergonomics in the car is whether to present information as a substitute for conventional, electromechanical instruments in their usual 'head down' position, or to produce a 'head up display' (HUD) similar to aircraft cockpits. Near infinity collimation of a CRT would be expensive and is not recommended for this reason; considering also the fewer and different demands of a driver compared to a pilot [3.1]. Superposition of the image on a cluttered outside world could be distracting and of little benefit; apart from the possibility of having a pair of vertical lines to indicate likely width of a vehicle in front for safe stopping distance. The driver is far less likely to be viewing at true infinity and so superposition against the plain background of the vehicle's bonnet has greater appeal. Work at the UK Transport and Road Research Laboratory has shown only a marginal benefit for a HUD system in terms of saving time for the change of eye accomodation and head position. Instruments near the top of the dashboard are almost as fast to read.

Another problem for car HUDs is the need for high luminance multiple colours, not usually even a feature in aircraft.

In spite of the disadvantages and minimal gains, some manufacturers have proposed such displays. Originally, Smiths had a prototype system for police cars, and GEC considered high intensity LEDs for reflection from the windscreen. In 1978, Siemens also suggested a similar approach with liquid crystal, electrochromic or electroluminescent displays. Other patents at that time invoked the transparency of a liquid crystal display for use as a HUD, omitting collimation altogether as Delco demonstrated on the Oldsmobile Aerotech in 1987 [3.2]. Also, in 1987, Nissan announced that it was the first company to develop a practical, working, compact, low cost HUD for automotive use by windscreen reflection and that it would be introduced on its high line models in the autumn of 1988.

Tables 3.2 and 3.3 show the introduction of solid state displays by three USA car manufacturers in the decade up to 1984 [3.3]. Tables 3.4 and 3.5 show the positions in Europe and in Japan.

3.1.4. Commercial vehicle instruments

Most of the solid state display products have been fitted as optional extras and on high line cars. Commercial vehicles such as trucks, buses and off highway bulldosers or tractors pose particular problems and opportunities for electronic displays; so far fewer dedicated products have appeared.

Environmental conditions are much tougher as regards direct exposure to moisture and mechanical knocks in in agricultural vehicles; and the requirement of extended operating life of at least 20,000 hours for trucks. Usually, the instruments lie more nearly horizontal than vertical which can exaggerate problems of specular reflection as usually there is no cowl. Safety aspects of monitoring many more functions demands high reliability. Generally the volume of business is smaller even compared with high line cars so that certain custom designed displays such as VFDs might be too expensive in terms of tooling cost and thus modular use of standard parts becomes more advantageous.

In some ways, commercial vehicles could benefit most from electronic display technology. Multiple warning lights over a large area can be replaced by an alpha-numeric

Fig. 3.1 Programmable dot-matrix VFD for ISO warning symbols (Lucas Electrical, Electronics and Systems Ltd).

Table 3.2

USA Automotive displays

Model year	General Motors	Ford	Chrysler
1974			LED gauge warning system
1975	low fuel warning (LED)		
1976			
1977	cluster and trip computer (gas discharge display)		clock (VFD)
1978	cluster and trip computer (VFD)	miles to empty (gas discharge display)	
1979		electronic radio display (VFD) clock (VFD)	electronic radio display (VFD) trip computer (VFD)
1980	cluster and trip computer (VFD) fuel monitor (LED)	cluster (VFD) cluster w/message center (VFD)	
1981	clock (VFD)		cluster (VFD)
1982		trip computer (VFD) navigation/information centre (CRT concept)	
1983	full dashboard (LCD)		trip computer (VFD)
1984	trip computer + radio and ventilation controls (CRT)		Pontiac 6000 STE (LCD Information Centre)
1985			H-body series cluster (VFD)
1987	Buick Riviera control system and trip computer (CRT) Colour CRT using liquid crystal shutter		

Table 3.3

Functions displayed

Current product offerings	Speedo	tach	odom	temp	fuel	trip F'CNS	driver info monitor	radio tuning
Ford								
Continential 84	VF			VF	VF	VF	VF	VF
Thunderbird 84	VF			VF	VF	VF	VF	VF
General Motors								
Corvette 84	LCD	LCD		LCD	LCD	LCD	LCD	LED
Cutlass ciera 83	VF			VF	VF			LED
Buick 83 Rivera and Regal	VF				VF	VF		LED
Pontiac STE 83							LCD	LED
Cadillac 83 Eldorado and Seville	VF					VF		LED
Chrysler								
Dodge 83						VF		VF
Imperial 83	VF		VF		VF	VF		VF
Lebaron 83						VF		VF
Dodge 600 ES 84	VF	LED	VF	VF	VF			VF
Daytona and Daytona turbo						VF		VF
Laser XE 84	VF	LED	VF	VF	VF	VF	VF	VF
executive Sedan 84	VF		VF	VF	VF			VF
Chrysler New Yorker 84	VF		VF	VF	VF	VF		VF

Table 3.4

European automotive displays

Date	Technology	Manufacturer	Description
late '70s	LED	BMW	trip computer
early '80s	DCEL	Smiths Industries	aftermarket tacho, clock and battery voltage
	LCD	VW	car clock
	VFD	Austin Rover	car clocks trip computers
	VFD	Smiths Industries	aftermarket trip computer
1983	VFD	Audi (Quattro)	full dashboard
	VFD	Austin Rover (Maestro)	full dashboard
	VFD	Ford (Sierra)	vehicle condition monitor
	LCD (G-H)	Renault (Renault 11)	full dashboard
	CRT	BL Technology Ltd	concept car full dashboard
	LED	Fiat	vehicle map warning lights
1984	LCD	Opel (Senator and Monza)	full dashboard
	LCD	Austin Rover (Montego)	full dashboard
	VFD (alpha-numeric)	Austin Rover (Montego)	message panel and trip computer
	VFD	Citroën (BX TRS)	tachometer and minor gauge
	CRT (colour)	Daimler	monitor for mobile office
1985	LCD	Vauxhall (Astra)	full instrumentation
1986	CRT	Leyland Trucks (TX 450)	instruments, loads, maintenance
1987	VFD	Kemitron (Aston Martin Lagonda)	speedometer, tachometer, warning symbols and dot-matrix message
	LCD	Vauxhall (Carlton GSi and Senator CD)	full dashboard and trip computer

Table 3.5

Japanese automotive displays

Date	Technology	Model	Description
1982	LCD	Cordia LS	digital speedo analogue tacho fuel and oil bargraphs
1983	LCD	Mirage	
	LCD	Tredia	
	LCD	Eterna	
	LCD	Lancer	
	VFD and LED hybrid	Starion	
	LCD	Cedric/Gloria	full dashboard
	VFD and LED hybrid	Capella turbo	
	VFD and LED hybrid	Cosmo turbo	LED speedo, tacho; VFD clock
	VFD	Leone	includes trip computer and digital tacho
	LCD TN + dye	Laurel turbo	full dashboard
	LCD	Cultus GS	
	LED	Piazza	
	LCD double layer guest host	New Corona	full dashboard (optional)
	VFD and LED	Soarer Sprinter Corolla Celica Crown Corona Cresta Camry	full dashboard VFD speedo, fuel, temp LED tacho
	VFD and LED	Silvia	LED tacho
	VFD and LED	Bluebird	
		City turbo Vigor Accord Ballade Prelude	LCD and with electromechanical
	CRT mono	Vigor	navigation aid
	CRT colour	Eterna	

Concept packs

(late '80s)	LCD dot-matrix	Nippon Seiki	flexible format
	CRT colour	Toyota FX-1	6 inch diagonal instruments
	CRT colour	Yazaki	6 inch diagonal + touch switches
	CRT colour	Nissan NRV II	route guidance
	CRT colour	Nissan NX 21	transparent overlay touch switches
1987	CRT colour	Toyota Crown	6 inch diagonal, navigation

message panel or coloured dot-matrix graphics panel as shown in figure 3.1. Multi-function and navigation requirements could be met also with dot-matrices or CRT. Tachographs recording journey details on hard copy are a particular feature of commercial vehicles where electronic print-out and display is possible. Viewing angles horizontally are about half of those in cars being about ± 15°, sufficient for the driver alone. In theory this should allow easier use of LCD and narrow angle contrast enhancement filters on active displays.

Some work has begun using standard small VFDs and LEDs, or sometimes a custom designed LCD. The improvement of technology, especially in colour dot-matrices and CRTs, may see more use for this market area in the future.

3.1.5. Marine instruments

Another terrestrial transport application is equipment for small boats and ships. Like commercial vehicles, demands for reliability are high and moisture protection is of course an important aspect. In 1984, Marinex made available their Cetrek 7000 autopilot using two dot matrix LC alpha-numeric displays and a membrane touch pad alongside. Compass and rudder information could be processed, shown on the LCDs and controlling direction. As for some car systems, satellite navigation is available from some manufacturers such as Kelvin Hughes. Mobile office concepts apply to yachts as well as limousine cars. Shipcom by West Electronics is an on-board telex system and includes a Sharp microcomputer terminal for word-processing and weather fax reception.

More sophisticated marine displays include naval applications such as a semiautomatic plotting system from in which an ancillary keypad has 8 rows of 20 alpha-numeric characters as a DCEL dot-matrix display. Cathode ray tubes are used at sea for the traditional ships radar, and colour ones are under evaluation also for sonar use but need careful colour selection to avoid degradation by background noise. In some cases, the liquid crystal sequential colour filter and 'monochrome' mixed green/red CRT combination is offered for

quite large seaborne monitors used below deck where high luminance is not
vital. One application for conventional colour CRT displays on board ship is in
superimposing radar information onto nautical charts in a system supplied by
EASAMS to Raytheon Marine Sales and Service Company for use by tankers in the
very narrow channels of the Arabian Gulf.

3.2. Visual requirements

3.2.1. The environment

In table 3.6, some requirements on the luminous conditions of car instruments
are indicated. A particular feature of vehicle displays is the emphasis on
adequate legibility, to comply with legislative ruling on speedometers and
warning lights, as well as achieving this consistent with excellent styling
appeal in terms of colour and design. The car manufacturer and the ultimate
driver want improved performance, such as accuracy, reliability and extra
features, without giving way on good looks. In some cases there is a fashion to
be set or followed and even traditional colours favoured for a particular model
of car. Thus, the task of the ergonomist in this area is by no means easy.
The ambient sunlight on a vehicle dashboard may enter at a fairly steep angle
through the side windows and sunshine roof; it may also shine through the rear
window when low in the sky. Illumination under these conditions is generally
taken as about 60.000 lux (windows open), 48.000 lux (windows closed) and
20,000 lux (rear window) [3.4]. There is in addition an even more difficult and
common case of driving into the sun, or of sunlight scattered and reflected
from a road surface with little falling directly onto the display. The eyes
then adapt to a luminance of maybe 8500 cd m^{-2} and may require at least 70–140
cd m^{-2} minimum luminance from the display [3.5]. The tendency for high line
model cars to have tinted windows may reduce the ambient illuminance: it is
usually these types which are fitted with displays. This would not help of
course where a sliding sunroof exists, or no roof at all in convertibles.
At night, twilight, and in overcast or foggy weather only a few lux exist so
that instrument glare will become objectionable unless dimmed automatically
with a photocell or it can be under manual control; the total dimming ratio
required thus being ideally 50 to 1 although presently 16 to 1 is common [3.6].

Table 3.6

Car instrument specification

Luminance		blue	208–415
(maximum luminance, cd m^{-2}		green	1385
under direct sunlight)		yellow	415
		orange	346
		red	277–460

Halve values for end of life or off-state matching background.
Approximately 70–140 cd m^{-2} for light adapted viewing.
Two orders of magnitude less for night driving.

Contrast	2.5:1 at 50,000 lux
(on-off luminance ratio)	36:1 at 0 lux

Uniformity	1.25:1 to 2:1

Life	2000 hours (max drive)
	3000 hours (min drive)
	5000 hours (to half brightness or contrast)
	82600 hours storage

3.2.2. Brightness and contrast

The main factors affecting both the legibility and the visual appeal are
brightness (perceived luminance) and contrast ('on' to 'off' luminance ratio).
Although closely related they are distinct, and one is achieved at the expense
of the other making compromise necessary; e.g. contrast enhancement filters and
high efficiency polarisers have low transmission making the display appear
dull. Conversely, active displays often have body colour such as the white of
many CRT or VFD phosphors. This raises overall luminance but reduces contrast
(see example I.1.7); so that such displays easily "wash-out". Luminance can be
an important factor for legibility, particularly for the older drivers [3.7].
It can also help to give apparently more vivid colours and provide a balance
with bright surroundings. Luminance requirements for given levels of
illumination depend on display surroundings, filter characteristics, colour of
the display and immediate mask colour around individual characters. The facia

around each display should be matte black to avoid specular reflections. In fact, the background of the display should both in colour and in luminance match that of the surround. This is supported by the Chrysler study [3.8] in which the required VFD luminance was halved using an immediately adjacent mask of equal reflectivity to the phosphor. The lack of disturbing contrast is of great benefit to the eye. Error rate studies [3.9,3.10] have shown the minimum required monochrome contrast ratio to be 1.5 (compare section I.1.4.3) for acceptable legibility and 2.5 for comfortable readings, especially necessary when glancing away from the road.

Besides luminance contrast, colour contrast can be an overriding factor in legibility under high ambient illumination. Discriminability between two adjacent areas on the display face is not only determined by different luminousity but also by colour differences which exceed the JND. A useful concept to combine the two types of contrast was introduced by Galves and Brun [3.11] under the assumptions that the luminance and colour contrasts are not mutually dependent and the contrast distance can be measured on a linear scale: the Discrimination Index (DI), see section I.2.5. This index is obtained by taking the square root of the sum of the squared contrasts. For example: with broad band colour emissions such as that of green VFDs often used in car instruments, sunlight wash-out is best avoided by using a neutral density filter rather than a green filter because the sunlight does not produce a green background in the off-state as it would with the green filter! Here colour contrast is the more significant contribution to the Discrimination Index. Obviously the neutral density filter should be no more dense than absolutely necessary, otherwise the display will look too dim. Furthermore, there may be several colours present in an automobile dashboard so that neutral density then is obligatory.

The choice of filter should also take into account the problem of specular reflections from convex curved glass of some small VFD or CRT displays. Circular polarisers can overcome this but should be sloped or concave to direct its own front surface reflections away from the driver. Diffuse matte finish on the front of filters (example I.1.6) also can prevent specular reflection but increases diffuse reflection and hence makes luminance contrast worse. For VFDs it will also blur the image when the back luminous variety is used. With LCDs, the blurring effect is less due to the thinner glass and so a diffuse front finish is often used.

3.2.3. Colour

Vehicle instruments require only a few colours but suitably chosen across the spectrum to maximise DI from blue to red. They are needed to code warnings such

as red for exceeding limits, amber for caution and green for indicator signals
or general presentation. Blue is a legislative requirement for the main beam
headlamp. Yellow and white have been also for general information
including digital speedometers, odometers and minor gauges. Thus, in general,
there is conformity with recommendations on avoiding colour confusion by using
no more than six separate colours [3.12]. Usually, the black background used
and the fixed positions of instrument displays assist further in avoiding any
confusion. The advent of the colour CRT or dot-matrix panels might aggravate
this aspect but assuming no grey scale capability is used, the possibility of
mistaken colours is minimal.

Some car manufacturers have their own colour preferences for lighting at night
conventional instruments such as clocks, speedometers, bargraphs and message
panels. Often these do not match the criteria needed for easier legibility
under direct sunlight. There have been several studies [3.4,3.13,3.14,3.15,
3.16] of the display luminance necessary for each colour to be as readily
visible as another (highest wash-out illumination, most rapid interpretation
time, highest accuracy of reading). Generally, extremes of the spectrum such as
red or blue are better seen than green, even though green lies at the peak of
the eye's photopic sensitivity. Under 50,000 lux of sunlight and with optimised
contrast enhancement filters for a luminance contrast of 2.5 (before filter),
colour luminance values would be in the order green, yellow, orange and red
(1582, 474, 394, 419 cd m^{-2} respectively). In practice [3.8] values specified
in cars are often at least twice as high which allows for gradual dimming
during operation over time, as ageing (to half brightness called half life) or
at higher temperatures when VFD brightness can be half that at room
temperature. There is also up to 2:1 manufacturing tolerance in matching
brightness between different panels.

Desaturation of VFD and CRT displays with neutral density filters, due to wash-
out, can be reduced using fairly saturated phosphor emission colours;
especially for warning indications. For LCDs, even as pseudo-active displays
with backlit and coloured negative images, sunlight does not cause so much
wash-out; it can even make pastel colours more saturated. In this case,
slightly less vivid colour filters may be chosen for more relaxed viewing at
night.

Other factors of course also influence the colour choice. Extremes of the
spectrum and very saturated colours can be visually tiring and can produce
accommodation difficulties, although quite useful as warning indicators. In
this sense, the traditional green or perhaps yellow colours are recommended and
are most widespread in vehicle dashboards. The wide range of visual ability
among drivers includes those with varying colour blindness for which yellow can
be quite suitable.

Differences of colour are noticeable to different extents depending on the spectral region involved (section I.2.5). For green, several shades of hue are less well differentiated and so variations between panels of this colour should be more acceptable. In the past, matching VFD luminance values has been more of a problem but for LEDs varying hues have arisen as well.

A specific marine application of displays is in the detection of changes in a sonar pattern. Here colour contrast also can enhance the detection in terms of shorter reaction time (RT). Research has shown [3.17] that there is very little gain in RT for various illumination conditions: low level white, blue and red of the same illuminance and even for no illumination at all, when averaged over all possible foreground and background colours (RT about 440 ms). A similar averaging procedure for foreground colours and illumination conditions showed that for backgrounds of equal luminance, blue was superior (RT of 420 ms against 440 ms for red, yellow and green); however when the backgrounds were adjusted for equal brightness this advantage was lost over green, tho differences still being small. For the foreground colours it was found that opponent colour pairs (foreground-background) like green on red, blue and purple on yellow, and red/purple on green generally yielded the shortest reaction times as predicted by the discrimination index DI; like luminance, contrast improves detection.

3.2.4. Viewing angle

The design of a new car requires knowledge of the driver's physical measurements for correct relative positions of the seat, controls and the instruments. An average is taken for the population; for instrumentation the median oscular position is defined by a vertical ellipse (figure I.2.10) including the extreme percentiles likely to be encountered. Typically, the mean is $\pm 3^{\circ}$ with respect to the horizontal. This in turn defines the viewing angle required of the instrument display, although the passenger must be kept in mind especially for the clock or radio tuner. The other relevant factor is the position of windows and sun roof so as to avoid specular reflections and wash-out of the display. Allowance must be made for both left and right hand drive models of the same vehicle. As a general rule, angles are $\pm 30^{\circ}$ to $\pm 45^{\circ}$ sideways from the normal in the horizontal plane, and $+30^{\circ}$ to -5° in the vertical plane, with respect to the horizontal. Obstructions to viewing can include the steering wheel, the instrument cowl used to shield sunlight and prevent reflections in the windscreen, as well as parallax arising from display masks, graphic overlay legends or scales and any separate colour filters. The only real limitation imposed by the display itself is with twisted nematic LCDs, especially if multiplexed more than two ways (see section II.3.3.3 figures 3.6 and 3.7).

The advantage of dyed LCDs for wider viewing angle has been mentioned (section II.3.3.6 and further) but where double layers of liquid crystal have been used, glass substrates must be sufficiently thin to avoid parallax otherwise negative images appear to have shadows when viewed in reflection. Likewise, even single layers can suffer in this way if the rear glass substrate is too thick, or if the reflector is spaced too far from the back. Dyed phase change LCDs can have internal reflectors which would assist in this respect.

Front plastic overlays carrying scales and legends also can introduce unwanted parallax, although it is usual to print on the rear surface of these plates to avoid the problem. Sometimes a gap remains however depending on the frame used to support several small LCDs or to house the elastomer connectors rigidly.

Internal graphics can avert this by delineating the alignment layer causing 90° rotation in twisted nematic liquid crystals or physically masking over the alignment layer. An added advantage is the better luminance matching of graphics and the moving elements of the display.

3.2.5. Flicker

Unlike conventional electromechanical instruments, electronic displays often rely on the persistence of vision to form the total image. In the case of directly driven LEDs, VFDs and LCDs this is not a problem but when multiplexed then the critical frequency for fusion must be exceeded (section I.2.4). Generally speaking this is the case, although vibrations and rapid turns of the head can sometimes cause a strobing effect with LEDs where the decay time in pulsed operation is extremely short.

For CRTs the frame rate and phosphor persistence will be the determining factor and should be as high as acceptable to display dynamics, using a non-interlaced raster. The frequency should be at least 70 Hz at 100 cd m^{-2}. This is especially necessary in a vehicle active display, since the critical flicker frequency rises with display brightness which is high to combat sunlight wash-out. There is also the problem of flicker being more noticeable in peripheral vision which again is the predominant mode for a car driver. It is also necessary to offset apparent movements of jitter and wavering (swim) when moving the head relative to motion on the screen. Fortunately older drivers are less susceptible to flicker effects.

3.2.6. Font and graphic design

For fast and accurate legibility of and alpha-numeric characters, certain height, width and luminous area dimensions are necessary and to a lesser extent choice of font (section I.2.2 and III.1.5). For the typical driver to instrument distance of 813 mm the following parameters apply derived from the vast literature on aerospace and other displays [3.18].

Minimum character height, H, for subtense angle of 27': 6.3 mm

Character width	70	% H
Character spacing	75	% H
Stroke width	12.5	% H

Fonts such as NAMEL, Lincoln/Mitre, Leroy and Mackworth are possible [3.19, 3.20]. Slanting to the right offers little gain in reading accuracy but is used as a styling feature. Digital speedometers are usually nearly four times the minimum character height for greater appeal and ease of viewing by older drivers.

Dot-matrices should allow adequate representation in all languages in the case of message panels. Thus each character may require 7x9 dots for which the Lincoln/Mitre is recommended rather than the more limited 5x7; and supporting lower case, including descenders and ascenders, as well as upper case.

Segmented alpha-numerics are more common as they are easier to drive with lower multiplex rates (section I.1.5) down to 1:4 for LCDs instead of the 1:7 or 1:9 for the dot-matrix. Either 14 or 16 segment symbols can be used; although

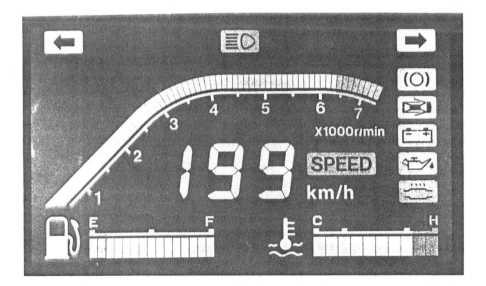

Fig. 3.2 Partly linear, partly curved bar graph for better quick glance estimation of analogue indication. Dyed TN LCD (EPSON).

legibility is said to be little different to dot-matrices. Reinig [3.21] does suggest that character height should be 30 % greater.

Graphic design is far less well defined, with greater artistic styling
contribution. The traditional circular electromechanical instrument has been
replaced by analogue bargraphs, usually partly linear and partly curved (see
figure 3.2). Some designers have made up for lack of 'texture' in the true
three-dimensional moving coil pointers, by producing perspective bargraphs and
digits with apparent depth as in a painting, originating mainly it seems in

Japan (figure 3.3). Stylists in
Europe however, have used printed
graphics, scales and legends to
give visual appeal even when the
display is not working to avoid the
'black hole' effect.

With the advent of CRTs and flat
dot-matrix panels, warning lamps
can be replaced by pictorial

Fig. 3.3 NEC VFD digits with apparent
depth [3.57].

representation of the ISO symbols
providing that it is shown in an
associated warning colour, or that
simultaneously the warning colour is generated in a general warning panel.

Electronic displays permit also animated symbols such as for the fan speed.
Although technologically possible, there may be ergonomic disadvantages as
motion in peripheral vision can be most distracting; indeed, as a method of
warning it might be used to better effect. Likewise, as an aid to navigation it
could be used well, both for showing the past, present and future route, and as
a compass in which the scale moves so that the pointer stays at North (the top
indicating direction of travel).

Too many individual instruments close together can give a cluttered, busy, look
to panels and may confuse the driver. Sensible presentation with limited
information at any one time on a need-to-know basis, perhaps as prioritised
warnings and multi-function gauges, would be more helpful. Moderate use of
supplementary audible signals or speech can reduce the load on the visual
channel.

An early consideration in design of dashboards was the debate on whether to use
digital, analogue or both styles for the speedometer and other functions.
Experiments have shown that digital representation is appreciated faster and
with higher accuracy, leading to a subjective preference [3.22]. Contrary to
expectations, older drivers are the more enthusiastic. Thus several dashboards
use only digital speedometers dispensing altogether with a backup analogue
bargraph and so reducing clutter and cost at the same time.

Conclusions

It is clear that many factors are present in automotive instrument ergonomics, [3.23,3.24,3.25,3.26,3.27]. The whole system must be considered with very careful compromise between ideal features for the driver and the cost involved. The instrument must blend in and be compatible with other aspects of the car. Mixed display technologies should not interfere with performance. Active light emitting displays are sometimes placed alongside passive LCDs or electro-mechanical gauges. Poor regard for this part of the design could lead to premature rejection of solid state display technology in its infancy, especially where there are high introductory costs.

3.3. Display technologies

The main contenders will be considered for their evaluation as vehicle instrument displays.

3.3.1. Incandescence

The oldest light emitting display technology used in vehicle instruments is the incandescent bulb. It is mentioned here merely for completeness and serves as a basis of comparison. In addition to power, thermal dissipation and life problems, even warning lights using incandescent bulbs can wash-out under direct sunlight when inadequately packaged and be too dim to attract attention, especially in peripheral vision. Older individual lights with lens mouldings could occassionally capture sunlight and glisten as though switched on: this is the converse of the wash-out of a genuine switched-on lamp. Also legislation demands that such lights shall not be dimmed; this can lead to excessive glare at night, particularly disturbing if the warning is of only moderate urgency. Thus, generally, these lights are themselves inadequate. Modern packaging methods have helped using semi-opaque secret-till-lit matte colour filter legends.

Application of incandescence continues both for warning lights and for backlight illumination of liquid crystal displays. Generally, the greater number of warning lamps favours the less expensive bulb. Tungsten bulbs used may last several hundred hours. Higher efficiency quartz halogen bulbs are finding favour for back lighting where greater brightness is needed without trading life or power.
Digital 7-bar or bargraph incandescent displays have been developed which have enormous luminances up to 50,000 cd m^{-2}, can accomodate colour filters and may

be multiplexed. Generally sufficiently robust for aircraft instruments they are too expensive and lack in styling flexibility for use in cars.

3.3.2. Gas discharge (chapter II.4)

The DC type of gas discharge was promoted as an early form of automotive display [3.28,3.29], in more recent times [3.30]. The original versions were the flat panel equivalent of neon bulbs and the later Nixie digital displays which represented the first of modern day display technology. Although used in some Chrysler cars and the Aston Martin Lagonda, this device had a number of disadvantages. These included high voltage drive, around 150 V, hence high cost of driver circuits unless using the Burroughs Self Scan principle. Other problems were colour, limited mainly to orange, while the low luminance is hardly sufficient for sunlight viewability. At low temperatures, mercury added to offset electrode sputtering could become less effective and even condense out to cause short circuits.

In one technique [3.30], some of these difficulties have been overcome, illustrated by a 120x270 mm panel at 130 V using only 5 or 7 drivers with self scan phase addressing whereby one segment primes the one next to it for subsequent activation. Special timing allows addressing to linger on the last addressed bar of a bargraph to enhance brightness. Up to 1400 cd m^{-2} luminance is claimed at a power consumptiom of 12 W; various colours are achieved using CRT phosphor activated by the UV from the negative glow. The main limitation seems to be the half-life of only 1000 hours.

3.3.3. Light emitting diodes (chapter II.6)

In some ways, LEDs appear to offer advantages for vehicle display use. They operate from only low voltage DC to give most colours required at low cost and with long life. On the other hand, until recently, efficiency was low giving low luminance and high currents. The advent of improved fabrication using transparent GaP substrates and the liquid phase epitaxial process overcomes some of these limitations, at least for small diodes used as indicators in trip computers. They may even replace incandescent warning lamps [3.31]. The very brightest can be relatively expensive however and have narrower viewing angles. The blue colour from SiC or GaN materials, required for main beam headlamps warnings, has a very much lower luminance than emission at the opposite end of the spectrum. LEDs are a small area display technology not readily given to styling flexibility possible with larger area technologies such as for LCDs. On the other hand, they are unique in being easily butt joined as pin connections protrude from the rear. Thus linear bargraphs are possible of varying lengths and closely stacked 7- or 16 segment alpha-numerics. In practice, they are

seldom used in this way since the luminance of segmented LEDs is much lower
than for narrow angle single diodes. Sunlight wash-out is then a more
significant problem than for vacuum fluorescent panels unless high current
ceramic mounting is used as in military or commercial vehicles where the
greater cost can be withstood.

Uniformity of both colour and luminance are a problem area (chapter II.6),
especially where a number of separate devices are used for high information
densities. Matching and selection to within a factor of 1.35 can add to the
overall cost [3.6]. Dimming is possible over a wide range without sacrificing
uniformity using pulse width modulation as LEDs are most efficient in the
strobed drive mode [3.31].

Bicolour variants can be obtained by reversing current in two anti-parallel
wired dice mounted within the same reflecting cavity. The red/green transition
through amber is particularly useful. The LED is an established electroluminescent technology of low voltage forward bias operation, with temperature,
humidity and vibration ranges conforming to automotive specifications; life can
exceed 500,000 hours MTBF.

3.3.4. Electroluminescence (chapter II.5)

Investigations into DCEL began in the mid '70s to determine the suitability of
applying 120 V pulses at 1¼ % duty cycle to ZnS:Mn phosphor thick films
[3.32,3.33].

Each particle of phosphor is coated with copper ions for high DC conductivity,
diffused away from the anode during an initial forming stage to leave a high
field region which is excited by hot electrons, impact exciting the activator
luminescent centres such as manganese for wide band yellow emission. The appeal
lay in its large area fabrication using paint deposition techniques on
patterned transparent anodes and an aluminium evaporated back electrode. Low
production cost seemed compatible with the automotive industry. Multiple
colours are possible using other activators and hosts; drive electronics were
minimised using the easy multiplexing ability of custom bipolar drive ICs in a
complementary 60 V mode.

By the early '80s, several disadvantages remained in spite of considerable
improvements and better appreciation of the mechanisms involved. Luminance of
around 350 cd m^{-2} was marginal given the reflectivity of the almost white
phosphor in sunlight. Furthermore, the displays aged to half-life within about
2000 hours from a variety of causes, but mainly because of the continued
diffusion of surface copper from anodes as further forming; causing migration
into the phosphor particles throughout the layer of phosphor at high
temperature, thus raising the series resistance. Softer I-V characteristics

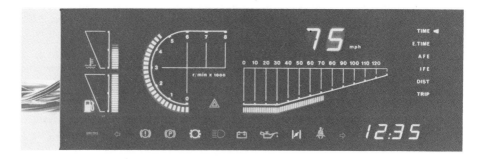

Fig. 3.4 DC EL prototype instrument panel (Smiths Industries PLC).

could arise as well and prevent satisfactory constant current operation.
Reproductivity of the phosphor, binder and encapsulation materials also added
to the problems experienced. Inevitably, as for ACEL powder electroluminescence
in the '50s, use has been restricted to what is most suitable: in this case
indoor dot-matrix alpha-numeric and graphic panels for indoor use where
temperature and ambient light levels are not too high.

The more expensive AC driven thin-film displays were investigated in 1987 by
Delco on their concept car, the Oldsmobile Aerotech. Transparency, and a black
background obtained using circular polariser filters, gave high contrast for
the yellow diagnostic panel of 512x256 dels. The luminance of such displays
suffers considerably however, since there is only a low duty cycle to cover so
many dels.

3.3.5. **Vacuum fluorescent displays (chapter II.2)**

As a cathodo luminescent system, modern flat panel VFDs have much in common
with CRTs and valve technology. Three major Japanese companies are the almost
exclusive suppliers. They concentrated their efforts in the automotive sector
once LCDs had taken over their market in portable calculators. The filaments
and grid electrodes are constructed of fine wires and originally mounted in
front of the phosphor segments under a shallow glass envelope (figure II.2.2a).
The filaments operate at such a low temperature, that they do not emit visible
light; although very fine they are carefully tensioned, leading to a
surprisingly robust device. Indeed, filament failure is most rare, although
local darkening due to evaporated particles is sometimes possible.

At first single, then multiple 7-bar characters digital displays (figure 3.5)
were put in a glass tube but a flatter structure became available by the mid

BRAKE PADS WORN

Fig. 3.5 Segmented VFD message panel (Lucas Electrical, Electronics and Systems Ltd).

'70s where graphite anode patterns and conductive metal tracks could be screen printed on the back substrate. Usually, zinc oxide is used as the brightest phosphor needing only 12 V direct drive.

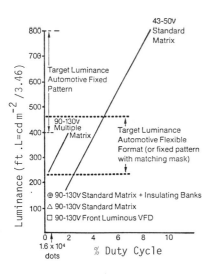

Fig. 3.6 VFD luminances vs duty cycle.

Other conductive phosphors followed to cover the entire spectrum, although at higher operating voltages up to 50 V. Multiplexing down to 10 % duty cycle is possible but also requires a proportionally higher voltage to maintain adequate luminance of about 2800 cd m^{-2} [3.34], compare figure 3.6.

In the early 80's, "front luminous" VFDs were produced [3.35,3.36] offering 40 % less luminance but enhanced life with phosphor placed on the front glass adjacent to rear mounted filaments and grid electrodes (figure II.2.2b). Better thermal characteristics are obtained, also reduced loss of luminance at high temperatures, which can be 50 % for a conventional VFD. Another advantage is the sharp image possible when matte finish filters are used, overcoming specular reflection without blur. Wider viewing angles are also possible.

Other improvements relevant to automotive applications include thin-film wiring patterns suitable for alpha-numeric and dot-matrix graphic panels at higher yield and lower cost. Metal substrates were suggested for heat sinking. Insulation colour to match the phosphor body colour avoids off-state elements (causing eye distraction) and helps legibility decreasing luminance required by 50 % [3.8]. Coplanar grids and anodes improve efficiency by around 50 %. Dot-matrix VFDs became brighter using multiple matrices [3.37] and insulating banks [3.38].

Differently coloured adjacent dots rather like colour CRTs have been demonstrated capable of showing many various warning graphics in different colours all in the same position.

Thin-film transistor drive at each pixel can restore DC addressed luminance levels even for high resolution matrices. Chip on glass circuits reduce the instrument connections down to only a handfull of wires. Good spacing and envelope manufacturing have allowed full dashboard size panels to be made at reasonable cost.

3.3.6. Liquid crystal displays (chapter II.3)

In many ways LCDs are a natural for car instruments as they modulate rather than emit light and so are not easily washed-out under direct sunlight. However adequate backlighting is still required for driving at night, twilight and in overcast or foggy weather, as well as when driving into the sun with light adapted viewing [3.18].

Early forms were unsuitable however for the car environment: dynamic scattering DC operated versions were of poor contrast and life.

Until the mid '70s AZO/Schiff based liquid crystal molecules were most subject to ageing effects of UV, oxygen and moisture, especially as seal materials were relatively poor unless glass frit was used. Positive image, purely reflective types of only moderate contrast also lacked the colour range required and did not add to styling appeal. Moreover, the temperature range of early materials was quite insufficient for vehicle use; with low clearing point temperature and excessive response time at low operating temperature. Multiplexing was considerably more difficult than with most of the active displays (contrast and viewing angle problems, see addressing, section II.3.4).

In the later '70s, dramatic advances were made in the materials used. Biphenyl liquid crystals overcome the vulnerability to environmental degradation and inexpensive plastic seals became the norm. Temperature range was widened to satisfy the automotive requirement, although a greater margin still is preferred. Multiplexing up to 1:7 became feasible, with steeper contrast – voltage characteristics using temperature compensated drive voltage. Certainly, 1/16 was a distinct posibility by the mid '80s when 400 line displays became available for indoor use. Polarisers were developed incorporating UV barriers with a choice of iodine for high contrast or dye for temperature and humidity resistance.

The later technology centered around negative image twisted nematic LCDs, usually backlit with incandescent bulbs and acrylic light guides. Thus, the sparkle and colour range of active displays could be achieved, usually with a

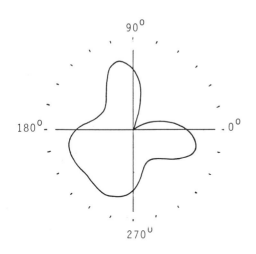

Fig. 3.7 Automotive backlit negative image TN LCD: contrast variation with azimuth viewing angle within 45° cone of the normal (right and drive and display located in front or to the right of driver).

transflector to partially transmit and partially reflect light for visibility at all times. In the off-state very little sunlight could penetrate through to the reflector, being absorbed twice by the poor transmission of polarisers and liquid crystal (< 1 %). By the same token, the on-state luminance was poor, usually with about 90 % being absorbed mainly by both polarisers, the transflector and the colour filters. At night there were also problems as the bleed-through in the off-state was then quite noticeable against the black mask delineating each character or legend.

Another difficulty concerns the restricted viewing angle of twisted nematic LCDs. There is a preferential viewing quadrant defined by the directions of both substrate alignment and orientation of polarisers (chapter II.3.3.3). Usually the top left and right is chosen if the panel tilts forward to avoid specular reflection, or the bottom left or right for backward tilted panels. The latter is more common in Japan and does have the benefit of better matte blackness from the underside of the cowl (figure 3.7). Sometimes the enclosing cowl is not black so as to blend in with the interior trim colour scheme but this does cause reflection problems for displays and from the windscreen. Matte finish coatings on the front polariser are used to avert some specular reflections and yet image blur is minimal as the display glass is very thin.

A top or bottom dead centre might be thought a more appropriate position for both left and right hand drive vehicles but this is incompatible with viewing using polaroid sunglasses and is also more wasteful in cutting sheets of polariser.

Windows made of moulded acrylic are common in instrument cowls as used for moving pointers to exclude dust, but are not so appropriate for LCDs. Although curved to avoid specular reflection, they can cause non-uniform colouration and

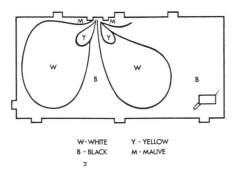

W-WHITE   Y - YELLOW
B - BLACK   M - MAUVE
3

Fig. 3.8 Concave-up plastic window.
Schematic of view through crossed
polarisers.

patchy brightness due to stress
birefringence when viewed with
polaroid sunglasses (figure 3.8).
They also further enclose an
instrument adding to the thermal
insulation where power dissipation
from backlights and driving
electronics plus sunlight radiation
can take the LCD above its clearing
point to the isotropic state.
Therefore, abrasion resistant anti-
glare matte coatings are often
applied to the display polariser or
a covering flat glass plate.

Contrast and luminance limitations were common in early examples. Contrast
ratios of 8 were obtained even for directly driven LCDs, especially if the less
efficient but more robust dyed polarisers were used. Optimisation of cell gap
(figure 3.9) and use of iodine polarisers (figure 3.10) raised the contrast
ratio to a more acceptable 36:1, largely by reducing the off-state bleed-
through. Non continuous exposure
to moisture allows water to dry
out before reaching the innermost
sensitive region of the laminated
polariser. Sometimes filter colours
can be chosen to block the natural
bleed-through colour to give
figures as high as 100:1 but this
restricts styling and functional
features possible with multiple
colours. At high temperature,
contrast falls as birefringence
rises increasing the bleed-through
with a range of colours (figure
3.11). Thus, a high contrast at
room temperature is adviseable to
allow for this effect, and also
for the fall associated with
a change from transmissive to

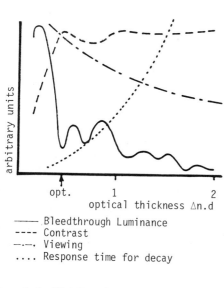

arbitrary units

opt.   1   2
optical thickness Δn.d

——— Bleedthrough Luminance
- - - - Contrast
—·—· Viewing
· · · · Response time for decay

Fig. 3.9 TN LCD cell gap optimisation.

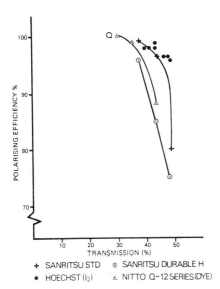

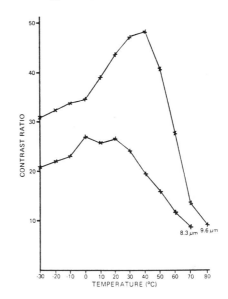

Fig. 3.10 Polariser properties.

Fig. 3.11 Contrast ratio vs temperature.

reflective mode where contrast may decrease to a minimum acceptable of 2.5 (figure 3.12).

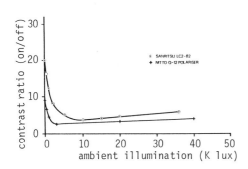

Fig. 3.12 TN LCD contrast vs ambient.

Luminance is a trade-off with contrast ratio, depending largely on polariser transmission as well as reflectors, backlight and colour filter. Well in excess of 70 cd m^{-2} should be available through the LCD when driving into the sun as the display must then work in the transmissive mode and viewing is with strongly light adapted eyes. High luminance can also enhance the perceived colour features, although excessive vividness and saturation can be irritating or tiring. Up to 340 cd m^{-2} may be required depending on contrast, requiring special fluorescent lighting techniques.

Liquid crystal displays operate at very low levels of current, $\simeq 2$ μA cm^{-2}, and at low voltage, 4-8 V rms. Apart from the low power dissipated, indium tin oxide used for the transparent front electrodes may be connected by very fine tracks to the output pads at the edge of the glass. Thus, considering the greater difficulty of multiplexing, many individual leads can be accommodated allowing maximum utilisation of the substrate for display contents. Very close to the operating segment they may thin down to 10 μm width invisible to the naked eye at the driver's distance. Thus, these display tracks can overlap the common backplane and tolerance in manufacture is easier. Elsewhere, a front mask can be used to delineate a character or digit, although such a front mask is often too noticeable against the incomplete off-state and, if possible, designs avoiding this would be preferred.

Connections to directly driven LCDs may be numerous and although there may be room for the tracks their terminations at contact pads often demands an especially small pitch. Elastomeric connector strips are then needed and associated rigid, heat resistant mounting frames. Sometimes these generate a parallax problem if front graphic legends are printed onto a transparent flat sheet overlapping the frame. They also make be backlighting difficult, requiring PCBs with holes for insertion of bulbs or a translucent PCB locally devoid of components, or extra deep elastomers to allow space for the bulb and light guide. Other types of connector include particularly long pins or directly attached metal-on-foil connectors by soldering, hot melt or conductive adhesive. Electronic addressing chips bonded to the glass substrate decrease the required number of external connections significantly, as shown in figure 3.13.

Multiplexing can reduce the number of connections necessary but degrades contrast and viewing angle (figure 3.14). The temperature range of LCDs is restricted since threshold voltage decreases as temperature rises, doing so non-linearly within 20 C of the clearing point. Present automotive practice restricts panels to only 2-way multiplexing without temperature compensation but 9-way for 9x7 dot-matrix alpha-numeric message panels is conceivable, using temperature compensation circuits. These need to be non-linear within more than 20 C of the clearing temperature.

Once this level of multiplexing is achieved more complex message panels will become available as they have for VFDs. Ultimately, much larger dot-matrix panels of at least 128x128 dots are conceivable in vehicles, for graphics use as warning lights and vehicle condition monitors (e.g. car outline with open doors and lights indication) or even as navigation aids (compass or route indication) [3.34,3.39,3.40]. This type of LCD will probably require quasi static addressing of individual dels for adequate viewing angle, contrast and temperature range. Thin-film transistors, or other means of active matrix addressing such as metal-insulator-metal devices (MIMs) or Schottky a-Si diodes, could achieve this but at a cost dependent on yield over large areas [3.41].

Fig. 3.13 Reduction of external connections by chips on glass technique (AEG).

Colour for these more complex graphic LCDs is also likely as already demonstrated from '83 to '87 in small colour LCD TVs by the Japanese [3.42]. Very fine stripes of red, green and blue filters can be produced by electro-deposition, but use in vehicles requires more saturated hues. In this way, universal panels are possible in various sizes for many models of car where tooling is only in the software control. The panel is thus very flexible and can present sequential information in a single space making the instrument compact laterally as well as in depth.

Response time of LCDs for vehicle instruments was unsatisfactory originally but new LC materials and small cell gaps with close tolerance can provide response times under 500 ms at – 30 C without the need for a heater. Again, trade-off is against contrast since too narrow a gap (compare figure 3.6) can lower contrast and introduce colour fringes for optical densities $\Delta n.d < 0.5$. (Mauguin limit; section II.3.3.3).

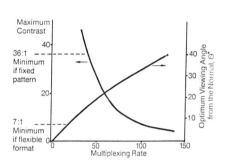

Fig. 3.14 TN LCD contrast and viewing angle dependence on multiplexing.

Finally, one can speculate on possible solutions for some of the problems raised so far. For instance, the lack of sufficient luminance because of backlight power, overheating and life; as well as polariser, filter and transflector absorption, could be overcome using improved guest–host LCDs. The preferred types include single or double LC layers of the dyed phase change cholesteric White-Taylor variety, or double layer dyed nematic LCDs. Additional advantages would be omission of the expensive and environmentally vulnerable polarisers and much wider viewing angles. Disadvantages include slower response time, lower contrast and greater difficulty of multiplexing. Extended response time could be alleviated with a heater at moderate on-cost if the perceived benefit were considered worthwhile. Combination with active matrix addressing could satisfy the addressing problem although some limited multiplexing has been reported also [3.43,3.44]. For a dot-matrix panel where there is no front mask differentiating from the background bleed-through, the relative lower contrast values of between 5:1 and 10:1 may be just acceptable in the case of uniform colour. Thin powder ACEL could provide total, rather than partial, reflection. This would contribute further to enhanced luminance, with the ACEL panel acting as a full reflector for daylight viewing and yet double as a backlight at night.

Examples of innovative LCD technology in the early '80s include the Nissan Cedric from Japan and the Renault 11 from France both providing a full instrument package.

Conventional TN LCDs were used on the Cedric, in combination with high intensity but low power fluorescent lighting. This allowed use of lower transmittance polarisers, passing 27 % of the light, which have higher polarising efficiency to provide a higher contrast of up to 30:1. Furthermore, the fluorescent tubes were encased in resin as insulation to retain heat for

working at low temperatures. They were driven with special circuitry for
luminance control over a wide range, with minimal radio frequency interference.
Compared to incandescent bulb lighting there is greater brightness, better
uniformity and near constant colour tone under varying ambient light, whereas
colour temperature shifts towards red on dimming incandescent bulbs. There is
also less contribution to ambient heating of the LCD beyond the upper
temperature limit. A 10 watt heater on the LCD glass substrate using a
transparent film of indium tin oxide ensured a response speed of only 100 ms
even down to − 30 C within 23 seconds.

The Renault 11 LCD used double layer dyed nematic guest host technology with
three pieces of glass as one single large complete dashboard. The absence of
polarisers allowed greater ruggedness and a brighter display in reflected light
although contrast was only 5.6:1 and incandescent back lighting for night use
was only 7 cd m^{-2}.

In both cases, the extra complexity of construction entailed higher cost not
easy to bear in the automotive market.

3.3.7. Electrochromic and electrophoretic displays (chapter II.7)

Both ECDs and EPDs provide a change of colour by application of a DC voltage in
which a changed particle is transported between electrodes. They offer superior
luminance and viewing angle compared to LCDs, as well as memory in the off-
state and have no vulnerable polarisers, but are worse for life, multiplexing
and general drive difficulty since there is no threshold voltage and reverse
polarity is required to complete erasure.

The particle in the ECD is an ion such as H^{+} in liquid electrolyte, often an
acid, or in a solid fast-ion electrolyte such as Na^{+} in sodium β-alumina. A
film of material such as tungsten trioxide then changes colour.

One method examined in the late method '70s [3.45] intended for vehicle
instruments used AgI and NaI salts in organic solvents with a silver back
electrode. Driven at around 0.8 V, the reduction of silver ions at the
transparent electrode constitutes electrolytic metallic deposition as a
specific example of redox reactions in electrochromic panels where usually just
a colour change occurs.

Response time is around 220 ms and satisfies the automotive requirement of
− 40 C to + 80 C. For 0.5 mA mm^{-2}, 10 nm of deposited silver could give a
contrast ratio of 10:1. Memory could last 10 minutes and measured life was 10^{7}
cycles. Care is needed to avoid reduction of the transparent oxide electrode,
using specially doped material and restricted write current. For uniformity,
high conductivity of the oxide electrode is needed for the current levels used
and an optical diffusing coating to mask silver inhomogenieties after extended
operation. Positive, near black on white, images are produced not readily

useful for the backlit colour mode usually wanted. Fast changing information, such as for seconds indication, is not possible given the limited number of cycles. Disadvantages outweighed the desirable features and no serious automotive applications emerged.

For electrophoretic displays a larger particle is used suspended as a non-settling colloid in a density matched opaque insulating fluid. White titanium dioxide particles of about 0.2 µm diameter have been used [3.46,3.47] for high contrast (15:1) and reflectance (23 %). Unlike the ECD example quoted, drive voltage is quite high at 100 V although current transients are only 15 nA mm^{-2} and response times are as low as 10 µs. Memory can last at least several hours. Special dispersant surfactants are needed to prevent particle agglomeration which can shorten operating life.

Hysteresis can occur, however, and response depends on duration of the open circuit state. Interactions must be avoided between the black dye in the fluid and the charge adsorbed onto the particles. As for the ECD, backlit negative image colour mode is not possible as the particles and dyed fluid would absorb. As reflective mode panels however, they most closely resemble the white on black point associated with electromechanical instruments. There is even a possibility of having dot-matrix displays using a grid array as column electrodes and a third common electrode [3.47].

Suspended particles displays can be backlit and have been evaluated for automotive displays [3.48,3.49]. Special dichroic colloidal dye molecules are used which appear blue/black in their random state but do not absorb when subjected to an electric field which aligns their axes parallel to the field and normal to the electrodes, so that the electric vector of incident light is at right angles to the molecular axis. In this way, they most closely resemble the dyed phase change LCD.

Optimised suspensions have been produced [3.49] for a good compromise between contrast ratio (38:1) and response time (250 ms) for operation over – 40 C to + 80 C and so compares advantageously with the dyed phase change LCD. It is possible also to operate with CMOS compatible voltage of 18 V driven AC to avoid electrophoretic and electrolytic effects. Storage tests over two years and accelerated life tests suggest compatibility with the automotive specification. Excess particle build-ups at electrode edges can remain on switch-off and deplete the zone immediately adjacent, but this effect has been reduced considerably to tolerable limits using modified suspensions. Other advantages include anticipated low cost of manufacture as there are no polarisers or alignment layers, and a large cell gap of 60 µm is easy to control with less chance of short circuits. Display reflectance can reach 20 % for the preferred suspension and voltage.

3.3.8. **Cathode ray tubes (chapter II.1)**

Although an older technology the CRT has become a later contender for the vehicle displays market [3.34,3.50,3.51]. Its greater depth, high voltage requirements and total system cost could make it unattractive. However, its capacity to display large amounts of data satisfies the need for display of information from multiple sou. .c. in military vehicles, luxury cars and in trucks where, under programme or manual control, different picture formats can

Fig. 3.15 CRT display in route guidance mode (Renault).

be presented. Formats include the usual dashboard gauges as well as trip computer data, diagnostics, navigation or route guidance (figure 3.15), alpha-numeric messages and animated graphics. It is also possible to display messages and images from external sources such as a mobile computer or local TV broadcasts and point to point communications. As such it could be invaluable for salesmen and business executives linked via a radio telephone to the home base and other networks.

The type of CRT technology to accomplish these multipurpose tasks needs careful choice for vehicle compatibility. Over a number of criteria, and compared to shadow-mask colour CRTs, single gun tubes with a wideband emission offer most chance of immediate use for higher luminance (7000 cd m^{-2}), greater ruggedness and resolution at lower system cost. Electrostatic unipotential lens focussing is common. Warm-up time is a feature of CRTs and 7 seconds is likely, but door or seat switched pre warming can reduce this delay down to 2 seconds.

Electromagnetic interaction can be a particular problem; proper screening would prove difficult and expensive. As no screening is provided, nearby motors and rear window demist currents can cause the image to shift. For navigation systems using an overlay map or for systems with overlay touch switches and menu presentation this could prove to be a weakness. In a mixed technology dashboard, where the CRT might be a secondary instrument and LCD a primary instrument, conversely an unscreened CRT might be a source of electromagnetic emission which is an even worse nuisance, and would also affect radio reception.

Resolution is a function of both the tube and the way in which it is addressed. The sharp outline of fixed image electromechanical, LCD or VFD instruments suggests a requirement of approximately 0.4 mm dot spacing (2.5 dots/mm) and typically 252x192 dots are conveniently obtained from CRT controller chips. At very high luminance with 1 mA beam current, some blur could become apparent, especially at edges where the spot size could be 80 % greater. Fortunately, 3500 cd m^{-2} before filtering should be sufficient which will reduce the tendency to blur in a smaller tube, assuming moderate contrast enhancement and colour filter transmission of say $\simeq$ 14 %.

The conventional shadow-mask colour CRT has barely adequate luminance unless driven in the stroke-written mode which would be too expensive (section II.1.5). Naturally, a variety of colours will be essential if the CRT is to act also as a primary instrument. The single gun CRT can provide this by using a wideband phosphor to which colour filters can be added locally. This can introduce parallax, limit the flexibility for total screen use and render colour warnings visible in sunlight even when they are ostensibly off. One way around this is to use specific colour phosphor in selected regions but this still limits screen flexibility. A promising technique under consideration by Delco is the use of a sequential colour filter placed in front of the faceplate transmitting red and green alternately at a sufficient frame rate to avoid flicker [3.52]. High contrast is possible from the temporal interruption of ambient light, but colour range is limited by the polarisers, liquid crystal and phosphor used. Temperature range has been too restricted for automotive use in spite of the significant reduction in switching time achieved with a π-twisted LCD (section II.3.3.7). Three colour versions are possible but at even greater sacrifice of brightness, already down to 12 % transmission, and with a need for more expensive higher frequency frame rate to avoid flicker (monochrome might require at least 60 Hz, the two and three colour options 120 and 180 Hz respectively).

Zenith Radio Corporation have produced two vehicle CRTs of 200x75 mm and 100x75 mm, both of the monochrome single gun variety with high luminance (of 3500 and 7000 cd m^{-2} respectively), and long life phosphor from which colours are possible using overlay filters. Magnetic deflection is used, phase locked horizontally and injection-locked vertically. The tube operates at 12 KV, 50 μA average beam current and its centre spot size is 0.9 mm for 1.5 mA beam current. A frame rate of 47-70 Hz is possible.

Operating life of the CRT is 6000 hours to half of the initial luminance if the photocell automatic dimming allows maximum luminance for only 5 % of the time, 75 % of maximum for 50 % of the time and 25 % of maximum for 45 % of the time. A particular feature of this system is the option of infra red touch switching using a number of infra red LEDs, mounted along the top and side opposite Fresnel facets which direct rays onto just two photodiodes thus saving on the number of receptors needed and so lowering the cost. Alternatively, transparent overlay touch switches can be fitted as in the 1984 Buick from General Motors.

By far the greatest potential use of the CRT is for navigation. Early examples used monochrome CRTs: Ford with a 5-inch diagonal tube and overlay switches, for use with satellite orientation; Honda with a product called an Electro-Gyro-Cator used in conjunction with a transparent map overlay under which the CRT image marked the vehicle's present position. Etak actually presented the map itself on the CRT [3.53]. Colour CRTs have been installed as computer and video displays in limousines, coaches and taxis. Prototype route guidance systems also used them in 1983 such as the demonstrations in Japan by Mitsubishi, Nissan, Toyota and Yazaki; likewise in Europe Philips discussed such a system in 1987 [3.54]. The shadow-mask colour CRT with circular holes absorbs 80 % of the electron beam producing rather too dim a picture for serious use in a car. In 1987 Sony developed a ruggedised version of their high brightness Trinitron CRT which has a stretched aperture grille of vertical bars and spaces aligned with vertical phosphor stripes [3.55]. This can produce a white light luminance of 400 cd m^{-2} at 0.6 mm spot size (or 1000 cd m^{-2} for boost brightness). A horizontal resolution of 300 dots is claimed for a useful screen area of 103 mm by 77 mm in the 6 inch diagonal tube. Compared to the shadow-mask CRT, the luminance of a Trinitron is greater because of the greater transparency to the electron beam (30 % better) and larger current density from better focussing with the larger electron lens diameter. Thus, it is an appropriate colour CRT technology for use in a high luminance application such as cars, but adequate ruggedisation has been a problem hitherto. A rival competition to the Trinitron is future might well be the beam index CRT (section I.1.4.6) which promises even more luminance and higher resolution, with white light at 3400 cd m^{-2} for a spot size of only 0.4 mm. This would be ideal for higher definition when used for maps and other requirements for high information density, especially where space is naturally limited and therefore only relatively small screen sizes can be accommodated. Ruggedisation is an inherent feature as there is no mask at all and only a single electron gun. Already, such tubes are being proposed for use in avionic applications [3.56].

REFERENCES

[3.1]   Swift, D.W. and Freeman, M.H., Application of head-up displays to cars, Displays 7, no. 3, (1986), pp. 107-110.

[3.2]   Train, M.H., Advanced instrumentation of the Oldsmobile Aerotech., SID Digest, (1987), pp. 37-40.

[3.3]   Grimm, R.A., Beyerlein, D.G., Engelman, J.C. and Carol, Jr. J.A., Electronic displays – automotive applications. SAE Congress, paper 830906, (1983).

[3.4]   Yamaguchi, T., Kishino, T. and Doris, J.M., The visual recognition of vacuum fluorescence displays under sunlight conditions, SAE Congress, paper 820013, (1982).

[3.5]   Ketchell, J., The effects of high intensity light adaptation on electronic display visibility, SID Symposium Digest, (1967), pp. 219-230.

[3.6]   Schultz, F.K., Display dimming and matching considerations for LED and VF Displays, SAE Congress, (1982).

[3.7]   Mourant, R.R. and Langolf, G.D., Luminance specifications for automobile instrument panels, Human Factors, 18, no. 1, (1976), pp. 71-84.

[3.8]   Kuzak, D.M. et al, An Objective measure of the readability of electronic display optical filter combinations. SAE Congress, paper 790050, (1979).

[3.9]   Beatty, P.H.J. and Shepherd, B., Active and passive displays: the case for both. Automotive Electronics Conference Proceedings IEE, (1981), pp. 299-309.

[3.10] Akeyoshi, K. and Terade, I., Consideration of LCD legibility for automobiles. Displays 4, no. 1, (1983), pp. 11-15.

[3.11] Galves, J.P. and Brun, J., Colour and brightness requirements for cockpit displays: proposal to evaluate their characteristics, AGARD Conference Proceedings 167, paper 6, (1975). See also ref. 3.15.

[3.12] Laycock, J., Selected colours for use on colour cathode ray tubes. Displays, 5, no. 1, (1984), pp. 3-14.

[3.13] Yoshida, Y., Miyazaki, T. and Maruyama, F., Multi-colour fluorescent indicator panel for automotive applications, SAE Congress, paper 820042, (1982).

[3.14] Ellis, B. Burrel, J., Wharf, J. and Hawkins, D.F., The format and colour of small matrix displays for use in high ambient illumination. SID Symposium Digest, (1974), p. 106.

[3.15] Christiansen, P., Design considerations for sunlight viewable displays. Proceedings of the SID, 24, no. 1, (1983), pp. 29-41.

[3.16] Tyte, R.N., Wharf, J.H. and Ellis, B., Legibility of a light emitting diode array in high illuminance. Proceedings of the SID, 21, no. 1, (1980), pp. 21-29.

[3.17] Van Cott, H.P. and Kinkade, R.G., Human Engineering Guide to equipment design. U.S. Government Printing Office, Washington D.C., USA. Library of Congress card no. 72-600054.

[3.18] Beatty, P.H.J., Automotive LCD instruments: design considerations, Automotive Electronics International Conference IEE., (1983).

[3.19] Semple. A. et al, Analysis of human factors data for electronic flight display systems, Report no. AD 884770, Wright Patterson Flight Dynamics Laboratory.

[3.20] Kolk van der, R.J., Dot-matrix display symbology study, Report no. AFFDL-TR-75-72, Wright Patterson Air Force Flight Dynamics Laboratory, (1975).

[3.21] Reinig, H., Legibility of characters with different segmentations and colours, Proceedings Eurodisplay, Munich (1981), pp. 122-125.

[3.22] Rolfe, J.M., Some investigations into the effectiveness of numerical displays for the presentation of dynamic information, Institute of Aviation Medicine report no. 470, Royal Aircraft Establishment, (1969).

[3.23] Fowkes, M. and Haslegrave, Ergonomics of Electronic displays in automobiles. MIRA Group sponsored project K485590, Report no. K4855, Motor Industry Research Association, (1984).

[3.24] Baines, P.A., Spicer, J., Galer, M. and Simmonds, G., Ergonomics in automotive electronics, Third Automotive Electronics Conference IEE, London (1981).

[3.25] Mackie, C., Vehicle condition monitoring and electronic instrument displays, paper C175, (1981), pp. 103-112.

[3.26] Greene, E.S. and Sendelbach, R., Definition of driver information instrumentation features. SAE Congress, paper 80035, (1980), pp. 59-63.

[3.27] Galer, M., Spicer, J., Geyer, T.A.W. and Holtum, C., The design and instrumentation of a trip computer and a vehicle condition monitor display, Third Automotive Electronics Conference IEE, London (1983), pp. 192-196.

[3.28] Tagg, F., Gas plasma display for automotive applications, Electronic Equipment News, (1977), p. 22.

[3.29] Cornell, P.J., Planar gas discharge displays for automotive applications. SAE Congress, paper 770276, (1977).

[3.30] Okomoto, Y. et al, A DC gas discharge display for automobiles. IEEE Transactions Electron Devices, vol. ED-30, no. 8, (1983), pp. 904–907.

[3.31] Greene, M. and Ross, K., LED display in distributed information systems, SAE Congress, paper 820014, (1982).

[3.32] Shepherd, B., Electroluminescent instrumentation, Displays techn. applications 6486, no. 7, vol. 2, (1981), pp. 337–340.

[3.33] Wilson, B.M., Microprocessor control of DCEL instrumentation, Third International Automotive Electronics Conference IEEE London, paper C199/81, (1981), pp. 275–281.

[3.34] Beatty, P.H.J., Comparison of CRTs and dot-matrix flat panels for vehicle instrumentation. Displays, 5. no. 4, (1984), pp. 196–202; Electro optics/laser conference, Brighton, April 1981; Automotive electronics conference, Toulouse, May 1981.

[3.35] Morimoto, K., Watanabe, H. and Pykosz, T.L., Graphic front luminous vacuum fluorescent display, SAE Congress, paper 840155, Detroit (1984).

[3.36] Morimoto, K. and Dorris, J.M., Front luminous vacuum fluorescent display, SAE Congress, paper 83004, Detroit (1984).

[3.37] Yamaguchi, T., Dorris, J.M. and Hasegawa, M., Graphic display for automobile, SAE Congress, paper 840150, Detroit (1984).

[3.38] Uchiyama, M. et al, Two colour 320x240 vacuum fluorescent image panel, SAE Congress, paper 840154, (1984).

[3.39] Terada, I. and Akeyoshi, K., Dot-matrix LCD for automotive application. SAE Congress, paper 840146, (1984).

[3.40] Talaki, K.A. and Schlax, T., Reconfigurable automobile instrument cluster. SID Digest, (1984), pp. 200–203.

[3.41] Morozumi, S. et al, 4.25 in and 1.51 in B/W and full-colour LC video displays addressed by poly-Si TFTs, SID Digest, (1984), pp. 316–319.

[3.42] Ugai, Y., et al, A 7.23 in diagonal colour LCD addressed by a-Si TFTs, Sid Digest, (1984), pp. 308–311.

[3.43] Waters, C.M., Brimmell, V. and Raynes, P., Highty multiplexable dyed LCDs. Japan Display, (1983), pp. 396–399.

[3.44] Clerc, J.F., Muller, F., Perrin, A. and Vinet, F., Multiplexed dot-matrix displays with guest-host CNPT mixtures in homeotopic orientation, SID Digest, (1984), pp. 130–132.

[3.45] Duchene, J., Meyer, R. and Delapierre, G., Electrolytic display. Biennial Display Research Conference SID, (1978), pp. 1–4.

[3.46] Novotny, V. and Hopper, M.A., Optical and electrical characterisation of electrophoretic displays. J. Electrochem., Soc., 126, no. 12, (1979), pp. 2211-2216.

[3.47] Maru, P., Characteristics of an X-Y addressed electrophoretic image display (EPID), SID Digest, (1984), p. 141.

[3.48] Saxe, R.L., Thompson, R.I. and Forlini, M., Suspended particle display with improved properties, International Display Research Conference, SID/IEEE, (1982), pp. 175-179.

[3.49] Rachner, H. and Morrissy, J.H., New results in colloid display technology, SAE Congress, paper 830036, (1983), pp. 1-8.

[3.50] Dietch, L. and Zeidler, H.R., The genesis of the cathode ray tube as a vehicle display device, SAE Congress, paper 810306, (1981), pp. 89-96.

[3.51] MacKie, C. and Williams, M., An experimental CRT instrument display, Automotive Electronics Conference IEE, London (1983), pp. 231-236.

[3.52] Vatne, R., Johnson, Jr. P.A. and Bos, P.J., A LC/CRT field sequential colour display SID Digest, (1983), pp. 28-29.

[3.53] Electronic display world, 6, no. 9, (1986), pp. 28-35.

[3.54] Fernhout, H., The CARIN car information and navigation and the extension to carminat, Proceedings Sixth International Conference on Automotive Electronics IEE, London. Conference publication no. 280, (1987), pp. 139-143.

[3.55] Okada, M. et al, A 6 inch high resolution Trinitron CRT for automotive applications, SID Digest, (1987), pp. 45-48.

[3.56] Saito, S. et al, A 6.3x6.3 inch avionics beam index CRT, SID Digest, (1987), pp. 178-181.

[3.57] Yoshida, et al, A new design concept for the graphics of fluorescent indicator panels for automotive applications, SAE Congress, (1985).

SUBJECT INDEX

AUTHOR INDEX